Phyllis B. Boynton

Structured C for Technology

Text   29.95

49.95

Merrill Publishing Company
Columbus   Toronto   London   Melbourne

To my brother, John Dusky Adamson—An expert programmer and a good friend

Published by Merrill Publishing Company
Columbus, Ohio 43216

This book was set in Garamond Light

Administrative Editor: David Garza
Production Coordinator: JoEllen Gohr
Cover Designer: Russ Maselli

Library of Congress Catalog Card Number: 89-62577
International Standard Book Number: 0-675-20993-5
Printed in the United States of America
1 2 3 4 5 6 7 8 9—93 92 91 90

Preface

Why This Text?

This text is designed to be an introductory course in the programming language called C. The text material is written for students in technically oriented programs at community or junior colleges, trade and technical schools, and colleges that offer four-year engineering technology programs. No prior programming experience is required.

Almost all texts written about the C programming language are oriented toward mathematical or computer science applications. This text is specifically directed toward the technology student and includes the solutions to technical problems. No mathematics beyond elementary algebra and introductory trigonometry are required in the technology problems used here.

Why C?

The programming language called C was developed at Bell Labs in 1972 by Dennis Ritchie. C was created as a programming tool for working programmers. The main goal of C is to be a useful language, which is not true for other programming languages such as BASIC and Pascal. (BASIC is a computer language that makes it easy for a beginner to learn programming. Pascal is a computer language that was developed to teach proper programming principles by forcing the programmer into a specific program structure.)

C was developed to create the most powerful and flexible way of giving instructions to a computer—any computer! Unlike BASIC or Pascal, where control of the computer is of secondary importance, C offers the user programming power and options never before available in any single computer language. It's not surprising to those who know C that its popularity has spread into the microcomputer field as well as larger and more sophisticated systems.

Why This Text Can Work for You

This is a good text for a first course in programming for the following reasons:

1. Each new programming concept is presented in an environment of a required structure using the programming language called C.
2. C is used to teach the beginning programming student the good programming habits desired by industry.
3. Programming is presented in an arena that requires consistent program structure and utilizes top-down design.
4. This text uses Turbo C by Borland as well as Microsoft Learn C. Those features that are unique to each environment are pointed out so the student may also use other C environments.
5. The ANSII C standard is used throughout this book. This includes function prototyping as well as **enum** and **void**.
6. All examples and problems are presented in an environment of technical applications. Each area of technology is clearly defined to stimulate student interest.
7. No mathematics beyond elementary algebra and right-angle trigonometry are required.
8. The simplest possible examples are used to illustrate new concepts. This prevents the student from getting bogged down in complex mathematical or technical details that may distract from learning the programming material.
9. The practical applications aspect of programming is presented so that the technology student sees this as another important tool for the solution, analysis, and design of technical problems.
10. Additionally, two purchasing options are available: (1) the text, or (2) the text with two programming diskettes offering Microsoft's powerful Quick C compiler in a Learn C environment with built-in debugger and help. Also included at the end of the text are a notice from Borland and a coupon from Microsoft regarding their special educational pricings for their C compilers.

This text is intended for use as the major text for a comprehensive three-credit course. There are many outstanding learning features available for users of this book:

1. Chapter objectives are clearly stated at the beginning of each chapter.
2. Review questions are presented at the end of each chapter section. These are open-ended questions that are designed to stimulate class discussion on key topics.
3. Each chapter contains a unique *interactive exercises* section. These exercises encourage student participation at the computer console. Programming

activities are selected to give immediate feedback and quickly demonstrate the unique properties of the computer being used by the student.

4. Each chapter contains a *self-test* which questions the student about a particular C program. Program debugging and the careful analysis of a practical program may be required here.

5. All end-of-chapter problems are carefully selected to fit into representative areas of technology. This is done so the technology student may work in an area of interest and see the programming commonalities of other technology disciplines.

6. All answers to Section Reviews, Self-Tests, and selected end-of-chapter problems are contained in the text.

Acknowledgments

My heartfelt thanks to the many reviewers who helped make this a quality text book. They spent many hours carefully reading the original manuscript, which included the important details of programming code. Their suggestions and comments resulted in a final text that will be appreciated by instructors and students.

Dion Benes	DeVry Institute of Technology Phoenix, Arizona
John Blankenship	DeVry Institute of Technology Decatur, Georgia
Gary Boyington	Chemeketa Community College Salem, Oregon
Bill Champion	DeVry Institute of Technology Irving, Texas
Charles Goodspeed	Hartnell Community College Salinas, California
Robert Hilles	DeVry Institute of Technology Calgary, Alberta (Canada)
Daniel Morrow	Shasta College Redding, California
Jim Pearson	DeVry Institute of Technology Irving, Texas
Iran Scheer	DeVry Institute of Technology Columbus, Ohio
Jim Stewart	DeVry Institute of Technology Woodbridge, New Jersey
Todd Walker	Gwinnett Technical Institute Lawrenceville, Georgia
Tom Wheeler	DeVry Institute of Technology Kansas City, Missouri

I would also like to thank the professional staff at Merrill Publishing, specifically, Catherine M. Parts, Editorial Assistant—Electronics Technology. I am very fortunate to have worked with Ms. Parts, whose commitment to quality textbooks is well respected. Also, I would like to thank Stephen Helba, executive editor, David Garza, administrative editor; JoEllen Gohr, production manager.

Contents

1 C Fundamentals   1

1-1 **The C Environment   3**
Overview   3 | The C Environment   3 | Conclusion   3 | Section Review   5

1-2 **Why C?   5**
Background   5 | Disadvantages of C   5 | What C Looks Like   6 | The Different Parts   6 | Conclusion   6 | Section Review   7

1-3 **Program Structure   7**
Discussion   7 | Structured vs. Unstructured Programming   8 | Program Blocks   9 | The Programmer's Block   10 | Advantages and Disadvantages of Structured Programming   11 | Conclusion   12 | Section Review   12

1-4 **Elements of C   12**
What C Needs   12 | C Statements   14 | Conclusion   15 | Section Review   15

1-5 **The printf( ) Function   15**
What **printf( )** Does   16 | More Than One   18 | Conclusion   18 | Section Review   18

1-6 **Identifying Things   19**
Discussion   19 | What Is a Function?   19 | What Needs Identification?   19 | Creating Your Own Identifiers   19 | Case Sensitivity   20 | Key Words   21 | Conclusion   21 | Section Review   21

1-7 **Declaring Things   21**
What Is a Variable?   22 | Declaring Variables   22 | Initializing Variables   23 | More of the Same Type   24 | Why Declare?   24 | Section Review   24

1–8   **Introduction to C Operators   24**
What Are C Operators?   25 | Arithmetic Operators   25 | Important
Considerations   26 | Precedence of Operations   26 | Compound
Assignment Operators   27 | Conclusion   29 | Section Review   29

1–9   **More printf( )   29**
Escape Sequences   29 | Field Width
Specifiers   29 | Conclusion   30 | Section Review   31

1–10   **Getting User Input   31**
Discussion   31 | The **scanf( )** Function   31 | Format Specifiers   32 |
A Real Technology Problem   32 | Using the E Notation   34 |
Conclusion   34 | Section Review   34

1–11   **Program Debugging and Implementation:**
Common Programming Errors   34
Discussion   34 | Types of Error Messages   35 | Case
Sensitivity   35 | The Semicolon   36 | Incomplete or Nested
Comments   37 | Conclusion   39 | Section Review   40

1–12   **Case Study   40**
Discussion   40 | First Step—Stating the Problem   40 | Blocking Out the
Program   41 | The Next Step   42 | Completing the Study   43 |
Checking the Output   46 | Conclusion   47 | Section Review   47 |
Interactive Exercises   48 | Self-Test   50 | End-of-Chapter Problems   52

2   **Structured Programming   57**

2–1   **Concepts of a Program Block   58**
Discussion   58 | A Blocking Example   59 | Blocking the Structure   59 |
Defining a Block Structure   61 | Some Important Rules   61 | Types of
Blocks   62 | A Theorem   63 | Conclusion   63 | Section Review   63

2–2   **Using Functions   63**
Discussion   63 | What Is a C Function?   63 | Making Your Own
Functions   64 | What Makes a Function?   64 | Program Analysis   66 |
Conclusion   70 | Section Review   70

2–3   **Inside a C Function   70**
Discussion   70 | Basic Idea   71 | Program Analysis   72 |
Conclusion   75 | Section Review   76

2–4   **Functioning with Functions   76**
Discussion   76 | More Than One Argument   76 | Program Analysis   77 |
Calling More Than One Function   78 | Program Analysis   80 | Calling
Functions from a Called Function   81 | Program Analysis   84 | How
Functions May Be Called   85 | Conclusion   86 | Section Review   86

2–5   **Defining Things   86**
Discussion   86 | Basic Idea   86 | Program Analysis   87 | Defining
Constants   88 | Program Analysis   88 | Defining Operations   89 |
Forms for **#defines**   90 | Revisiting an Old Program   90 | Program
Analysis   91 | Expanding the Concept   91 | Program Analysis   92 |
Including the **#define**   93 | Conclusion   93 | Section Review   93

2–6 **Program Debugging and Implementation:**
Making Your Own Header Files **94**
Discussion 94 | An Example 94 | Program Analysis 94 | Saving Your
#defines 94 | A Sample Program 96 | Program Analysis 96 | Making
the Header File 98 | Using Your Header File 98 | Token Pasting and
String-izing Operators 99 | Conclusion 100 | Section Review 100

2–7 **Case Study** **100**
Discussion 100 | The Problem 101 | First Step—Stating the
Problem 101 | Developing the Algorithm 102 | First Coding 103 |
Program Analysis 104 | Entering the First Code 104 | Formula
Decision 105 | Adding the Calculate and Display Block 105 | Program
Analysis 108 | Final Program 109 | Conclusion 109 | Section
Review 110 | Interactive Exercises 110 | Self-Test 112 | End-of-
Chapter Problems 112

3 Branching and Logic **117**

3–1 **Relational Operators** **118**
Discussion 118 | Relational Operators 118 | Program Analysis 120 |
Equal To 121 | Conclusion 122 | Section Review 122

3–2 **The Open Branch** **123**
Discussion 123 | Basic Idea 123 | The **if** Statement 123 | Program
Analysis 124 | A Compound Statement 125 | Program Analysis 127 |
Calling a Function 129 | Program Analysis 130 | Conclusion 131 |
Section Review 131

3–3 **The Closed Branch** **131**
Discussion 131 | Basic Idea 132 | The **if...else** Statement 132 |
Program Analysis 133 | Compound **if...else** 135 | Program
Analysis 136 | Another Way 138 | Program Analysis 140 |
Conclusion 142 | Section Review 142

3–4 **Logical Operation** **142**
Discussion 142 | Logical AND 143 | Program Analysis 144 | The
OR Operation 145 | Program Analysis 146 | Relational and Logic
Operations 147 | Program Analysis 149 | Compounding the
Logic 149 | Program Analysis 150 | Application Program 150 |
Conclusion 152 | Section Review 152

3–5 **Conversion and Type Casting** **152**
Data Types 153 | Type Casting 153 | Lvalue 153 |
Conclusion 153 | Section Review 154

3–6 **The C switch** **154**
Discussion 154 | Decision Revisited 154 | Program Analysis 155 |
How to Switch 155 | Program Analysis 157 | Compounding the
switch 158 | Program Analysis 161 | Functional Switching 161 |
Program Analysis 163 | What May Be Switched 164 | Switching within
Switches 164 | Conclusion 164 | Section Review 164

3-7   One More switch and the Conditional Operator   164
Discussion   164 | One More **switch**   164 | Program Analysis   166 | The
Conditional Operator   167 | Program Analysis   169 | Conditional
Operator Application   169 | Program Analysis   171 | Conclusion   171 |
Section Review   172

3-8   **Program Debugging and Implementation   172**
Discussion   172 | Conditional Directives   172 | Conclusion   173 |
Section Review   173

3-9   **Case Study: A Robot Troubleshooter   174**
Discussion   174 | The Hypothetical Robot   174 | First Step   174 |
Developing the Algorithm   175 | First Development Stage   175 | Program
Analysis   177 | Adding Some Structure   177 | Program Analysis   180 |
Final Program   182 | Program Analysis   185 | Conclusion   185 | Section
Review   186 | Interactive Exercises   186 | Self-Test   190 | End-of-
Chapter Problems   190

4   Loops   195

4-1   **The for Loop   196**
Discussion   196 | What the **for** Loop Looks Like   196 | Program
Analysis   197 | **for** Loop Facts   198 | More Than One Statement   199 |
The Increment and Decrement Operators   200 | Program Analysis   202 |
The Comma Operator   202 | Conclusion   202 | Section Review   203

4-2   **The while Loop   203**
Discussion   203 | Structure of the **while** Loop   203 | Program
Analysis   204 | **while** Application   205 | Program Analysis   206 |
Functions Inside the **while**   207 | Conclusion   207 | Section Review   207

4-3   **The do while Loop   207**
What the **do while** Loop Looks Like   208 | Using the C **do while**   208 |
Program Analysis   209 | Loop Details   209 | Sentinel Loops   211 |
Program Analysis   212 | Loop Comparisons   212 | Conclusion   212 |
Section Review   213

4-4   **Nested Loops   213**
Discussion   213 | Basic Idea   213 | Program Analysis   214 | Nested
Loop Structure   215 | Nesting with Different Loops   216 | Program
Analysis   217 | Conclusion   218 | Section Review   218

4-5   **Program Debugging and Implementation   219**
Run Time Errors   219 | A Sample Problem   219 | Auto Debug   221 |
Auto Debug Function   222 | Conclusion   223 | Section Review   223

4-6   **Case Study   223**
Discussion   223 | Background Information   223 | The Problem   225 |
The First Step—Stating the Problem   225 | Developing the
Algorithm   226 | Program Analysis   228 | Designing the get_values
Function   229 | Designing the calculate_and_display Function   230 |
Designing the calculate_below and calculate_above Functions   231 | Final
Program Phase 232 | Program Execution   234 | Conclusion   235 | Section

Review   235 | Interactive Exercises   236 | Self-Test   239 | End-of-Chapter Problems   240

5   Pointers, Scope, and Class   245

5–1   **Internal Memory Organization   246**
Discussion   246 | Basic Idea   246 | Storing a Program   247 | Another Way   247 | Conclusion   250 | Section Review   250

5–2   **How Memory Is Used   251**
Discussion   251 | Storage Size   251 | **char** Type   251 | **sizeof**   253 | Representing Signed Numbers   254 | Binary Addition   254 | Complementing Numbers   255 | Binary Subtraction   256 | Conclusion   261 | Section Review   262

5–3   **Pointers   262**
Discussion   262 | Basic Idea   262 | Program Analysis   264 | Using the Pointer   265 | Passing Variables with Pointers   266 | Program Analysis   267 | Some Examples   268 | Conclusion   269 | Section Review   270

5–4   **Passing Variables   271**
Discussion   271 | Basic Idea   271 | Using Pointers   272 | Pointer Application   274 | Program Analysis   277 | Key Points   279 | Conclusion   280 | Section Review   280

5–5   **Scope of Variables   280**
Discussion   280 | Local Variables   280 | Global Variables   281 | Caution with Global Variables   282 | Conclusion   284 | Section Review   285

5–6   **Variable Class   285**
Discussion   285 | Constants   285 | Automatic Variables   286 | External Variables   286 | Static Variables   286 | Register Variable   287 | Overview   288 | Conclusion   289 | Section Review   289

5–7   **Program Debugging and Implementation   289**
Problems in C Programming   289 | Pointers   289 | Not Initializing Pointers   290 | Assignment and Equality   290 | Forgetting to Use the Address Operator   291 | Conclusion   291 | Section Review   291

5–8   **Case Study   292**
Discussion   292 | Bit Manipulation   292 | Bitwise Complementing   292 | Bitwise ANDing   294 | Bitwise ORing   295 | Bitwise XORing   296 | Shifting Bits   297 | Conclusion   298 | Section Review   299 | Interactive Exercises   299 | Self-Test   301 | End-of-Chapter Problems   302

6   Strings and Arrays   309

6–1   **Characters and Strings   310**
Discussion   310 | Storing Strings   310 | Program Analysis   311 | An Inside Look   312 | Program Analysis   313 | Where Are the Elements?   313 | Conclusion   316 | Section Review   316

6-2   **More about Arrays   316**
Discussion   316 | Basic Idea   316 | Type **int** Array   317 | Inside
Arrays   318 | Putting in Your Own Values   320 | Passing Arrays   321 |
Program Analysis   322 | Passing Arrays Back   323 | Revisiting
Strings   324 | Passing Strings   326 | Conclusion   327 | Section
Review   328

6-3   **Multidimensional Arrays   328**
Discussion   328 | Basic Idea   328 | Program Analysis   330 | Array
Applications   332 | Program Analysis   333 | Adding More Columns   334 |
Program Analysis   336 | Conclusion   338 | Section Review   338

6-4   **Introduction to Array Applications   339**
Discussion   339 | Working with the Array Index   339 | Changing the
Sequence   340 | Program Analysis   341 | Finding a Minimum
Value   342 | Program Analysis   343 | Conclusion   344 | Section
Review   344

6-5   **Sorting   344**
Basic Idea   344 | Sample Program   346 | Program Analysis   348 |
Conclusion   350 | Section Review   350

6-6   **More about Strings   350**
Discussion   350 | An Array of Characters   351 | Program Analysis   351 |
Another Way   352 | A Simple Way   353 | A Closer Look   354 | Pointing
to Pointers   355 | Program Analysis   357 | Manipulating the Pointer
Indexes   358 | Program Analysis   360 | Doing It with Pointers   361 |
Conclusion   362 | Section Review   362

6-7   **Program Debugging and Implementation   362**
I/O Problems   362 | **scanf( )** Again   363 | The **gets( )**   363 | Long
Strings   364 | A Character at a Time   365 | Checking Characters   366 |
More Character Checking   367 | Program Analysis   369 | Looking for
Numbers   370 | Conclusion   371 | Section Review   371

6-8   **Case Study   371**
Discussion   371 | The Problem   372 | First Step—Stating the
Problem   372 | Developing the Algorithm   372 | Built-in C String
Functions   372 | Program Development   373 | Function Analysis   375 |
Final Program Phase   376 | Conclusion   379 | Section Review   379 |
Interactive Exercises   380 | Self-Test   381 | End-of-Chapter Problems   382

7   **Structuring and Saving Data   387**

7-1   **Enumerating Types   388**
Discussion   388 | Expressing Data   389 | Program Analysis   390 |
Another Form   391 | Enumeration Example   392 | Assigning **enum**
Values   392 | Conclusion   393 | Section Review   393

7–2   **Naming Your Own Data Types**   **394**
Discussion   394 | Inside **typedef**   394 | Program Analysis   397 | Other
Applications   398 | Conclusion   398 | Section Review   398

7–3   **Introduction to Structures**   **398**
Discussion   398 | The C Structure   399 | Putting Data into a
Structure   401 | Program Analysis   403 | Conclusion   404 | Section
Review   404

7–4   **More Structure Details**   **405**
Discussion   405 | The Structure Tag   405 | Program Analysis   406 |
Naming a Structure   407 | Structure Pointers   408 | A Function as a
Structure   410 | Program Analysis   412 | Conclusion   413 | Section
Review   413

7–5   **The union and Structure Arrays**   **413**
Discussion   413 | The C **union**   413 | Initializing the C Structure   415 |
Program Analysis   416 | Structure Arrays   417 | Program Analysis   420 |
Conclusion   422 | Section Review   422

7–6   **Ways of Representing Structures**   **422**
Discussion   422 | Structures within Structures   422 | Arrays within
Structures   423 | Multidimensional Structure Arrays   424 |
Conclusion   424 | Section Review   424

7–7   **Disk Input and Output**   **425**
Discussion   425 | Creating a Disk File with C   425 | A DOS
Review   425 | A First File Creation Program   426 | Program
Analysis   426 | Putting Data into the Created File   427 | Program
Analysis   427 | Reading Data from a Disk File   427 | Program
Analysis   428 | Saving a Character String   429 | Reading a Character
Stream   430 | Conclusion   431 | Section Review   431

7–8   **More Disk I/O**   **431**
Discussion   431 | The Possibilities   432 | File Format   432 | Example
File Program   434 | Program Analysis   436 | Mixed File Data   438 |
Program Analysis   439 | Retrieving Mixed Data   439 | Text vs. Binary
Files   440 | Record Input   441 | Program Analysis   443 | Record
Output   444 | Conclusion   445 | Section Review   445

7–9   **Program Debugging and Implementation**   **446**
Discussion   446 | Inside I/O   446 | The Buffered Stream   446 | Standard
Files   448 | Redirecting Files   448 | Piping Files   449 | Filtering   449 |
I/O Levels   449 | A Word about File Handles   450 | Random Access
Files   450 | File Pointers   450 | Program Analysis   452 | Command
Line Arguments—The Basic Idea   454 | Developing Command Line
Arguments   454 | Conclusion   456 | Section Review   456

7–10   **Case Study**   **456**
Discussion   456 | The Problem   456 | First Step—Stating the
Problem   457 | Developing the Algorithm   457 | Program
Development   457 | Program Analysis   464 | Conclusion   464 |
Interactive Exercises   465 | Self-Test   472 | End-of-Chapter
Problems   472

8   Color and Technical Graphics   477

　　8–1   C Text and Color   478
　　　　Discussion   478 | What Your System Needs   479 | The Color
　　　　Monitor   479 | Display Capabilities   479 | The **mode** Function   480 |
　　　　Text Color   482 | Changing the Background   486 | Conclusion   487 |
　　　　Section Review   487

　　8–2   **Starting Turbo C Graphics   488**
　　　　Discussion   488 | Basic Idea   488 | What You Need to Know   489 |
　　　　Selecting Your Graphics Driver   490 | The Graphics Screen   490 |
　　　　Lighting up a Pixel   492 | Using Color   493 | Making an Autodetect
　　　　Function   495 | Conclusion   496 | Section Review   496

　　8–3   **Knowing Your Graphics System   496**
　　　　Discussion   496 | Your System's Screen Size and Colors   496 | Your
　　　　System's Modes   497 | Creating Lines   500 | Conclusion   501 | Section
　　　　Review   502

　　8–4   **Built-in Shapes   502**
　　　　Discussion   502 | Changing Graphics Modes   502 | Line Styles   504 |
　　　　Making Rectangles   506 | Creating Circles   508 | Area Fills   510 |
　　　　Aspect Ratios   512 | Conclusion   512 | Section Review   512

　　8–5   **Bars and Text in Graphics   512**
　　　　Discussion   512 | Creating Bars   513 | Filling in the Bars   514 |
　　　　Text with Graphics   519 | Changing Text Fonts   520 | Conclusion   523 |
　　　　Section Review   523

　　8–6   **Graphing Functions   523**
　　　　Discussion   523 | Fundamental Concepts   524 | Scaling   525 |
　　　　Coordinate Transformation   526 | Putting It Together   528 | Analyzing
　　　　the Program   531 | Conclusion   532 | Section Review   532

　　8–7   **Programming Style   532**
　　　　Discussion   532 | C Reserved Words   533 | Naming Things   533 |
　　　　Your Structure   533 | Conclusion   536 | Section Review   536

　　8–8   **Case Study   536**
　　　　Discussion   536 | The Problem   537 | First Step—Stating the
　　　　Problem   537 | Developing the Algorithm   537 | Arithmetic
　　　　Functions   537 | Program Development   537 | Program Analysis   542 |
　　　　Conclusion   542 | Interactive Exercises   542 | Self-Test   543 |
　　　　End-of-Chapter Problems   543

9   Hardware and Language Interfacing   549

　　9–1   **Inside Your Computer   551**
　　　　Discussion   551 | Basic Architecture   551 | General Purpose
　　　　Registers   551 | Pointer and Index Registers   554 | Memory
　　　　Segmentation   556 | Calculating the Address   557 | Flag
　　　　Register   557 | Conclusion   557 | Section Review   558

9–2   **Assembly Language Concepts   558**
Discussion   558 | Basic Idea   559 | What Assembly Language Looks
Like   559 | Types of Assembly Language Instructions   559 | Converting C
to Assembly Language Code   559 | C at the Machine Level   560 | Saving
Space for Data   561 | Saving μP Contents   561 | C Arguments   562 |
Storing Static Variables   563 | Conclusion   563 | Section Review   564

9–3   **Memory Models   565**
Discussion   565 | Basic Idea   565 | Memory Models   567 | Near and
Far Pointers   567 | Conclusion   568 | Section Review   569

9–4   **C Source Code to Assembly Language   570**
Discussion   570 | Doing the Conversion   570 | Meaning of the
Instructions   573 | Conclusion   573 | Section Review   573

9–5   **Pseudo Variables and Inline Assembly   573**
Discussion   573 | Introduction to Pseudo Variables   574 | Using Pseudo
Variables   575 | Introduction to Inline Assembly   576 | Inline Machine
Code   577 | Conclusion   578 | Section Review   578

9–6   **Programming Utilities   578**
Discussion   578 | Utility Overview   578 | The MAKE Utility   578 |
How MAKE Works   582 | Date/Time Utilities   584 | Library Utility   584 |
Using the Turbo C Library Utility   584 | Using the Quick C Library
Utility   585 | File Search Utilities   585 | Conclusion   586 | Section
Review   586

9–7   **BIOS and DOS Interfacing   586**
Discussion   586 | BIOS and DOS   587 | Some BIOS Functions   587 |
DOS Calls   588 | BIOS Application   588 | DOS Call Application   590 |
Conclusion   591 | Section Review   591

9–8   **Program Debugging and Implementation: Interrupts   592**
Discussion   592 | Basic Idea   592 | What an Interrupt Does   592 |
Interpreting a Vector Table   598 | Typical Interrupts   599 | Creating
Interrupt Handlers   601 | Conclusion   604 | Section Review   604

9–9   **Case Study   604**
Discussion   604 | The Problem   604 | First Step—Stating the
Problem   605 | Developing the Algorithm   605 | Program
Development   605 | The Parallel Printer Port   605 | Defining
Data   606 | The Completed Program   607 | Program Analysis   608 |
Conclusion   608 | Interactive Exercises   609 | Self-Test   610 | End-of-
Chapter Problems   611

Appendices   615

Appendix A—Introduction to IBM and MS-DOS   615
Appendix B—Learn C Programming Environment   620
Appendix C—Using the Learn C Help Menu   622
Appendix D—The Learn C Debugger   626
Appendix E—Learn C Color and Graphics   629

Appendix F—ASCII Character Set   634

Appendix G—ANSI C Standard Math Functions   635

Answers   639

Index   677

Summary of Keywords and Statements   686

1   C Fundamentals

Objectives

This chapter gives you the opportunity to learn:

1. Why C is used as a programming language and what you need to use C in a computer.
2. Some beginning C commands and how to use them to develop a **structured program**.
3. How to recognize block structure and use it in program development.
4. What a programer's block is and why it is useful.
5. The basic elements needed to write a program in C.
6. How the **printf( )** function operates.
7. What identifiers are and how to use them.
8. The keywords used in C.
9. The importance of declaring variables and how to do this.
10. The methods C uses to perform arithmetic operations.
11. Ways of formatting your output using the **printf( )** function.
12. The development of the **scanf( )** function for getting user input.
13. Some common reasons for bugs in a C program.
14. The steps used to develop a technology program using the C language.

Key Terms

Structured Program	LEARN C
Environment	QUICK C
TURBO C	Editor

Compiler
Library Files
Linker
Source Code
Program Structure
Programmer's Block
Remarks
Top-down Design
Computer Memory
Token
Keyword
Character
String
Integer
Float
Data Type
Statement
Expression

Single Statement
Compound Statement
Standard Output
Format Specifier
Argument
Field
Function
Declaration
Identifier
Type Specifier
Assignment Operator
Arithmetic Operators
Precedence
Compound Assignment Operator
Escape Sequence
Field Width Specifier
E Notation

Outline

1-1   The C Environment
1-2   Why C?
1-3   Program Structure
1-4   Elements of C
1-5   The printf( ) Function
1-6   Identifying Things
1-7   Declaring Things

1-8   Introduction to C Operators
1-9   More printf( )
1-10 Getting User Input
1-11 Program Debugging and
      Implementation
1-12 Case Study

Introduction

This chapter introduces you to some of the fundamentals of the C language. You will learn how to display values on the monitor screen and how to get values from the program user. This chapter also shows you how to do basic arithmetic operations with C.

When you complete this chapter, you will be able to write some of your first technology programs in C. The chapter concludes with the design of an actual technology program.

1-1   The C Environment

Overview

This section presents what is called the **C environment**. An environment in programming means all the different programming tools you will need in order to work with a particular programming language.

When you program in the programming language called BASIC, the environment is usually built into the microcomputer (in its ROM). In this case, all that is usually necessary is to turn the computer on, start typing in BASIC commands, and type in a command to instruct the computer to execute the program. This is not the case with other programming languages.

The C Environment

This book emphasizes two C environments: **TURBO C** by Borland International and **LEARN C** by Microsoft Inc. LEARN C is actually a condensed version of **QUICK C**. The main difference is that LEARN C will not store a stand-alone executable C program. All of these C environments contain an **editor, compiler, library files, linker,** and much more. Each of these does the following:

Editor   Allows you to enter and modify your C source code.

Compiler   A program that converts the C program you developed into a code understood by the computer.

Include Files   Disk files that consist of many separate definitions and instructions that may be useful to a programmer in certain instances.

Library Files   These are previously compiled programs which perform specific functions. These programs can be used by you to help you develop your C programs. For example, the C function that allows you to display text on your monitor screen (the **printf( )** function) is not in the C language. Instead, its code is in a library file. The same is true of may other functions, such as graphics, sound, and working with the disk and printer, just to name a few. You can also create your own library files of routines developed by you that you use over and over again in different C programs. By doing this, you can save hours of programming time and prevent programming errors.

Linker   Essentially the linker combines all of the necessary parts (such as library files) of your C program to produce the final executable code. Linkers play an important and necessary role in all of your C programs. In larger C programs, it is general practice to break the program down into smaller parts each of which is developed and tested separately. The linker will then combine all of these parts together to form your final executable program code.

You will also need some kind of a disk operating system to assist you in saving your programs. The main parts of the C environment are shown in Figure 1–1.

Conclusion

In this section, you were introduced to the C environment. Here, you got an idea of what it takes to enter a C program into your computer and what else is needed in order to get the program to do what you want it to do (executing the program).

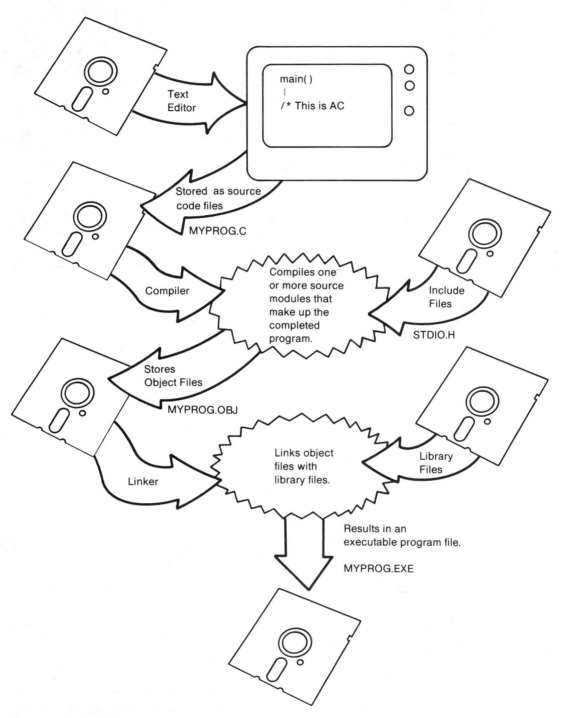

Figure 1-1   The C Environment

In the next section, you will learn some of the reasons why C is such an important programming language and how you can use it in the field of technology.

1-1 Section Review

1   State the purpose of an editor. Why is it needed?
2   Give the reason for using a compiler.
3   Explain the action of a compiler.

1-2   Why C?

Background

C is an increasingly popular programming language in industry and schools and for personal use. The many reasons for this are outlined in Table 1-1.

Disadvantages of C

Since C is such a versatile language, its code can be written with such brevity that it becomes almost unreadable. This style of programming is not encouraged and is not presented in this book. Because of its flexibility, C will allow you to write programs that could wind up containing very difficult-to-find "bugs." Programming in C is like having a sports car with a 500 mph top speed and very few highway regulations—you have to be very careful how you handle such unrestricted power.

Table 1-1   The Power of C

Advantage	What This Means to You
Designed for top-down programming	Your programs will be easier to design.
Designed for structure	Your programs will be easier to read and understand.
Allows modular design	Enhances the program's appearance so others can easily follow and modify. Makes it easier for you to debug.
An efficient language	More compact, quicker-running programs.
Portability	A program you write on one computer will operate on another computer system with few if any changes.
Computer control	You exercise almost absolute control over your computer.
Flexibility	You can easily create other languages and operating systems.

What C Looks Like

Here is a C program:

Program 1–1

```
#include <stdio.h>
/* This is a C program. It will print a message on the */
/* computer screen. */

main ( )
{
printf("Send me a 10 ohm resistor.");
}
```

Note several things about this program. It is easy to read. No one doubts what the program will do. The program has a structure that makes it easy to read. There are some "strange" symbols (like the /* that will be explained shortly). What you see in this program is the **source code**. This is what you, the source, would type in.

In the above program, the items required by C are in boldface. The optional items you may put in are not in boldface. From this, you can conclude that the word **main** immediately followed by **( )** is required in a C program. When the program is executed it displays:

```
Send me a 10 ohm resistor.
```

on the monitor screen.

The Different Parts

As outlined in Table 1–2, all C programs must have certain parts. Don't worry about the other items (such as the **printf**) used in the above program. These were used to assist C in displaying the sentence on the monitor. You will learn the meaning of these and where to use them shortly.

Conclusion

Congratulations! You have just seen your first C program. It just displayed a message on the screen, something any computer can easily do without using C. It was used here just to keep things as simple as possible so you could concentrate on the main items of a C program.

In the next section, you will learn some important points about structure and why you should use it.

Table 1–2   Major Parts of a C Program

Item	Purpose
#include <stdio.h>	Tells the compiler to *include* the standard input/output file.
main	This marks the point where the C program begins execution. Required for all programs.
( )	Must appear immediately after **main**. Usually information that will be used by the program is contained within these parentheses.
/* */	These symbols are optional and are used to enclose comments. Comments are remarks used by you to help clarify the program for people. They are ignored by the compiler.
;	Each C statement ends with a semicolon. For now, think of a C statement as consisting of a C command.
{ }	The braces are required in all C programs. They indicate the beginning and the end of the program instructions.

1–2 Section Review

1   State three advantages of the C programming language.
2   Explain what is meant by portability in a programming language.
3   What C command must all C programs start with?
4   Give an example of a comment in a C program.
5   What is the purpose of the braces { } in a C program?

1–3   Program Structure

Discussion

In this text, all C programs will conform to a specific structure. You can think of a **program structure** as the format you use when entering the program. The first C program had a structure:

```
#include <stdio.h>

/* This is a C program. It will print a message on the */
/* computer screen. */

main( )
{
printf("Send me a 10 ohm resistor.");
}
```

When compiled and executed, the above program would cause the monitor to display (on the first line starting at the left of the screen):

`Send me a 10 ohm resistor.`

The above program could just as well have be written with a different structure:

Program 1-2

```
#include <stdio.h>

/* This is a C program. It will print a message on the */
/* computer screen. */
main( ) { printf("Send me a 10 ohm resistor."); }
```

When compiled and executed, the above Program 1-2 would still do exactly the same thing. The difference is that the structure has been changed. The program is a little harder to read. You could have written the same program as:

Program 1-3

```
#include <stdio.h>

main( ) { printf("Send me a 10 ohm resistor."); }
```

And again, when compiled and executed, the program would have done the same thing. The point is that program structure is considered good structure when the structure is easy for people to read and understand.

For the simple program introduced here, program structure may seem to make very little difference. However, in larger programs, program structure becomes very important.

Structured vs. Unstructured Programming

One of the measures of a "good" computer program is that it can be read and understood by anyone—even if that person does not know how to program.

For example, the first C program presented in this book was easy to understand. Admittedly it didn't do much, but the point is if good structure is observed, the program will be easier to understand, modify, and debug when necessary. An unstructured program is one where little or no effort is given to helping people read and understand it. Remember, the structure of a program makes no difference to the computer, only to the people who have to work with it.

Program Blocks

You can think of a structured program as having certain parts of the program located at particular positions in the program document. These positions can be thought of as *blocks* of information. A letter contains blocks of information:

> John Student
> 123 Page Mill Road
> Programville, USA
> 01234
>
> The Resistor Company
> 321 Mill Page Road
> Sourcecode, USA
> 43210
>
> Dear Sir:
>
> Send me a 10 ohm resistor.
>
> Sincerely,
>
> John Student
>
> P.S. Thanks for the prompt delivery of my last order.

The above letter could just as well have been written as follows:

> John Student 123 Page Mill Road Programville, USA 01234
> The Resistor Company 321 Mill Page Road Sourcecode, USA 43210
> Dear Sir: Send me a 10 ohm resistor. Sincerely, John Student
> P.S. Thanks for the prompt delivery of my last order.

The difference between the two letters is that one is structured and the other isn't. The unstructured letter would take less time to write, less time to print, and use less paper. Over a period of time, if all your letters were done this way, you might even save some postage. But they would be difficult to read, not because of the writing style but because of their structure.

The same letter is shown again below, but this time its block structure is emphasized:

> John Student
> 123 Page Mill Road
> Programville, USA
> 01234

← Return address block.

> The Resistor Company
> 321 Mill Page Road
> Sourcecode, USA
> 43210

← Recipient address block.

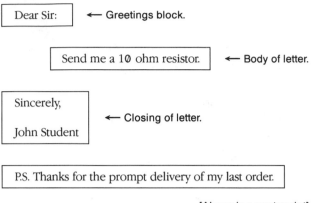

[Above is a post script]

The block structure of the letter makes the letter easy to read. In a similar fashion, when block structure is used in programs, they too are easier to read. Just as there is an agreed-upon block structure for letters, in this text, there will be an agreed-upon block structure for writing your C programs.

The Programmer's Block

All C programs will start with a block called the **programmer's block**. It consists of **remarks** that contain the following information:

1. Program name
2. Developer and date
3. Description of the program
4. Explain all variables
5. Explain all constants

You already know enough about C to do this now. Suppose you need to develop a C program that would solve for the voltage drop across a 10 ohm resistor with a specified amount of current. The formula for this relationship is

$$V = I \times R$$

Where

$V =$ The voltage across the resistor measured in volts.
$I =$ The current in the resistor measured in amps.
$R =$ The value of the resistor measured in ohms.

The C program would start with the programmer's block. This is illustrated in Program 1–4.

Program 1–4

```
#include <stdio.h>

/***************************************************/
/*                 Voltage Solver                  */
/***************************************************/
/*          Developed by: A. Good Structure        */
/*                  September 1993                  */
/***************************************************/
/*     This program will solve for the voltage drop */
/* across a 10 ohm resistor. The program user must  */
/* enter the value of current.                       */
/***************************************************/
/*                 Variables used:                  */
/* _____ */
/*          V = Voltage across resistor.            */
/*          I = Current in resistor.                */
/***************************************************/
/*                 Constants used:                  */
/* ------------------------------------------------ */
/*          R = 10 (A 10 ohm resistor)              */
/***************************************************/

main( )

{
  /* Body of the program to do the above. */
}
```

The above program will compile and execute, but nothing will happen because no commands were put into the program. The program only consists of remarks and the essential elements of a C program. But the programmer's block is complete. It tells you exactly what the program will do, and, just as important, it defines all of the variables (V and I) and the value of the constant R that will be used in the program. When doing **top-down** design (presented later in this chapter), stating the programming problem in words is the essential first step in programming design. Selecting the variables and defining them is another important step. For now, just know what must be included in a programmer's block and know the mechanics of how to write such a block using C.

Advantages and Disadvantages of Structured Programming

The main disadvantage of structured programming affects primarily those who have been programming in an unstructured way. Old habits die hard. If you've never programmed before, then structured programming will not have any disadvantages for you. The only other disadvantage of structured programming is that it makes short unstructured programs longer.

In the past, **computer memory** was relatively expensive and limited. Thus, at one time it did pay to conserve computer memory space by making programs as brief as possible. Today, this is no longer the case. Even pocket computers have more memory than many of the older large machines. Thus, now there is no necessity to be brief in programming. There is, however, the necessity to be clear in programming and to develop good programming habits that result in completed programs that are easy to understand, easy to modify, and easy to correct. That's the purpose of this book.

What you learn in this text can be applied to any structured language such as Pascal and even BASIC. When you complete this text you will be proficient in creating programs using structured programming techniques.

Figure 1–2 illustrates the structure of a complete C program using ANSI prototyping standards. The programming details will be covered in the following chapters. For now, look through the program to get the general idea.

Conclusion

This section presented a general idea of the difference between structured and unstructured programming. You will learn much more about these areas, including the power of C as a programming language. Test your new skills in the following section review.

1–3 Section Review

1   Define the term structure as used in programming.
2   Is it necessary to give a structure to C in order for it to compile without errors?
3   Describe a block structure and give an example.
4   Explain the reasons for structured programming.
5   Describe a programmer's block and state what it must contain.

1–4   Elements of C

This section lays the ground rules for the fundamental elements of all C programs. In this section you will see many new definitions. These definitions will set the stage for the remainder of this chapter as well as the remainder of this text.

What C Needs

To write a program in C, you use a set of characters. This set includes the uppercase and lowercase letters of the English alphabet, the ten decimal digits of the Arabic

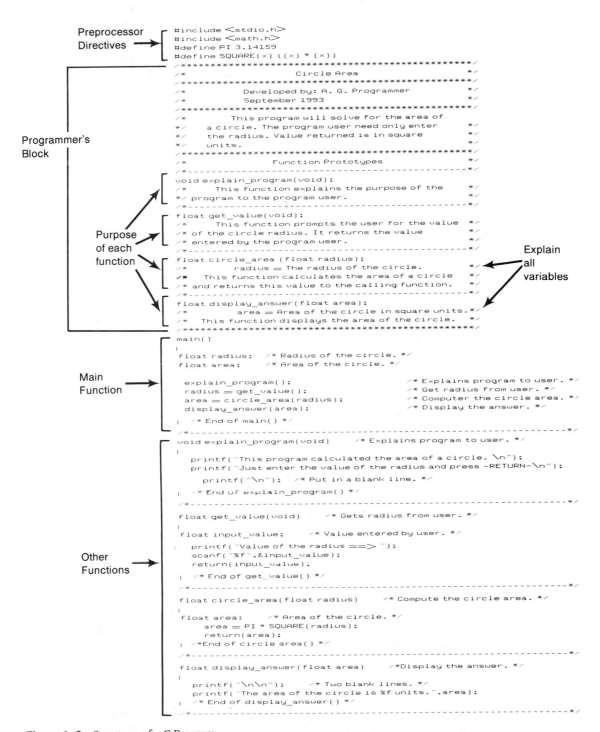

```
                        #include <stdio.h>
Preprocessor            #include <math.h>
Directives              #define PI 3.14159
                        #define SQUARE(x) ((x) * (x))
                        /*****************************************************/
                        /*                    Circle Area                  */
                        /*****************************************************/
                        /*          Developed by: A. G. Programmer         */
                        /*                September 1993                   */
                        /*****************************************************/
                        /*        This program will solve for the area of  */
                        */      a circle. The program user need only enter  */
                        */      the radius. Value returned is in square     */
                        */      units.                                      */
                        /*****************************************************/
                        /*                Function Prototypes              */
                        /*-------------------------------------------------*/
                        void explain_program(void);
                        /*      This function explains the purpose of the  */
                        */  program to the program user.                   */
                        /*-------------------------------------------------*/
                        float get_value(void);
                        /*      This function prompts the user for the value */
                        /* of the circle radius. It returns the value       */
                        /* entered by the program user.                     */
                        /*-------------------------------------------------*/
                        float circle_area (float radius);
                        /*      radius = The radius of the circle.         */
                        /*   This function calculates the area of a circle  */
                        /* and returns this value to the calling function.  */
                        /*-------------------------------------------------*/
                        float display_answer(float area);
                        /*      area = Area of the circle in square units.*/
                        /*   This function displays the area of the circle. */
                        /*****************************************************/
                        main()
                        {
                         float radius;   /* Radius of the circle. */
                         float area;     /* Area of the circle. */

                          explain_program();                /* Explains program to user. */
                          radius = get_value();             /* Get radius from user. */
                          area = circle_area(radius);       /* Computer the circle area. */
                          display_answer(area);             /* Display the answer. */
                        }  /* End of main() */
                        /*-------------------------------------------------------------*/
                        void explain_program(void)    /* Explains program to user. */
                        {
                          printf("This program calculated the area of a circle. \n");
                          printf("Just enter the value of the radius and press -RETURN-\n");
                            printf("\n");   /* Put in a blank line. */
                        }  /* End of explain_program() */
                        /*-------------------------------------------------------------*/
                        float get_value(void)     /* Gets radius from user. */
                        {
                         float input_value;       /* Value entered by user. */
                          printf("Value of the radius ==> ");
                          scanf("%f",&input_value);
                          return(input_value);
                        }  /* End of get_value() */
                        /*-------------------------------------------------------------*/
                        float circle_area(float radius)     /* Compute the circle area. */
                        {
                         float area;       /* Area of the circle. */
                            area = PI * SQUARE(radius);
                            return(area);
                        }  /*End of circle area() */
                        /*-------------------------------------------------------------*/
                        float display_answer(float area)     /*Display the answer. */
                        {
                          printf("\n\n");     /* Two blank lines. */
                          printf("The area of the circle is %f units.",area);
                        }  /* End of display_answer() */
                        /*-------------------------------------------------------------*/
```

Labels: Programmer's Block; Purpose of each function; Explain all variables; Main Function; Other Functions

Figure 1-2   Structure of a C Program

number system, and the underscore (_) character. Whitespace characters (such as the spaces between words) are used to separate the items in a C program, much the same as they are used to separate words in this book. These whitespace characters also include the tab and carriage return, as well as other control characters that produce white spaces.

Tokens   In every C source program, the most basic element recognized by the compiler is a single or group of characters known as a **token**. Essentially, a token is source program text that the compiler will not break down any further—it is treated as a fundamental unit. As an example, in C **main** is a token; so is the required opening brace { as well as the plus sign (+).

ANSI C Keywords   **Keywords** are predefined tokens that have special meaning to the C compiler. Their definition cannot be changed; thus they cannot be used for anything else except for the intended action they have on the program in which they are used. These keywords are

auto	double	int	struct	break	else
long	switch	case	enum	register	typedef
char	extern	return	union	const	float
short	unsigned	continue	for	signed	void
default	goto	sizeof	volatile	do	if
static	while				

Types of Data   The C language allows three major types of data: numbers, characters, and strings. A **character** is any item from the set of characters used by C. A **string** is a combination of these characters.

Numbers Used by C   C uses a wide range of numbers. Numbers used by C fall into two general categories: **integer** (whole numbers) and **float** (numbers with decimal points). These two main categories can be further divided as shown in Table 1–3.

As you can see from Table 1–3, C offers a rich variety of **data types**. Generally speaking, the larger the value range of the data type, the more computer memory it takes to store it. As a general rule, you want to use the data type that conserves memory and still accomplishes the desired purpose. As an example, if you needed a data type for counting objects—such as the number of resistors for a parts order— the type **int** would probably do the job. However, if you were writing a program for as much precision as possible you may consider the **double** which can give you 15-digit accuracy.

As you may note from Table 1–3, some of the data types produce the same range of values (such as **int** and **short**). This is because on other computer systems, the **short** and **int** may actually have different ranges.

C Statements

A C **statement** controls the flow of the execution of a program. It consists of keywords, **expressions**, and other statements. In C, an expression is a combination of operands and operators that express a single value (such as **answer = 3 + 5 ;**).

Table 1-3 Subdivisions of C Data Types

Type Identifier	Meaning	Range of Values (IBM PC)
`char`	character	−128 to 127
`int`	integer	−32,768 to 32,767
`short`	short integer	−32,768 to 32,767
`long`	long integer	−2,147,483,648 to 2,147,483,647
`unsigned char`	unsigned character	0 to 255
`unsigned`	unsigned integer	0 to 65,535
`unsigned short`	unsigned short int	0 to 65,535
`unsigned long`	unsigned long int	0 to 4,294,967,295
`enum`	enumerated	0 to 65,535
`float`	floating point	3.4E +/− 38 (7 digits)
`double`	double floating point	1.7E +/− 308 (15 digits)
`long double`	long double floating point	1.7E +/− 308 (15 digits)

There are two types of statements in C. These are **single statements** and **compound statements**. A *compound statement* is delimited by braces ({ }) while a single statement ends with a semicolon (;). As you progress through this book, reference will be made to C statements.

Conclusion

This section presented the elements of a C program. The rest of this chapter will show you how to use them. Check your understanding of this section by trying the following section review.

1-4 Section Review

1 State what is used to write a C program.
2 Explain what is meant by a token in C.
3 What are the major data types used by C?
4 Explain what is meant by a keyword. Give an example.
5 What data type handles the largest number?

1-5   The printf( ) Function

This section introduces you to the powerful **printf( )** function used in C. This is actually a separate C function (like **main( )** is a function) and is contained in the standard library that comes with your C system.

What **printf( )** Does

The **printf( )** function is used to write information to **standard output** (normally your monitor screen). You saw the **printf( )** used in the last chapter. The structure of this function is:

printf(characters, strings, format specifiers);

      Characters are set off by single quotes (such as **'a'**), and strings are set off by double quotes (such as **"This is a string."**). A **format specifier** instructs the **printf( )** function how to convert, format, and print its **arguments**. For now, think of an argument as the actual values that are within the parentheses of the function. A format specifier begins with a percent (**%**) character. As an example,

printf("This is a C statement.");

when executed produces

This is a C statement.

With a format specifier and an argument:

printf("The number 92 in decimal is %d.",92);

when executed produces

The number 92 in decimal is 92.

Another way of producing the same output is

printf("The number %d in decimal is %d.",92,92);

      Table 1–4 lists the various type fields used by the **printf( )** function in format specifiers.

      Program 1–5 illustrates the use of the different **field** type format specifiers.

Program 1–5

```
#include <stdio.h>

main( )
{
  printf("The value 92 using field type d is %d. \n", 92);
  printf("The value 92 using field type i is %i. \n", 92);
  printf("The value 92 using field type u is %u. \n", 92);
  printf("The value 92 using field type o is %o. \n", 92);
  printf("The value 92 using field type x is %x. \n", 92);
  printf("The value 92 using field type X is %X. \n", 92);
  printf("The value 92.0 using field type f is %f. \n", 92.0);
```

(continued)

Program 1–5 *(continued)*

```
printf("The value 92.0 using field type e is %e. \n", 92.0);
printf("The value 92.0 using field type E is %E. \n", 92.0);
printf("The value 92.0 using field type g is %g. \n", 92.0);
printf("The value 92.0 using field type G is %G. \n", 92.0);
printf("The value 92 using field type c is %c. \n", 92);
printf("The character '9' using field type c is %c. \n", '9');
printf("The string 92 using field type s is %s. \n", "92");
}
```

Execution of this program results in the following output:

```
The value 92 using field type d is 92.
The value 92 using field type i is 92.
```

Table 1–4 Field Type Format Specifiers Used by **printf( )**

Character	Argument	Resulting Output
d	integer	Signed decimal integer.
i	integer	Signed decimal integer.
o	integer	Unsigned octal integer.
u	integer	Unsigned decimal integer.
x	integer	Unsigned hexadecimal integer using lowercase letters.
X	integer	Unsigned hexadecimal integer using upper-case letters.
f	floating point	Signed floating point number.
e	floating point	Signed floating point number using e notation.
E	floating point	Signed floating point number using E notation.
g	floating point	Signed decimal number in either e form or f form, whichever is shorter.
G	floating point	Signed decimal number in either E form or f form, whichever is shorter.
c	character	A single character.
s	string	Prints character strings.
%	none	Prints the % sign.

Note: Use an l prefix with %d, %u, %x, %o, to specify long integer (for example %ld).

```
The value 92 using field type u is 92.
The value 92 using field type o is 134.
The value 92 using field type x is 5c.
The value 92 using field type X is 5C.
The value 92.0 using field type f is 92.000000.
The value 92.0 using field type e is 9.20000e+01.
The value 92.0 using field type E is 9.20000E+01.
The value 92.0 using field type g is 92.
The value 92.0 using field type G is 92.
The value 92 using field type c is \.
The character '9' using field type c is 9.
The string 92 using field type s is 92.
```

For now, don't worry about the \ you see in each **printf( )** function. You will learn more about these in this chapter. What they do here is to cause a carriage return and a new line of output. Without them in the above program, all of the output would run together.

Note that a character argument used the single quote while the string argument used a double quote. Also note that the numerical value of the backslash (\) is 92. (Refer also to Appendix F which contains the ASCII Table.)

More Than One

You can use more than one format specifier in a **printf( )** function. However, you must have at least as many arguments as format specifiers; if not, the results will be unpredictable. You can have more arguments than format specifiers; however, the extra arguments will just be ignored.

An example of using more than one format specifier is

```
printf("A character is %c and a number is %d.",'a',53);
```

Note that the arguments are separated by commas.

Conclusion

This section introduced you to the **printf( )** function in C. You saw how to use this function to display strings, characters, and numbers. You also saw what a format specifier is and how to use various field type format specifiers. Check your understanding of this section by trying the following section review.

1–5 Section Review

1   State the purpose of the **printf( )** function.
2   How are characters distinguished from strings?
3   What is the purpose of a format specifier in the **printf( )** function?
4   What is an argument?
5   State the rule concerning the number of format specifiers and arguments.

1-6   Identifying Things

Discussion

In this section you will learn how to use words to identify specific parts of your C program. Here you will also get an introduction to functions. In this section you will see how to give names to strings, values, and parts of your program. Knowing how to do this will make your programming tasks in any programming language much easier.

What Is a Function?

A **function** is an independent collection of **declarations** and statements. A declaration states the relationship between the name and type of a variable or other functions. You will see more about this later in this chapter. What is important for now is that you realize that a function is usually designed to perform one task. All C programs must have at least one function called **main( )**. Dividing tasks into separate parts in a program makes programs easier to design, correct, understand, and modify.

What Needs Identification?

When you create a function, it's best to give it a descriptive name. As an example, a function that would calculate the total power dissipated in a resistor could be named

resistor_power( )

This is more descriptive than naming it:

function_1( )

These two examples both use **identifiers** to distinguish one function from the other. An identifier is nothing more than the name you give to parts of your C program. Identifiers can be used to name parts of a formula such as:

total = resistor_1 + resistor_2;

Identifiers can be used to assign a constant value that describes the value and can then be used in your program:

PI = 3.14159;
circle_area = PI * radius * radius;

As you can see from the above examples, identifiers play an important role in a C program.

Creating Your Own Identifiers

There are some rules to follow when making your own identifiers. First, all identifiers must start with a letter of the alphabet (uppercase or lowercase) or the underscore _.

The remainder of the identifier may use any arrangement of letters (uppercase or lowercase), digits (0 through 9), and the underscore—and that's it—no other characters are allowed. This means spaces are not allowed in identifiers. Most C systems will recognize the first 32 characters of an identifier. The following example illustrates.

EXAMPLE 1-1

Which of the following are legal C identifiers?

A. This _1   E. One for the road
B. _This1   F. The_Following:
C. 1_for_the_road   G. E_=_IR
D. One_for_the_road

Solution
Keep in mind the rules for identifiers explained prior to this example.

A. This_1   A legal C identifier.
B. _This1   A legal C identifier.
C. 1_for_the_road   Not a legal C identifier.
                       It does not start with a letter
                          of the alphabet or underscore.
D. One_for_the_road   A legal C identifier.
E. One for the road   Not a legal C identifier.
                       Spaces are not allowed.
F. The_Following: Not a legal C identifier.
                       The : symbol is not allowed.
G. E_=_IR   Not a legal C identifier.
                       The = sign is not allowed.

Case Sensitivity

Identifiers in C are case sensitive. This means that C makes a distinction between uppercase and lowercase letters in an identifier. Thus, as far as C is concerned, the following identifiers are not equal:

```
pi    PI    Pi    pI
```

Neither are the following:

```
This_One    THIS_ONE    this_one
```

All of the above identifiers are legal in C; it is just that they are not equal. What this means is that you must be careful with your use of identifiers in C. As an example, if you define the identifier **pi** to be equal to 3.14159, you must use the lowercase letters **pi** anywhere in the program you expect the identifier to equal 3.14159. If you used **PI** or **Pi**, neither of which you assigned a value to, then the program would contain errors.

Keywords

An identifier cannot have the same spelling and case as a C keyword. The following example illustrates.

EXAMPLE 1-2

State which of the following identifiers are legal and which are reserved, and state if any are equal.

A. Ohms_Law   D. reconstruct
B. continue   E. Reconstruct
C. Ohms Law   F. _continue

Solution

Keep in mind the discussions presented prior to this example.

A. Ohms_Law   Legal C identifier, not reserved.
B. continue   Legal C reserved word.
C. Ohms Law   Not a legal C identifier.
                   Spaces are not allowed.
D. reconstruct   Legal C identifier, not reserved.
E. _continue   Legal C identifier, not reserved.
                   (However, it is so close to
                       looking like a reserved word
                       that this practice is discouraged.)

Conclusion

This section presented the important concepts for naming parts of your C program. You learned what an identifier is and how it can be used to name (identify) parts of your C program. Check your understanding of this section by trying the following section review.

1-6 Section Review

1 What is a C function?
2 What is meant by an identifier?
3 What are the rules for creating your own identifiers?
4 How many characters of an identifier are recognized by C?

1-7   Declaring Things

In this section you will learn that all variables must be declared. This means that you must let the compiler know ahead of time, before you use them, the identifier that will be used for each variable as well as the type of variable you will be using.

At first, this may seem like a lot of extra work. But you will find that by doing this you actually reduce the chances for program errors.

What Is a Variable?

You can think of a variable as a specific memory location set aside for a specific kind of data and given a name for easy reference. Essentially you use variables so that the same memory space can hold different values of the same type at different times. As an example, if you were calculating the voltage across a fixed value of resistance as the current changed, the voltage variable would have a different value each time the current changed.

Declaring Variables

In C, you must declare all variables before using them. To declare a variable, you must declare its type and identifier. Table 1–5 presents the fundamental **type specifiers** that will be used for variables.

Table 1–5   Fundamental C Type Specifiers

Integers	Floating Point	Other Types
char	double	const
enum	float	void
int	long double	volatile
long		
short		
signed		
unsigned		

Program 1–6 illustrates how to declare variables.

Program 1–6

```
#include <stdio.h>

main( )
{
  char a_character;      /* This declares a character. */
  int an_integer;        /* This declares an integer. */
  float floating_point;  /* This declares a floating point. */
```

(continued)

Program 1-6 *(continued)*

```
    a_character = 'a';
    an_integer = 15;
    floating_point = 27.62;

    printf("%c is the character.\n",a_character);
    printf("%d is the integer.\n",an_integer)
    printf("%f is the floating point.\n",floating_point);
}
```

When executed, Program 1-6 produces

```
a is the character.
15 is the integer.
27.620000 is the floating point.
```

In the program, the variable declarations are

```
char a_character;        /* This declares a character. */
int an_integer;          /* This declares an integer. */
float floating_point;    /* This declares a floating point. */
```

Observe that declaring a variable consists of stating its type and its identifier. In the body of the program, each variable was assigned a value:

```
a_character = 'a';
an_integer = 15;
floating_point = 27.62;
```

This was done with the **assignment operator** (=). You will learn more about this operator in the next section.

Initializing Variables

You can combine a variable declaration with the assignment operator thus giving the variable a value at the same time that it is declared. This is shown in Program 1-7.

Program 1-7

```
#include <stdio.h>

main( )
{
char a_character = 'a';         /* This declares/assigns a character. */
int an_integer = 15;            /* This declares/assigns an integer. */
float floating_point = 27.62;   /* This declares/assigns a floating point */
```

(continued)

Program 1-7 *(continued)*

```
    printf("%c is the character.\n",a_character);
    printf("%d is the integer.\n",an_integer);
    printf("%f is the floating point.\n",floating_point);

}
```

Note that in each case, the variable declaration is followed by a comment that states the purpose of each variable. It's good to get in the habit of doing this kind of program documentation.

More of the Same Type

In C, if you have variables of the same data type, you can declare them as shown:

int number_1, number_2, number_3;

Even though this is legal in C, this practice will not be used in this text because it discourages you from commenting on each variable used in the program.

Why Declare?

When you declare variables, you gather all the information about them at one place in the program. This allows everyone reading your source code to quickly identify the data that will be used (assuming that you add comments that give a good explanation). Doing this also forces you to do some planning before leaping into the program code. Another important reason for doing this is that it prevents you from misspelling a variable within the program. If there were no requirements to declare, then you could create a new identifier without knowing it. This could cause disastrous problems that are very difficult to find and correct. The more you program, the more complex and practical your programs are, the more thankful you will be that variables must be declared before you can use them.

1-7 Section Review

1   How can you think of a variable in terms of a C program?
2   Explain what it means to declare a variable.
3   Name three fundamental C type specifiers.
4   Explain what is meant by initializing a variable.
5   What is one good reason for declaring variables?

1-8   Introduction to C Operators

In this section you will discover how to cause a C program to act on variables. This means you will learn about arithmetic operations as well as other operations that are unique to C.

What Are C Operators?

A C operator causes the program to do something to variables. Specifically, an **arithmetic operator** allows an arithmetic operation (such as addition, $+$) to be performed on variables. This section will introduce you to the most commonly used C arithmetic operators.

Arithmetic Operators

The common arithmetic operators used by C are listed in Table 1-6.

Program 1-8 illustrates the use of the first four arithmetic operators in C.

Program 1-8

```
#include <stdio.h>

main( )
{
 float number_1 = 15.0;     /* First arithmetic operator. */
 float number_2 = 3.0;      /* Second arithmetic operator. */
 float addition_answer;     /* Answer to addition problem. */
 float subtraction_answer;/* Answer to subtraction problem. */
 float multi_answer;        /* Answer to multiplication problem.*/
 float division_answer;     /* Answer to division problem. */

 addition_answer = number_1 + number_2;
 subtraction_answer = number_1 - number_2;
 multi_answer = number_1 * number_2;
 division_answer = number_1/number_2;

 printf("15 + 3 = %f\n",addition_answer);
 printf("15 - 3 = %f\n",subtraction_answer);
 printf("15 X 3 = %f\n",multi_answer);
 printf("15/3 = %f\n",division_answer);

}
```

Table 1-6   Common Arithmetic Operators

Symbol	Meaning	Example
+	Addition	answer = 3 + 5 (answer → 8)
−	Subtraction	answer = 5 − 3 (answer → 2)
*	Multiplication	answer = 5 * 3 (answer → 15)
/	Division	answer = 10/2 (answer → 5)
%	Remainder	answer = 3/2 (answer → 1)

Execution of the above program produces

```
15 + 3 = 18.000000
15 − 3 = 12.000000
15 × 3 = 45.000000
15/3 = 5.000000
```

Note that all of the numbers used in Program 1-8 were of type **float**. It's important to note that in C, division of type **int** variables will truncate the answer. This means that $5/2 = 2$ and that $2/3 = 0$. The remainder operator (%) requires the use of whole numbers. This operator is also called the *modulo* operator.

Important Considerations

C does not allow a reverse type of assignment structure. This means that you cannot have an assignment to an expression:

```
6 + 3 = answer;    ← Not allowed!
```

Nor can you have an assignment to a constant.

```
3 = answer;    ← Not allowed!
```

Both of the above attempts will produce compile time errors.

Precedence of Operations

Precedence of operations is simply the order in which arithmetic operations are performed. As an example, consider the expression:

```
X = 5 + 4/2;
```

Precedence of operations requires that division be done before addition. If this were not the case, then the above operation could be interpreted two different ways. If the division is performed first it would yield $(5 + 2 = 7)$. If addition is performed first 5 is added to 4 and then divided by 2 $(9/2 = 4.5)$. The interpretation of any expression must be consistent for reliable and predictable program results. For example, if you want to indicate that 5 is to be added to 4 first and the result then divided by 2, parentheses must be used.

```
X = (5 + 4)/2
```

Table 1-7 indicates the precedence of operations for C.

Table 1-7   Precedence of Operations

Priority	Operation
First	( )
Second	Negation (assigning a negative number)
Third	Multiplication *, Division /
Fourth	Addition +, Subtraction −

In all cases operations proceed from left to right. The following example illustrates.

EXAMPLE 1-3

Determine the results of the following operations:

A. $Y = 6 + 12/6 * 2 - 1$ C. $Y = 6 + 12/6 * (2 - 1)$
B. $Y = (6 + 12)/6 * 2 - 1$ D. $Y = 6 + 12/(6 * 2) - 1$

Solution
Using precedence of operations, the results are:

A. $Y = 6 + 12/6 * 2 - 1$
   $Y = 6 + 2 * 2 - 1$ (Operations from left to right. Division precedence.)
   $Y = 6 + 4 - 1$ (Multiplication precedence.)
   $Y = 9$ (Operations from left to right.)
B. $Y = (6 + 12)/6 * 2 - 1$
   $Y = 18/6 * 2 - 1$ (Work inside parentheses done first.)
   $Y = 3 * 2 - 1$ (Operations from left to right.)
   $Y = 6 - 1$ (Multiplication precedence.)
   $Y = 5$
C. $Y = 6 + 12/6 * (2 - 1)$
   $Y = 6 + 12/6 * 1$ (Work inside parentheses done first.)
   $Y = 6 + 2 * 1$ (Operations from left to right.)
   $Y = 6 + 2$ (Multiplication precedence.)
   $Y = 8$
D. $Y = 6 + 12/(6 * 2) - 1$
   $Y = 6 + 12/12 - 1$ (Work inside parentheses done first.)
   $Y = 6 + 1 - 1$ (Division precedence.)
   $Y = 6$ (Operations from left to right.)

Compound Assignment Operators

In C, **compound assignment operators** combine the simple assignment operator with another operator. For example, look at the following statement:

```
answer ⁻ answer + 5;
```

What the above means is that the memory location called answer will be assigned the new value of its old value plus 5. It does *not* mean "answer equals answer plus 5." The = sign does *not* mean equal in C; it means *assigned to* and is called the assignment operator. Thus, in the above example, if **answer** had the value of 10, then

```
answer = answer + 5;
```

would cause the new value of **answer** to be 15.

The above expression can be shortened in C by using the compound assignment:

```
answer += 5;
```

Table 1–8 lists the compound assignments of the arithmetic operators presented at the beginning of this section.

Table 1–8   Common Compound Assignments in C

Symbol	Example	Meaning
+=	X += Y;	X = X + Y;
−=	X −= Y;	X = X − Y;
*=	X *= Y;	X = X * Y;
/=	X /= Y;	X = X / Y;
%=	X %= Y;	X = X % Y;

Use of the compound assignment operators is illustrated in Program 1–9.

Program 1–9

```c
#include<stdio.h>

main( )
{
 int number = 10;     /* Value of number for example. */

 number += 5;
 printf("Value of number += 5 is %d\n",number);

 number −= 3;
 printf("Value of number −= 3 is %d\n",number);

 number *= 3;
 printf("Value of number *= 3 is %d\n",number);

 number /= 5;
 printf("Value of number /= 5 is %d\n",number);

 number %= 3;
 printf("Value of number %%= 3 is %d",number);

}
```

Execution of the above program produces

```
Value of number += 5 is 15
Value of number −= 3 is 12
Value of number *= 3 is 36
```

```
Value of number /= 5 is 7
Value of number %= 3 is 1
```

Observe the last **printf( )** function in the program. In order to get a printout of the **%** sign, a double **%%** was used.

Conclusion

This section presented the basic arithmetic operations in C. Here you learned the meaning of an assignment operator as well as compound assignments and order of precedence. Check your understanding of this section by trying the following section review.

1-8 Section Review

1   List the common arithmetic operators used in C.
2   In C what does integer division do to the remainder? What is the significance of this?
3   In C is the following allowed: 3 − 2 = **result**;? Explain.
4   What is meant by precedence of operation?
5   Give an example of a compound assignment used in C.

1-9   More printf( )

This section presents some more details about the **printf( )** function. You have already been using the **\n** as a part of this function and were briefly told that this causes a carriage return and a new line. In this section you will learn more about the power of the **printf( )** function.

Escape Sequences

The **\n** is an example of an **escape sequence** that can be used by the **printf( )** function. The backslash symbol (\) is referred to as the escape character. You can think of an escape sequence used in the **printf( )** function as an escape from the normal interpretation of a string. This means that the next character used after the \ will have a special meaning as listed in Table 1-9.

The Interactive Exercises for this chapter will give you some practice in using the **printf( )** escape sequences in your system.

Field Width Specifiers

The **printf( )** function allows you to format your output. Recall from the previous programs that when you printed the output of a type **float** it would appear as: 16.000000; even though you didn't need the six trailing zeros, they were still printed.

Table 1-9   Escape Sequences

Sequence	Meaning
\n	Newline
\t	Tab
\b	Backspace
\r	Carriage return
\f	Formfeed
\'	Single quote
\"	Double quote
\\	Backslash
\xdd	ASCII code in hexadecimal
\ddd	ASCII code in octal

Note: The double quote and the backslash can be printed by preceding them with the backslash.

The **printf( )** function provides **field width specifiers** so that you can control how printed values will appear on the monitor. The syntax is:

%W.DF

Where

% = Indicates the format specification.
W = The width of the field.
D = Number of digits to the right of the decimal.
F = The format specifier.

As an example, the statement

printf("The number %5.2f uses them.",6.0);

would display

The number   6.00 uses them.

Note that the width first indicates how the number is justified (five spaces), and then the number of digits following the decimal point are specified.

Conclusion

This section presented some more important details about the **printf( )** function. Here you saw the other escape sequences that could be used with the **printf( )** as well as the format specifiers. Test your understanding of this section by trying the following section review.

1 Explain the use of an escape sequence in the **printf( )** function.
2 What is the backslash character (\) sometimes called when used in a **printf( )** function?
3 Give three escape sequences that are used in the **printf( )** function.
4 What is a field width specifier as used by the **printf( )** function?

1-10 Getting User Input

Discussion

The real power of a technical C program is its ability to interact with the program user. This means that the program user gets to input values of variables. As you might guess, there is a built-in C function that lets you make this happen.

The scanf( ) Function

The **scanf( )** function is a built-in C function that allows your program to get user input from the keyboard. You can think of it as doing the opposite of the **printf( )** function. Its use is illustrated in the following program.

Program 1-10

```
#include <stdio.h>

/*              Getting user input              */
 main( )
 {

   float value;        /* A number inputted by the program user. */

   printf("Input a number => ");
   scanf("%f", &value);

   printf("The value is => %f", value);
   }
```

When the above program is executed, the output will appear as follows (assuming the program user inputs the value of 23):

Input a number => 23
The value is => 23.000000

Note that the **scanf( )** function has a similar format to that of the **printf( )**. First, it contains the **%f**, enclosed in quotation marks. This tells the program that a value will be entered which will be a floating point type. Next, it indicates the variable identifier where this value will be stored. It indicates this by using a comma outside the quotes and then an **&** (ampersand sign) immediately followed by the name of the variable identifier (**&value**). Now the value that the user inputs will be the value of the variable **value**. Also note that the result of using the **scanf( )** function is a carriage return.

Format Specifiers

The format specifiers for the **scanf( )** function are similar to those of the **printf( )** function. This is illustrated in Table 1–10.

It should be noted that either the **%f** or the **%e** format specifier may be used for accepting either exponential or decimal notation.

The **scanf( )** function can accept more than one input with just one statement as shown below:

```
scanf("%f %d %c",&number1, &number2, &character);
```

In the above case, the variable **number1** will accept a type **float**, the variable **number2** a type **int**, and **character** a type **char**. In this case, the program user would have to type in three separate values separated by a space. As an example:

```
52.7 18 t
```

Since it is easy for the program user to make errors entering data in this manner, multiple inputs with the **scanf( )** function will not be used in the text.

A Real Technology Problem

You now know how to input data, do a basic computation, and display the results. Program 1–11 solves for the voltage across a resistor when the value of the current and resistance are known. The mathematical relationship is:

$$voltage = current \times resistance$$

Table 1–10 scanf( ) Format Specifiers

Specifier	Meaning
%c	A single character.
%d	Signed decimal integer.
%e	Exponential notation.
%f	Floating point notation.
%o	Unsigned octal integer.
%u	Unsigned decimal integer.
%x	Unsigned hexadecimal integer.

Program 1-11

```
#include <stdio.h>
/*              Ohms Law              */

  main( )
  {
   float voltage;        /* Value of the voltage. */
   float current;        /* Value of the current. */
   float resistance;     /* Value of the resistor. */

   printf("Input the current in amps => ");
   scanf("%f", &current);

   printf("Input the resistance in ohms => ");
   scanf("%f", &resistance);

      voltage = current * resistance;     /* Compute the voltage. */

    printf("The value of the voltage is %f volts. ", voltage);

  }
```

Assume that the program user will enter the value of 3 for the current and 4 for the resistor value. Then execution of the above program would yield

```
Input the current in amps => 3
Input the resistance in ohms => 4
The value of the voltage is 12.000000 volts.
```

There are several key points to note concerning this program:

- All variables were declared, and a comment was made about each one.
- Each **scanf( )** function used **%f** to indicate that the input would be a floating point and the **&variable** to indicate which variable would store the user input value.
- The actual calculation was commented.
- The **printf( )** function used the **%f** to indicate that the numeric output would be floating point and the variable identifier whose value was to be displayed was indicated at the end of the quotes (**",voltage);**.

What you just observed was a fundamental problem in technology solved by the C language. This was a very simple problem that you could have easily solved on your pocket calculator. But, for now the point is to keep the problems simple so that they don't get in the way of understanding the C language. As you progress through the text, your programs will become much more powerful.

Using the E Notation

Some mention should be made of using **E (exponential) notation** with C. As pointed out earlier, C will accept E notation numbers for floating point types. You can do this for input as well as output. For example, in the last program, the program user could have inputted the more practical values of 0.003 amps for the current and 2000 ohms for the resistance. This could have been done using E notation:

```
Input the current in amps => 3E-3
Input the resistance in ohms => 2E3
The value of the voltage is 6.000000 volts.
```

It should be pointed out that C will accept either an uppercase E or a lowercase e for this kind of data representation.

Conclusion

This section brought you to the point where you can now develop some very basic technology programs using the C language. In the next section, you will learn how to perform more arithmetic functions using the C language. This will be an important step enabling you to handle almost any type of technology formula. Check your understanding of this section by trying the following section review.

1–10 Section Review

1 State the purpose of the **scanf( )** function.
2 How does the **scanf( )** function know what variable identifier to use for inputting the user data?
3 Does the the result of using the **scanf( )** function produce a carriage return to a new line?
4 How may floating point values be entered by the program user in a C program?
5 State what you must do in order to have values outputted to the screen in E notation.

1–11   Program Debugging and Implementation: Common Programming Errors

Discussion

The material in this section is designed to help you minimize some of the most common errors encountered by beginning C programmers.

Note: If you have the LEARN C option, also read Appendix D, The LEARN C Debugger.

Types of Error Messages

Depending upon the type of compiler you are using, you will receive different kinds of error messages. Some compilers, such as the Microsoft LEARN C ®, produce three types of error messages:

1. Fatal error messages
2. Compilation error messages
3. Warning messages

A *fatal error* message will immediately terminate the compiling process. It will not check for any further errors and stops with a message display of what may be the most likely problem. Most C systems will move the cursor to the line of source code where it "thinks" the error is located. You may find, depending on the nature of the error, that where the cursor is placed is not the actual location of the error. You must remember that no error location scheme is perfect, and not all possible errors are predictable. Your compiler is giving you its best guess of what is causing the error. It's up to you to determine exactly where the error is located and to correct it accordingly.

A *compilation error message* is a result of a less severe error. In these cases, the compiler will try to keep compiling and produce other compilation error messages if it encounters more errors of this type. In some cases, it may not be able to continue the compilation process and a fatal error will result. In any case, no object code is produced, and you will wind up with a list of error messages. You can then use this list to help determine where the problems are in your source code.

Warning messages are the results of programming errors that will allow your program to compile and link, but your program will not be executed. These warning messages are listed for you as a reference to use in determining the problems with your source code.

Case Sensitivity

Recall that an important aspect of C programming is that its identifiers are case sensitive. This means that C does make a distinction between uppercase and lowercase letters. The following program will not execute because of the uppercase M used in the function **main( )**.

Program 1–12

```
/*      Case sensitivity example.        */
{
   Main( )

   /*  This program will not execute!  */
}
```

Case sensitivity applies to all identifiers. As an example, the following program will not execute because the programmer capitalized the first letter of the identifiers when they were declared but forgot the capitalization when they were used in the program.

Program 1-13

```
/*      Another case of case sensitivity.      */
 main( )
{
 float This_One; = 2; /* Program constant.*/
 float That_One;       /* Program variable. */

    That_one = 2 * This_one;
    /*     This program will not execute!    */
}
```

There are two bugs in the above program, both of the same type. The declared variables `This_One` and `That_One` use two capital letters each. However, in the body of the program, the programmer forgot to use a capital O:

```
That_one = 2 * This_one;
```

Therefore, the compiler will think these are new identifiers which have not been declared!

The Semicolon

Another error common to beginning C programmers is the omission of the semicolon. In C, the semicolon identifies the end of a statement or instruction. The semicolon is actually a part of the C statement and must be included. A missing semicolon will always prevent your program from executing. However, a missing semicolon can really confuse the compiler as to what is causing the problem. You may get error messages from missing semicolons that make no mention of them. A good rule to remember is that if the warning message doesn't make sense, check first for missing semicolons.

As an example, both of the following programs have missing semicolons; however, each produces different error messages (the kind you get will depend upon your system). In both cases, the error messages will not recognize that the problem is a missing semicolon.

Program 1-14

```
/*      Missing semicolon.      */
 main( )
{
 float    This_One = 2     /* Program constant.*/
 float    That_One;        /* Program variable. */
    That_One = 2 * This_One;
   /*      This program will not execute!     */
}
```

Program 1-15

```
/*      Another missing semicolon.     */
 main( )
{
 float    This_One = 2;    /* Program constant.*/
 float    That_One         /* Program variable. */
    That_One = 2 * This_One;
   /*      This program will not execute!     */
}
```

Note that Program 1-14 has a semicolon missing from the constant declaration, while Program 1-15 has a semicolon missing from the variable declaration.

The best rule to follow is to always make a good visual inspection of your C program before attempting to compile it. Look for the missing semicolons used to terminate a C statement.

Incomplete or Nested Comments

Another area for program bugs is in the use of comments. These usually are one of two types. The first is to produce the starting comment delimiter /* but forget to include the ending one. This error is shown in the following program.

Program 1-16

```
/*      Omitting comment delimiters.    */
main( )
{
 float constant_1 = 2; /* Program constant.   */
 float variable_1;     /* Program variable.
       variable_1 = constant_1 + constant_1;
}
```

The above program will not execute because of the missing ending delimiter */ for the comment: /* Program variable. However, this so confuses the compiler that when a warning message is displayed, it does not indicate that the problem is with the comment. This brings up another good rule. When you get an error message you don't understand, after checking to ensure all statements have their required semicolons, then check to make sure each ending comment delimiter is present.

The next program gives an example of nested comments.

Program 1-17

```
           /*      Nested comments      */
    main( )
    {
    float constant_1 = 2; /* Program constant. */
    float variable_1;      /* Program variable.  */
     /* This is a comment...
    variable_1 = constant_1 + constant_1 /* Another comment...*/
        end of nested comment*/
    }
```

Normally, the above program will not execute because one comment is contained inside the other. Some compilers do allow you to set the operating environment so that you can nest comments. However, this practice is not recommended and will not be used in this book.

The following example gives you practice using your visual inspection skills to spot some common program bugs.

EXAMPLE 1-4

Determine if there is a bug in any of the following three programs that will prevent execution of the program. If so, state what you would do to correct them.

Program 1-18

```
    /*    Is there a bug here?        */
    main( )
    {
        {
        float This_Value = 15;  /* Program constant.  */
        float That_Value;       /* Program variable.  */
        That_Value = this_Value + this_value;
        }
    }
```

Program 1–19

```
/*    Any bugs here?      */
  main( )
    {
    float this_value = 15;    /* Program constant.  */
    float that_value;         /* Program variable. */
     that_value = this_value + this_value
     }
```

Program 1–20

```
/*    Any bugs here?      */
  main( )
  { float this_value = 15;      /* Program constant. */
    float that_value;           /* Program variable. */

    that_value = this_value + this_value;  }
```

Solution

Always perform a good visual observation of your source code before attempting to compile it. Doing this will train you to catch your own mistakes quickly. The ability to find your errors comes with a great deal of practice.

Program 1–18

The case of the declared constant identifier was changed.

```
That_Value = this_Value + this_value;
```

Program 1–19

Missing semicolon at end of a C statement.

```
that_value = this_value + this_value
```

Program 1–20

There are no bugs in this program. However, the style of having the opening and closing braces on the same line with program code is discouraged because they are not as easy to see. Good programming practice in C devotes a whole line for each one.

Conclusion

This section presented some of the most common program bugs encountered by beginning C programmers. Here you saw that good programming practice requires that you carefully check your programs for required semicolons, case sensitivity, and comment delimiters. Check your understanding of this section by trying the following section review.

1–11 Section Review

1   State the three types of error messages found in C.
2   Which type of error message will terminate the compilation process?
3   Explain how case sensitivity will cause your C program not to execute.
4   Why is it helpful to look for missing semicolons if you do not understand how an error message applies to your program?
5   What is meant by a nested comment? Is this legal?

1–12   Case Study

Discussion

The case study for this chapter is to design a program that converts a temperature reading from degrees Fahrenheit to degrees centigrade. This is an easy conversion to perform on most pocket calculators. It is used here to illustrate some of the fundamental elements of program design. The problem is intentionally kept simple so as not to dominate the programming development; yet the problem is rigorous enough to require most of the new material presented in this chapter.

First Step—Stating the Problem

The first, most important step in program design is to state the problem in writing. This problem statement must include some specifics about the requirements of the computer program itself. The specifics are:

- Purpose of the Program
- Required Input (source)
- Process on Input
- Required Output (destination)

Stating the case study in writing gives:

- Purpose of the Program: Convert a temperature reading from degrees Fahrenheit to degrees centigrade.
- Required Input: Temperature in degrees Fahrenheit. (Source: Keyboard)
- Process on Input: Temperature centigrade = 5/9 × (temperature Fahrenheit − 32).
- Required Output: Temperature in degrees centigrade.

From the above four areas you (and others who may be working with you) clearly know the purpose of the program. It is evident that the information for the program will be received from the keyboard and that it will consist of a single temperature reading in degrees Fahrenheit. The program will then process this input according to a very specific formula. The output will simply be displayed on the screen. It will be the temperature expressed in degrees centigrade.

For this simple problem, this information may seem obvious. However, for more complex programs where some information will be received from the keyboard while other information may be received from the disk, the processes may not be so obvious. Stating them in writing in this manner ensures that you and the person you are designing the program for are in agreement as to what is to be done, and just as important, what is not to be done.

Blocking Out the Program

Once you have completed the problem statement, the next step is to comment out your C program. This is illustrated in the following program.

Program 1–21

```
/*****************************************/
/*         Fahrenheit to centigrade      */
/*****************************************/
/*         Developed by: Able Programmer */
/*             September 15, 1992         */
/*****************************************/
/*    This program converts a temperature */
/* reading in degrees Fahrenheit to its   */
/* value in degrees centigrade.           */
/*****************************************/

main( )
{
/*    Explain program to user.    */

/*    Get Fahrenheit value from user.    */

/*    Do computations.    */

/*    Display the answer.    */

}
```

The above program outline gives the program a title and tells who the programmer is, the date of the program, and a description of what the program does. The rest is an outline of the major sections of the program. This will act as your guide as you develop the required code. Most important, your outline will actually become a part of your program, the comments for each of the major parts of your program. Once the outline is completed, it is saved to the disk, and a printed copy is made of it. This printed copy can verify that it fits the original intent of the program and serve as a source of documentation.

Note that the essential **main( )** and { and } are included so that the program will compile. The compiling should be done at this point to ensure that the commented sections do not contain any bugs (such as nested comments as explained in the last section).

The Next Step

Once you have the program outlined, the next step is to develop each section, one section at a time. This process reduces the amount of program debugging time. For this case study, the first step is to code the program explanation, compile it, and observe the output. This is shown in the following program. A new C function is introduced in this program called the **puts( )**. This function produces an automatic carriage return to a new line. It is a convenient function to use for just displaying text.

Program 1–22

```
#include <stdio.h>
    /*****************************************/
    /*        Fahrenheit to centigrade       */
    /*****************************************/
    /*        Developed by: Able Programmer  */
    /*           September 15, 1992          */
    /*****************************************/
    /*    This program converts a temperature */
    /* reading in degrees Fahrenheit to its   */
    /* value in degrees centigrade.           */
    /*****************************************/

    main( )
    {
    /*    Explain program to user.    */

     puts("");
     puts("This program will convert a temperature reading in");
     puts("degrees Fahrenheit to its equivalent temperature");
     puts("in degrees centigrade.");

     puts(" ");
     puts("You only need to enter the temperature in Fahrenheit");
     puts("and the program will do the rest.");

    /*    Get Fahrenheit value from user.    */

    /*    Do computations.    */

    /*    Display the answer.    */

    }
```

When executed, the above program will display

```
This program will convert a temperature reading in
degrees Fahrenheit to its equivalent temperature
in degrees centigrade.
```

```
You only need to enter the temperature in Fahrenheit
and the program will do the rest.
```

The **puts** requires the **#include** <**stdio.h**> added to the beginning of the program.

The point here is that the program is being developed in a step-by-step manner. This reduces the potential frustration of trying to discover program bugs when developing a large program. If there were a bug at this early stage, the programmer knows that it is confined to this part of the program. Once the bugs are worked out here, then the next section is developed, compiled, and executed. Again, if there are bugs they too must be confined to the most recent code entered (the exception to this is if the most recent code interacts with prior code—still, this method is the preferred method).

Completing the Study

The next step, of course, is to code and compile the next part of the program outline.

Program 1–23

```c
#include <stdio.h>

        /********************************************/
        /*        Fahrenheit to centigrade          */
        /********************************************/
        /*        Developed by: Able Programmer     */
        /*            September 15, 1992            */
        /********************************************/
        /*    This program converts a temperature   */
        /* reading in degrees Fahrenheit to its     */
        /* value in degrees centigrade.             */
        /********************************************/

        main( )
        {
        /*    Explain program to user.    */

         puts("");
         puts("This program will convert a temperature reading in");
         puts("degrees Fahrenheit to its equivalent temperature");
         puts("in degrees centigrade.");

         puts(" ");
         puts("You only need to enter the temperature in Fahrenheit");
         puts("and the program will do the rest.");

        /*    Get Fahrenheit value from user.    */
```

(continued)

Program 1-23 *(continued)*

```
    puts("");
    printf("Enter temperature value in Fahrenheit => ");
    scanf("%f'',&fahrenheit_temp);

    /*   Do computations.   */
    /*   Display the answer.   */

}
```

Execution of the above program yields

```
This program will convert a temperature reading in
degrees Fahrenheit to its equivalent temperature
in degrees centigrade.

You only need to enter the temperature in Fahrenheit
and the program will do the rest.

Enter the temperature value in Fahrenheit =>
```

If you enter a value at this point in the development of the program, no further processing will take place. But again, you have ensured that there are no compile time bugs up to this point.

Next, the processing part of the program is developed. Note the declaration of the two temperature variables.

Program 1-24

```
#include <stdio.h>

/******************************************/
/*        Fahrenheit to centigrade        */
/******************************************/
/*        Developed by: Able Programmer    */
/*            September 15, 1992            */
/******************************************/
/*    This program converts a temperature  */
/* reading in degrees Fahrenheit to its     */
/* value in degrees centigrade.             */
/******************************************/

main( )
{

float   fahrenheit_temp;     /* Temperature in degrees Fahrenheit */
float   centigrade_temp;     /* Temperature in degrees centigrade */

/*    Explain program to user.    */
```

(continued)

Program 1–24 *(continued)*

```
        puts("");
        puts("This program will convert a temperature reading in");
        puts("degrees Fahrenheit to its equivalent temperature");
        puts("in degrees centigrade.");

        puts(" ");
        puts("You only need to enter the temperature in Fahrenheit");
        puts("and the program will do the rest.");

    /*    Get Fahrenheit value from user.    */

        puts("");
        printf("Enter temperature value in Fahrenheit => ");
        scanf("%f",&fahrenheit_temp);

    /*    Do computations.    */

        centigrade_temp = 5/9 * (fahrenheit_temp - 32);

    /*    Display the answer.    */

    }
```

Note an important addition to this program. Since variables had to be used, they first had to be declared. Thus, the declaration block had to be added to the program so that the type of each variable could be stated. As before, these variables will be of type **float**. Each variable is commented as is required by good programming practice. Note that the * sign is used in the formula to indicate that multiplication is taking place.

Note: An error has been introduced into the program. You will be shown where it is shortly. Can you spot the error now?

The last programming step is to cause the results to be displayed.

Program 1–25

```
#include <stdio.h>

    /******************************************/
    /*       Fahrenheit to centigrade         */
    /******************************************/
    /*       Developed by: Able Programmer    */
    /*            September 15, 1992           */
    /******************************************/
    /*   This program converts a temperature  */
    /* reading in degrees Fahrenheit to its    */
    /* value in degrees centigrade.            */
    /******************************************/
```

(continued)

Program 1–25 *(continued)*

```
main( )
{

float fahrenheit_temp;    /* Temperature in degrees Fahrenheit */
float centigrade_temp;    /* Temperature in degrees centigrade */

/* Explain program to user.  */

puts("");
puts("This program will convert a temperature reading in");
puts("degrees Fahrenheit to its equivalent temperature");
puts("in degrees centigrade.");

puts(" ");
puts("You only need to enter the temperature in Fahrenheit");
puts("and the program will do the rest.");

/*   Get Fahrenheit value from user.   */

puts("");
printf("Enter temperature value in Fahrenheit => ");
scanf("%f",&fahrenheit_temp);

/*   Do computations.   */

centigrade_temp = 5/9 * (fahrenheit_temp - 32);

/*   Display the answer. */

puts("");
printf("A temperature of %f degrees Fahrenheit \n",fahrenheit_temp);
printf("is equal to %f degrees centigrade \n",centigrade_temp);

}
```

Good programming practice would now require the variable definitions to be placed in the programmer's block.

The last part of the program, the display the answer section, is coded, and the program is compiled again to test for any compile time errors. Note that the **%f** is used so that a float type may be displayed and the **\n** is used to indicate a new line of display. All this being done, there is one final and very important step left.

Checking the Output

Up to this point in the case study, you have confirmed that there are no compile time errors. However, to ensure that there are no run time errors, the program is checked for several different input values. The output results are then checked for accuracy. Several temperatures were entered (negative values, as well as positive values and zero), and the results were checked by a calculator by more than one person.

Doing this ensures that the program process was correctly coded to produce what was intended by the program designer.

A sample execution of the program yields:

```
This program will convert a temperature reading in
degrees Fahrenheit to its equivalent temperature
in degrees centigrade.

You only need to enter the temperature in Fahrenheit
and the program will do the rest.

Enter the temperature value in Fahrenheit => 212


A temperature of 212.000000 degrees Fahrenheit
is equal to a temperature of 0.000000 degrees centigrade.
```

As seen from the above execution of the program, there is a run time error! This is why it is so important to test the program with different values and check the answers. Just because there are no compile time errors does not mean the program is doing what you intended it to do. No compile time errors simply means that your coding is passable; it says nothing about your design of the program. The reason why you are getting 0.000000 for the centigrade temperature is because you are using integers $(5/9)$ in a division problem. Recall from the discussion on integers in this chapter that division by integers does not leave a remainder (it can't, they are integers). Thus $5/9$ will return a value of 0. This 0 value is then multiplied by the term (`fahrenheit_temp` $-$ 32) which again results in zero (any number multiplied by 0 is 0). Thus, no matter what value you put into the program for the Fahrenheit temperature, the result will always be zero. In order to correct this, the conversion constant $5/9$ needs to be changed from an integer. This is done by simply adding a decimal point followed by a zero:

```
centigrade_temp = 5.0/9.0 * (fahrenheit_temp - 32);
```

Conclusion

This section presented your first real case study for the development of a technical C program. You were introduced to the concept of first describing the program requirements in writing. You then saw how to develop an outline of the actual program using comments. From this, each comment section was developed and tested separately. Once the program was completely coded, its accuracy was tested for several different values.

Check your understanding of this section by trying the following section review.

1–12 Section Review

1 What is the first step in the development of a program?
2 State the items that should be included in the program problem statement.
3 Give the first step in the actual coding of the program.
4 Explain the process used to develop the final program.

Interactive Exercises

DIRECTIONS

These exercises require that you have access to a computer and software that support C. They are provided here to give you valuable experience and, most important, immediate feedback on what the concepts and commands introduced in this chapter will do. They are also fun.

Exercises

1   Predict what the output of the following program will be and then try it.

Program 1–26

```
#include <stdio.h>

main( )
{

    printf("The number is %d\n",15);
    printf("The number is %i\n",15);
    printf("The number is %x\n",15);
    printf("The number is %o\n",15);

}
```

2   The next program now uses **e** and **E**. What is different about the outputs?

Program 1–27

```
#include <stdio.h>

main( )
{

    printf("The number is %e\n",15.0);
    printf("The number is %E\n",15.0);

}
```

3   The next program is a fun one. See if you can predict what the output will be before you try it.

Program 1-28

```
#include <stdio.h>

main( )
{
  char letter;

     letter = 'b';

     printf("This is %d or %c or %x.",letter,letter,letter);

}
```

4 Figure the next program out with pencil and paper. Then try it. Do your results agree
with the computer?

Program 1-29

```
#include <stdio.h>

main( )
{

  float result = 10;

     result = 2*(3+5)/8 - 3;

     printf("The result is %f.\n",result);

}
```

5 The next program is a good test question. Make sure you try it!

Program 1-30

```
#include <stdio.h>

main( )
{

     printf("What is this ==> \\\n");
     printf("and this ==> \"");

}
```

6 For this last program see if you can predict what the output will be in each case. Then enter and execute the program. Be sure to write what your system gave you in your notes.

Program 1-31

```
#include <stdio.h>

   float number_1 = 125.738;

   main( )
   {
      printf("In decimal notation 125.738 = %d\n",number_1);
      printf("In float notation 123.738 = %f\n",number_1);
      printf("In E notation 123.738 = %e\n",number_1);

   }
```

Self-Test

DIRECTIONS

The following program was developed by a beginning C student. It may contain some errors. Answer the questions that follow by referring to this program.

Program 1-32

```
#include <stdio.h>
/*************************************************/
/*              Ohms Law Program.               */
/*************************************************/
/*         Developed by: A. Good Programmer.    */
/*            Date: Sept 12, 1992               */
/*************************************************/
/*    This program will solve for the circuit   */
/*  current. The program user must input the    */
/*  value of the circuit voltage and the        */
/*  circuit resistance.                         */
/*************************************************/
/*              Variables used:                 */
/*---------------------------------------------*/
/*       current = Circuit current in amps.     */
/*       voltage = Circuit voltage in volts.    */
/*     resistance = Circuit resistance in ohms. */
/*************************************************/
```

(continued)

Program 1–32 *(continued)*

```
main( )
{

/* Declaration block.   */

    float current;        /* Circuit current */
    float voltage;        /* Circuit voltage */
    float resistance;     /* Circuit resistance */

/* Explain program to user. */

    puts("This program will compute the value of the");
    puts("circuit current in amps.");
    puts("");
    puts("You must input the value of the circuit");
    puts("voltage in volts and the circuit");
    puts("resistance in ohms.");
    puts("");
    puts("");

/* Get input from user. */

    printf("Value of the circuit voltage = ");
    scanf("%f",&voltage);

    printf("Value of the circuit resistance = ");
    scanf("%f",&resistance);

/* Do the calculations. */

    current = voltage/resistance;

/* Display the answer. */

    puts("");
    puts("The current in a circuit with a total resistance");
    printf("of %e ohms and a total voltage of %e\n",resistance, voltage);
    printf("volts is %e amps.\n",current);

}
```

Questions

1 Will the above program compile and execute on your system? If not, why not?

2 Explain what the program does. How did you determine this?

3 How many variables are there in the program? What type are they? How did you determine this?

4 State why the **puts( )** function was used to explain the program to the user.

5   State how the user may input the values of the voltage and the current. How did you determine this?

6   In what manner will the output values be displayed? How did you determine this?

End-of-Chapter Problems

General Concepts

Section 1–1

1   What type of a program is used in order to enter C source code?

2   State the program that converts your source code into something the computer understands.

Section 1–2

3   What must all C programs start with?

4   State the purpose of the /*   */ used in C.

5   What indicates the beginning and the end of program instructions in the C language?

Section 1–3

6   What, in a program, makes it easy to understand, modify, and debug?

7   State the purpose of the programmer's block.

8   What is contained in the programmer's block?

9   Is it necessary to have a programmer's block for a C program to compile?

Section 1–4

10   What is a token in C?

11   Name the symbol other than the English alphabet and the ten decimal digits of the Arabic number system that can be used to develop a C program.

12   State what is meant by a predefined token.

Section 1–5

13   What is the main difference between how strings and characters are represented in a **printf( )** function.

14   State what in a **printf( )** function specifies how it is to convert, print, and format its arguments.

15   What is the name given to the actual values within the parentheses of a function?

16   How many arguments must a **printf( )** function contain?

Section 1–6

17   What is the name given to an independent collection of declarations and statements in C?

18   State what all identifiers must start with in C.

19   How many characters of an identifier are recognized by C?

20   What is an identifier?

Section 1–7

21   Name the three fundamental C type specifiers.

22   How do you set aside a specific memory location to later receive a value in C?

23   In C, what prevents a new variable from coming into your program as a result of a typing error?

24   What is it called when you combine a variable declaration with an assignment operator?

Section 1–8

25   State the unique characteristic of integer division in C.
26   Is there an order in which arithmetic operations must be performed? What is this called?
27   What does the C statement: `result *= 5;` mean?
28   What kind of statement is the statement of problem 27 called?

Section 1–9

29   What in a **printf( )** function causes a vast departure from the normal interpretation of a string?
30   Name the character in the **printf( )** function that is referred to as the escape character.
31   State what determines the number of digits to the right of the decimal point in a displayed value when using the **printf( )** function.

Section 1–10

32   Give an example of using the **scanf( )** function to enter the value of a declared constant `constant_1` in floating point.
33   Explain the purpose of the **scanf( )** function as described in this chapter.
34   Give an example of outputting the value of a declared constant `constant_1` in E notation.
35   For the C statement

$$\texttt{scanf("\%f",\&value);}$$

in what numerical format may the program user enter values?

Section 1–11

36   What are the three types of compile time error messages used in C?
37   State the action of a fatal error message.
38   What are the three most common errors made by beginning C programmers as presented in this chapter?
39   Do compiler error messages always identify the problem in your program? Explain.

Section 1–12

40   List the four specific areas, as presented in this chapter, that should be stated in writing when designing a program.
41   State the first step in the coding of a program.

Program Design

You now have enough information to actually develop all of the source code for the following programs. As with all programs you design, you are expected to use the steps presented in the Case Study section of this chapter.

Electronics Technology

42   Develop a C program that solves for the power dissipation of a resistor when the voltage across the resistor and the current in the resistor are known. The relationship for resistor power dissipation is

$$P = I \times E$$

Where

P = Power dissipated in watts.
I = Resistor current in amps.
E = Resistor voltage in volts.

43   Create a C program that would solve for the current in a series circuit consisting of three resistors and a voltage source. The program user must input the value of each

resistor and the value of the voltage source. The relationship for total current is

$$I_t = V_t / (R_1 + R_2 + R_3)$$

Where

I_t = Total circuit current in amps.
V_t = Voltage of voltage source in volts.
R_1, R_2, R_3 = Value of each resistor in ohms.

44   Develop a C program that will compute the inductive reactance for a particular frequency. The program user must input the value of the inductor and the frequency. The relationship for inductive reactance is

$$X_L = 2\pi fL$$

Where

X_L = Inductive reactance in ohms.
f = Frequency in hertz.
L = Value of inductor in henrys.

45   Create a C program that will solve for the capacitive reactance of a capacitor. The program user must input the value of the capacitor and the frequency. The relationship for capacitive reactance is

$$X_c = 1/(2\pi fC)$$

Where

X_c = Capacitive reactance in ohms.
f = Frequency in hertz.
C = Value of capacitor in farads.

46   Write a C program that will find the total impedance of a series circuit consisting of a capacitor and an inductor. Program user is to input the value of the inductor, capacitor, and applied frequency. The relationship for series circuit impedance consisting of a capacitor and inductor is

$$Z = X_L - X_c$$

Where

Z = Circuit impedance in ohms.
X_L = Inductive reactance in ohms.
X_c = Capacitive reactance in ohms.

Manufacturing Technology
47   Create a C program that will compute the number of items made in an eight-hour day assuming that the same number of items are made each hour. User input is the number of pieces manufactured in one hour.

Drafting Technology
48   Develop a C program that will compute the area of a circle. User input is the radius of the circle.

Computer Science
49   Write a C program that will allow the program user to determine the maximum and minimum size of a **float** for their computer system.

Construction Technology

50   Write a C program that will compute the volume of a room. User inputs are the height, length, and width of the room.

Agriculture Technology

51   Create a C program that will compute the number of acres of land. The program user is to input the width and length (assume a perfect rectangle) of the land in feet. The program is to display the answer in acres, without E notation.

$$\text{Formula: 1 acre} = 43{,}560 \text{ ft}^2$$

Health Technology

52   Develop a C program that will convert from degrees centigrade to degrees Fahrenheit. User input is the temperature in Fahrenheit. The relationship is:

$$F = 9/5C + 32$$

Where

C = Temperature in centigrade.
F = Temperature in Fahrenheit.

Business Applications

53   Write a C program that will compute a 6% sales tax on a purchase. The program user is to input the total amount of purchase. The program is to return the original amount of purchase, the tax, and the total of the two.

2 Structured Programming

Objectives

In this chapter, you will have the opportunity to learn:

1 How to recognize a block structured program.
2 How to develop block structure using C.
3 A very important theorem about programming.
4 The use of C functions in the development of a block structured program.
5 How to pass values between functions.
6 The meaning and use of a formal parameter.
7 The meaning and use of an actual parameter.
8 The meaning of preprocessing and preprocessor commands.
9 The use of the **#define** directive.
10 How to develop and save your own header file.
11 How to use top-down design in the design of a technology problem-solving program.

Key Terms

Program Block   Completeness Theorem
Block Structure   Function
Block Separators   Function Prototype
Action Block   Formal Parameter List
Loop Block   Actual Parameter
Branch Block   Formal Parameter

Value Passing	Macro
Multiple Arguments	Header File
Calling Functions	Token Pasting
Recursion	String-izing
Defining Functions	Prologue

Outline

2–1 Concepts of a Program Block

2–2 Using Functions

2–3 Inside a C Function

2–4 Functioning with Functions

2–5 Defining Things

2–6 Program Debugging and Implementation

2–7 Case Study

Introduction

The last chapter got you started developing your own C programs. You learned how to use C to get values from the program user, perform operations on these values and then display the results.

From this point on, you will learn information that will help you develop technical programs in C that are not so easily solved with the simple scientific pocket calculator. This means that your programs will become longer and perform many kinds of useful operations. Because of this, it is very important that you are introduced to the concept of structured programming. Doing this now will help you develop good programming skills early in your learning of the C language. These good programming habits will then be used in the remaining chapters of this book.

As you will see, it is the purpose of this chapter to help you design programs that are easy to read, understand, debug and modify. This is a very important chapter.

2–1   Concepts of a Program Block

Discussion

You were introduced to the use of block structure in programming back in Chapter 1. There you saw that dividing the program up into distinct blocks makes the program easier to read and modify. Remember the analogy to the structure of a business letter. The structure makes the letter easier to read.

This section will present more detailed information on how to develop block structured programs using C.

A Blocking Example

Consider Program 2–1. It does little more than put information out to the screen. However, it does illustrate one of the features of a block structure.

Program 2–1

```
/************************************************/
/*               Typical C Program              */
/************************************************/
/*       Developed by: A. Typical Programmer.   */
/************************************************/
/*                 September 20, 1991           */
/************************************************/
/*       This program illustrates nothing more  */
/* than a typical unstructured C program.       */
/************************************************/
main( )
{
/*    Explain program to user.    */
puts("This is a C program that illustrates the");
puts("typical unstructured approach to writing ");
puts("a program in C.");
puts("When the program is executed, the program");
puts("user can't tell that the program is unstructured");
puts("only the programmer can.");
puts("Hence a structured program in C is useful only");
puts("to the programmer, his boss, his teacher");
puts("those who need to modify it and to those");
puts("who are living with him while he is trying");
puts("to find the bugs in the program.");
}
```

Blocking the Structure

You probably had no problem in understanding what the above program will do. It simply prints a bunch of strings out to the monitor. So why structure it?

You should consider several points. First, the program looks boring. Every program line starts at the same place—at the same column on the left, one **puts** function after the other. The structure could be made more readable and a little more exciting if paragraphs could be used. This would help distinguish one part of the program from another. Each paragraph can be thought of as a **program block**. Each program block can be thought of as having one main idea needed in the program. Recall that the case study of Chapter 1 presented this same concept. It shall be used again here.

Program 2–2 shows a simple way of making Program 2–1 a little more interesting.

Program 2-2

```
/***********************************************/
/*                Typical C Program            */
/***********************************************/
/*      Developed by: A. Block Programmer.      */
/***********************************************/
/*                September 20, 1991            */
/***********************************************/
/*      This program illustrates the most       */
/* simple form of block structured programming */
/***********************************************/

main( )
{

/*----------------------------------------------------------*/

/*  First paragraph of explanation.    */
    puts("This is a C program that illustrates a");
    puts("typical unstructured approach to writing ");
    puts("a program in C.");
*/  End first paragraph of explanation.    */

/*----------------------------------------------------------*/
/*  Second paragraph of explanation.    */
    puts("When the program is executed, the program");
    puts("user can't tell that the program is unstructured");
    puts("only the programmer can.");
/*  End second paragraph of explanation. */

/*----------------------------------------------------------*/

/*  Third paragraph of explanation.    */
    puts("Hence a structured program in C is useful only");
    puts("to the programmer, his boss, his teacher");
    puts("those who need to modify it and to those");
    puts("who are living with him while he is trying");
    puts("to find the bugs in the program.");
/*  End third paragraph of explanation. /*

/*----------------------------------------------------------*/

/* End of program.    */
}
```

Note the use of the dashes to emphasize each block of the program. Also note the comments at the beginning and end of each block. These tell you the purpose of the block you are entering and state it again when you are leaving it. For the simple program here, the added structure did nothing to help you understand it, but it does serve as a model for what is to follow.

Defining a Block Structure

Block structure means that the program will be constructed so that it consists of a few groups of instructions rather than one continuous listing of instructions. This is the way the longer programs in the previous chapters were presented.

Each group or block of instructions starts with a remark statement telling what the block is to do. An example from the above program is

```
/* Third paragraph of explanation. */
```

Each program block is defined by separating spaces called **block separators.*** An example of a block separator is

```
/*-----------------------------------------------------------*/
```

The body of every program block is highlighted by being indented from the left column. The exact number of spaces from the left is not critical. What is important is that you make a distinction between the body of the block and its beginning and ending comments. An example from the above program is

```
/*   Third paragraph of explanation.   */
     puts("Hence a structured program in C is useful only");
     puts("to the programmer, his boss, his teacher");
     puts("those who need to modify it and to those");
     puts("who are living with him while he is trying");
     puts("to find the bugs in the program.");
/*   End third paragraph of explanation. /*
```

Some Important Rules

What is important in a structured program is that there is no jumping around. If you don't know how to do this (if you have never used a **GOTO**) then consider yourself lucky! People can understand things a lot easier if they can follow them logically from one step to another. Hence, you should always follow these important rules when doing block structured programming:

1. All blocks are entered from the top.
2. All blocks are exited from the bottom.
3. When the computer finishes one block, it goes on to the other or ends.

Can you design every program this way? Yes you can—with absolutely no exceptions. There is no excuse for writing a program in C that doesn't contain block structure.

---

*Use of block separators is optional. To conserve space, they are not always used in the sample programs.

Types of Blocks

No matter what programming language you use, there are only three necessary types of blocks:

1. Action block
2. Loop block
3. Branch block

What does each of these blocks do? An **action block** is the simplest kind of programming block. It is nothing more than a straight sequence of action statements, one following the other. A **loop block** can cause the program to go back and repeat a part of the program over again, while a **branch block** gives the option of performing a different sequence of instructions. The concepts of these three different kinds of blocks are shown in Figure 2–1.

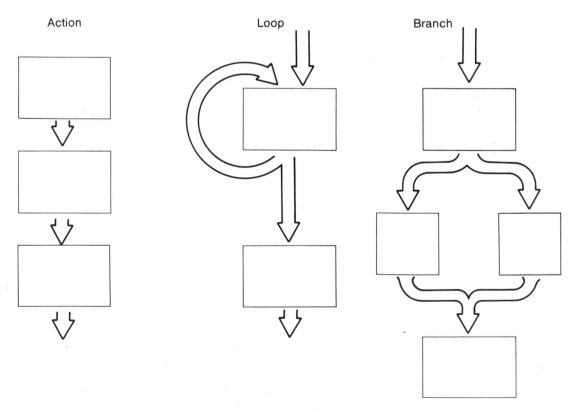

Figure 2–1   Concepts of the Three Kinds of Programming Blocks

A Theorem

In 1966, two programmers named Boehm and Jacopini proved that any computer program, no matter how complex, can be written using the following **completeness theorem**:

Any program logic, no matter how complex, can be resolved into action blocks, loop blocks and branch blocks.

You will soon be able to take on any programming problem, using structured programming techniques in any programming language that is or will be used in the foreseeable future.

Conclusion

This section emphasized the importance of block structure in programming and presented a specific example. You were also introduced to a method of breaking a C program into blocks. The important completeness theorem was also presented. You will come across this throughout the rest of this book. For now, test your understanding of this section by trying the following section review.

2-1 Section Review

1 Does a C compiler require that programs be structured? Explain.
2 Define the term block structure.
3 State one way of beginning a program block.
4 Explain how program blocks may be separated.
5 Explain how the body of a program block is highlighted.
6 Name the three types of blocks.
7 State the completeness theorem.

2-2   Using Functions

Discussion

In the last section you were reintroduced to the concept of block structured programming. In this section, you will see how a C program is designed for this kind of structure. This section will serve as an introduction to this important concept of C. You will have an opportunity to increase your understanding of it as you progress through the text.

What Is a C Function?

A C function is an independent collection of source code designed to perform a specific task. All C programs have at least one function called `main( )`.

You have already used other functions that are built into your C library such as **puts( )**, **printf( )** and **scanf( )**. All of these built-in functions actually do something.

Making Your Own Functions

You can also make your own functions in C. Doing this allows you to create a function and tell C what it is the function is to do Then you can use it over and over again just like the built-in functions of C. This means that you could create a function to solve for an electrical series circuit, one to solve for a parallel circuit or one to solve for the electrical characteristics of a transistor amplifier—to name just a few. Define it only once, give it a name and then call on it any time you want—just like you call on **puts( )** or **printf( )** any time you want. As you can see from this, to call a function, you simply use its name. You could even create your own library of functions, save them on your disk and then invoke them into your program the same as you do **puts( )**, **printf( )** and **scanf( )**.

This concept is shown in Figure 2–2.

What Makes a Function?

In C, when you create a **function** other than **main( )** you first declare it and then you define it. When you declare a function, you code in what is called the **function**

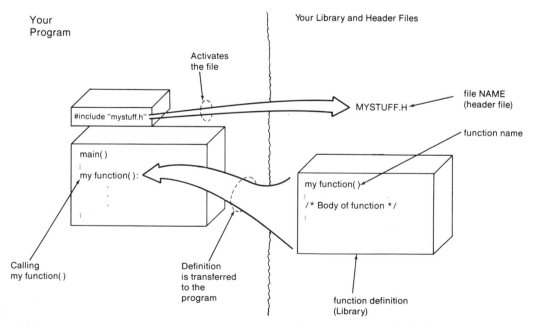

Figure 2–2   Concept of Using Your Own C Functions

prototype. A function prototype gives the function name plus other important information concerning the function. This appears at the beginning of the program before **main( )**.

When you define a function, you again give the name of the function and other information about it (just like you did in the prototype), and you also produce the body of the function which contains all of the source code to be used by that function. Everything said here is shown in the following example.

An Example   Take a look at the program from the last section. It has been restructured to use C functions. Its output to the screen will be exactly the same as it was before. Since you know exactly what the program will do, you can now concentrate on how it is going to do it.

Program 2–3

```
#include <stdio.h>

    /***********************************************/
    /*                Typical C Program            */
    /***********************************************/
    /*      Developed by: A. Block Programmer.     */
    /***********************************************/
    /*                September 20, 1991           */
    /***********************************************/
    /*      This program illustrates the most      */
    /* simple form of block structured programming */
    /***********************************************/
    /*           Function prototype                */
    /*-------------------------------------------*/
      void first_paragraph(void);
    /* This function presents the first paragraph  */
    /* for this program.                           */
    /*-------------------------------------------*/
      void second_paragraph(void);
    /* This function presents the second paragraph */
    /* for this program.                           */
    /*-------------------------------------------*/
      void third_paragraph(void);
    /* This function presents the third paragraph  */
    /* for this program.                           */
    /***********************************************/

    main( )
      {
      first_paragraph( );  /* First paragraph of explanation. */
      second_paragraph( ); /* Second paragraph of explanation. */
      third_paragraph( );  /* Third paragraph of explanation. */
      exit(0);
      }
```

(continued)

Program 2-3 *(continued)*

```
/*------------------------------------------------------------*/
void first_paragraph( ) /* First paragraph of explanation. */

     {
     puts("This is a C program that illustrates a");
     puts("structured approach to writing ");
     puts("a program in C.");
     }
/*  End paragraph of explanation.   */

/*----------------------------------------------------------*/

void second_paragraph( )  /* Second paragraph of explanation. */
     {
     puts("When the program is executed, the program");
     puts("user can't tell that the program is unstructured");
     puts("only the programmer can.");
     }

/*  End second paragraph of explanation. */
/*----------------------------------------------------------*/

void third_paragraph( ) /*  Third paragraph of explanation. */
     {
     puts("Hence a structured program in C is useful only");
     puts("to the programmer, his boss, his teacher");
     puts("those who need to modify it and to those");
     puts("who are living with him while he is trying");
     puts("to find the bugs in the program.");
     }
/* End third paragraph of explanation. */

/*----------------------------------------------------------*/

/* End of program. */
```

Program Analysis

The key to analyzing the above program is to note what the function **main( )** now
contains:

```
main( )
{
 first_paragraph( );   /* First paragraph of explanation. */
 second_paragraph( );  /* Second paragraph of explanation. */
 third_paragraph( );   /* Third paragraph of explanation. */
 exit(0);
}
```

The **main( )** function now consists of nothing more than a series of calls to other functions. The first function is called **first_paragraph( )**;. What it does is go to the definition of the function by the same name:

```
void first_paragraph( )   /* First paragraph of explanation. */
{
  puts("This is a C program that illustrates the");
  puts("typical unstructured approach to writing ");
  puts("a program in C.");
}
```

It is here that the function **first_paragraph( )** is defined. The definition consists of the program code put between the { and the }. This is exactly the same thing you have been doing with the C reserved function **main( )**!

This process of calling each function, in the order needed, is done by **main( )**. This is illustrated in Figure 2–3.

There are some other important things to point out. First note the statement of what each function will do under the heading of **Function prototype**:

```
/*          Function prototype                       */
/*--------------------------------------------------*/
  void first_paragraph(void);
/* This function presents the first paragraph  */
/* for this program.                           */
/*--------------------------------------------------*/
  void second_paragraph(void);
/* This function presents the second paragraph */
/* for this program.                           */
/*--------------------------------------------------*/
  void third_paragraph(void);
/* This function presents the third paragraph  */
/* for this program.                           */
/****************************************************/
```

Note that all three of the functions defined in the program (**first_paragraph**, **second_paragraph**, and **third_paragraph**) are not commented out (do not have **/*** and ***/** around them). This means that they represent some kind of instruction to the C system. The instruction actually tells the compiler that there will be a function by a certain name to be defined later in the program. This is the function prototype. The function has a type (just like variables and constants have a type), and for this program, the type is **void**. **void** means that the function will not return a value. This is so because the definition of the function only causes some words to appear on the monitor; it isn't computing anything or doing anything else with values. The second thing it states is that there will not be anything contained in its **formal parameter**

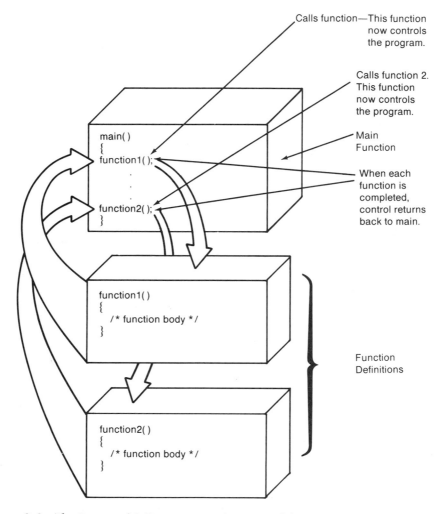

Calls function—This function now controls the program.

Calls function 2. This function now controls the program.

Main Function

When each function is completed, control returns back to main.

Function Definitions

Figure 2-3 The Process of Calling Functions from **main( )**

list (between the ( ))*. This is important because doing this instructs the compiler how much and what kind of memory to save for these functions. This is also declared as **void** because no values will be placed here.

The concept of a function prototype is illustrated in Figure 2-4.

Note that each function prototype ends with a semicolon. Also note that when the function was called in **main( )** it also ended with a semicolon. Now note—

---

*Information contained within the parentheses are parameters. The actual values used for this information are called *arguments*.

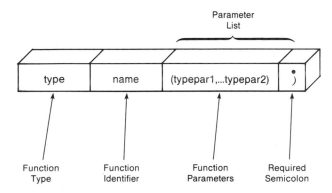

Figure 2-4   Concept of a Function Prototype

and this is important—when the function is defined, it does not end with a semicolon. This is no different from what you have been doing with **main( )**—when defining it, you don't end it with a semicolon.

Note the structure of the function definition:

```
void first_paragraph( )   /*  First paragraph of explanation. */
   {
   puts("This is a C program that illustrates a");
   puts("structured approach to writing ");
   puts("a program in C.");
   }
```

The definition states the function type (**void**—because it isn't returning any value to the calling function). It states its name, and it is immediately followed by the required ( )—which don't contain anything between them because it was already stated in the prototype declarations that they wouldn't (they were (**void**)). There is no semicolon (as stated before). The body of the function is defined by the opening brace { and then the closing brace }.

The structure of a function definition is illustrated in Figure 2-5.

There is one last important point. Look again at the function **main( )**:

```
main( )
   {
   first_paragraph( );  /* First paragraph of explanation. */
   second_paragraph( ); /* Second paragraph of explanation. */
   third_paragraph( );  /* Third paragraph of explanation. */
   exit(0);
   }
```

Its last function call is **exit(0);**. **exit** is a C function that causes normal termination of a C program.

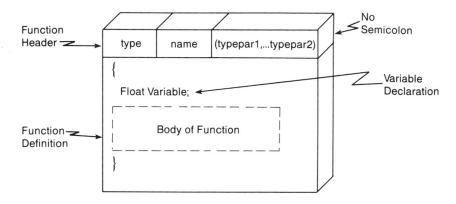

Figure 2-5   Structure of a Function Definition

Conclusion

This section contained a lot of new information. The importance of this section was not so much the exact details (although they will be important when it comes to programming) but more the concepts—the concept of how a program may be divided into discrete parts called functions, and how one function, the **main( )** function, can then call these other functions in an order that determines how the program is executed. Check your understanding of this section by trying the following section review.

2-2 Section Review

1   What is a C function?
2   Explain what is meant by a type **void**. Give an example.
3   State the meaning of a function prototype.
4   What is the purpose of using a function prototype?
5   State what is meant by calling a function.
6   Why is the **exit( )** used as the last function call in **main( )**?

2-3   Inside a C Function

Discussion

In this section, you will discover how functions may be used to actually return a value. This means you can now use functions to perform complex calculations for you and then return the results of these computations to the calling function.

Basic Idea

The following program gets a value from the program user and squares it.

Program 2-4

```c
#include <stdio.h>

  main( )
  {
    float number;    /* Number to be squared.   */
    float answer;    /* The square of the number. */

      printf("Give me a number and I'll square it => ");
      scanf("%f",&number);

      answer = number * number;

      printf("The square of %f is %f",number, answer);

    exit(0);

  }
```

When the program is executed and the user enters the value 3, the results are

```
Give me a number and I'll square it => 3
The square of 3.000000 is 9.000000
```

To illustrate how a C function can be used in the above program, consider Program 2-5.

Program 2-5

```c
#include <stdio.h>

  /*  Function prototype.  */
  float square_it(float number);
  /* This is the function that will square the number.  */

  main( )
  {
    float value;    /* The number to be squared. */
    float answer;   /* The square of the number. */
```

(continued)

Program 2-5 *(continued)*

```
        printf("Give me a number and I'll square it => ");
        scanf("%f",&value);

        answer = square_it(value);     /* Call the function. */

        printf("The square of %f is %f",number, answer);

        exit(0);
    }

  float square_it(float number)
     {
        float answer;    /* The square of the number.  */

        answer = number * number;

        return(answer);

     }
```

Program Analysis

The above program has created a separate function called `square_it`. This new function computes the square of a number. Where does it get this number? It gets it from the main function `main( )`. How does it get it? It is passed to it from `main( )`. How is it passed to it? It is passed to it through its parameter argument (`float number`). Figure 2-6 shows how **value passing** is done.

Note from the figure that the formal parameter list contains declarations of the function parameters. In this case, the function has only one, called `parameter_1`. This formal parameter is present in the function prototype as well as the head of the function definition. However, when the function is called (as called from `main( )`), its actual parameter need not have the same name as its formal parameter (it must still have the same type). For example, in the previous program, the **actual parameter** was `value` and the formal parameter was `number`. Thus the **formal parameters** define the type and number of function parameters, while the actual parameters are used when calling the function.

As you can see from the figure, the parameter of the function `square_it` is no longer void (it is not `square_it(void)` it is instead `square_it(float number)`).

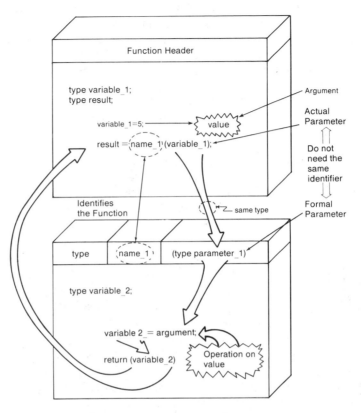

Figure 2-6   Passing a Value from One Function to Another

This was first done in the prototype part of the program:

```
/* Function prototype.   */
float square_it(float number);
/* This is the function that will square the number.   */
```

Note that the since this function will return a value, it is no longer of type
void. Hence the compiler must be told what type it is to be. Keeping with what
you have been doing up to this point, it is given the type **float**. This means that
the answer it gets after squaring the number will be of type **float**. The number
that it will square must also have a type. The compiler is also told this in the prototype
declaration:

```
(float number);
```

Thus, the prototype declaration for this function told the compiler to expect a function definition in the program that is identified as `square_it` which will return a value of type **float** and will do its operation (squaring in this case) on another number identified as **number** which is also of type **float**. All of this necessary information is given in the prototype declaration:

```
float square_it(float number);
```

Note that it ends with the necessary semicolon.

Next look at the function definition:

```
float square_it(float number)
   {
    float answer;    /* The square of the number.    */
    answer = number * number;
    return(answer);
   }
```

It starts out exactly like the prototype; the only important difference is that it *does not end with a semicolon!* This is important because it tells the compiler that what is to follow is the definition of this function. It also defines its type and the type of its parameter. This is exactly what was done in the prototype. Remember, the prototype tells the compiler what kind of memory to preserve; the function definition describes exactly what action is to take place.

There is a new C term used in this function definition. It is

return( );

This tells the computer which value is to be returned to the calling function. In this case, the value of the variable **answer** is to be returned. Thus, within the argument of **return**, the variable identifier **answer** is placed:

```
return(answer);
```

And this ends with the required semicolon. Thus, this function will get a value from the calling function. This value will be of type **float** and placed in the variable identifier **number**. A computation will be performed:

```
answer = number * number;
```

the results of which will be placed in the variable identifier called **answer**. The value contained in **answer** will then be passed back to the calling function by the command

```
return(answer);
```

Next, look at the function **main( )**:

```
main( )
  {
   float value;     /* The number to be squared. */
   float answer;    /* The square of the number. */
    printf("Give me a number and I'll square it => ");
    scanf("%f",&value);
```

```
    answer = square_it(value);        /* Call the function. */
    printf("The square of %f is %f",number, answer);
    exit(0);
}
```

The function **main( )** uses two variables. One variable is the value to be squared (**value**), the other is needed to contain the answer of the result (**answer**). The program simply asks the program user for a number to be squared and then gets the number from the keyboard. This is done with the **printf** and the **scanf** functions. But now look at what is different; to do the computation, the function that defines squaring need only be called. The value **value** is placed as a function parameter, and this function is made equal to the variable **answer**. The **exit(0);** terminates the program.

Table 2–1 illustrates some of the important function definitions:

Table 2–1 Important Function Definitions

Term	Example	Comments
Formal Parameters [Used when defining the function.]	**square_it(float number)**	**number** is a formal parameter. It states that a quantity of type float will be passed to the function called square_it (quantity).
Actual Parameters [Used when calling the defined function.]	**square_it(value)**	**value** is an actual parameter. The identifier used as the actual parameter may be different from that of the formal parameter, but it must be of the same type.
Arguments [Actual values passed to the called function].	**value = 5;** **square_it(value);**	The number 5 is the argument because it is the actual value passed to the function square_it (value).

Conclusion

This was an important section. Again, the details are less important than the concepts. Here the concept of passing a value from one function to another was introduced. You saw that this can be done using parameters that will represent a place holder for a value to be passed from one function to another. The second idea is using a special C command that will pass a value back to the calling function (**return( )**).

In the next section, you will see a more useful application of passing values between functions. For now, test your understanding of this section by trying the following section review.

2-3 Section Review

1   State what is meant by the parameter of a function.
2   What is the difference between a formal parameter and an actual parameter. How must these be the same? How may these be different?
3   Explain the meaning of passing values between functions.
4   What is the difference between the coding of a function prototype and the head of a function declaration?
5   How is a value passed back to the calling function?

2-4   Functioning with Functions

Discussion

In the last section you were introduced to the power of passing values between functions. In this section you will see how to do more of this as well as some more useful items about functions. You will find that the backbone of the C language is its functions. You will spend a lot of time learning about functions. This section presents some more important information.

More Than One Argument

The following program illustrates the passing of **multiple arguments** between functions. What the program does is to use a function to calculate the capacitive reactance of a capacitor given the value of the capacitor and the frequency. The mathematical relationship is

$$X_c = 1/(2\pi fC)$$

Where

X_c = The reactance of the capacitor in ohms.
f = The frequency in hertz.
C = The value of the capacitor in farads.

Program 2-6

```
#include <stdio.h>

/* Function prototype. */
float capacitive_reactance(float capacitance, float frequency);
/* This is the function that computes the capacitive reactance
   for a given value of capacitor and frequency. */

main( )
{
 float farads;        /* Value of the capacitor. */
 float hertz;         /* Value of the frequency. */
 float reactance;     /* Value of the reactance. */
```

(continued)

Program 2-6 *(continued)*

```
    printf("Input value of the capacitor in farads => ");
    scanf("%f",&farads);
    printf("Input value of the freqency in hertz => ");
    scanf("%f",&hertz);

    reactance = capacitive_reactance(farads, hertz);

    printf("The reactance of a %e farad capacitor\n", farads);
    printf("at a frequency of %e hertz is %e ohms.\n",hertz, reactance);

    exit(0);
}

float capacitive_reactance(float capacitance, float frequency)
{
  float reactance;    /* The capacitive reactance.  */
  float pi = 3.14159;

    reactance = 1/(2 * pi * frequency * capacitance);

    return(reactance);

}
```

Program Analysis

The above program uses two functions: **main( )** and `capacitive_reactance( )`. The feature of this program is that two values must be passed from **main( )** to `capacitive_reactance( )` (the value of the capacitor and the value of the frequency). This is accomplished by having two formal parameters for `capacitive_reactance( )`:

```
float capacitive_reactance(float capacitance, float frequency);
```

As you can see from the prototype, these formal parameters are identified as `capacitance` and `frequency`. Both are declared as type **float**. The function itself will return a value so it is also declared as a type **float**. Now look at the function definition:

```
float capacitive_reactance(float capacitance, float frequency)
  {
  float reactance;    /* The capacitive reactance.  */
  float pi = 3.14159;

    reactance = 1/(2 * pi * frequency * capacitance);

    return(reactance);

  }
```

Again, the function heading is exactly like its prototype—with the important exception that it does not end in a semicolon. The body of the function starts with the required { and then declares two identifiers. One, `reactance`, will represent the value of the computed capacitive reactance while the other is the value for pi. The calculation is then performed and the single value is returned back to the calling procedure.

Now, look at the function **main( )**:

```
main( )
{
  float farads;        /* Value of the capacitor. */
  float hertz;         /* Value of the frequency. */
  float reactance;     /* Value of the reactance. */

    printf("Input value of the capacitor in farads => ");
    scanf("%f",&farads);

    printf("Input value of the freqency in hertz => ");
    scanf("%f",&hertz);

    reactance = capacitive_reactance(farads, hertz);

    printf("The reactance of a %e farad capacitor\n", farads);
    printf("at a frequency of %e Hertz is %e ohms.\n",hertz,
           reactance);

    exit(0);
}
```

main( ) contains the type declarations of the three variables it will be using. Two, `farads` and `hertz`, will be entered by the program user though two separate **scanf( )** functions. Each one is preceded by the necessary **&**. The third, `reactance`, will be used to display the resulting answer. After the user input is received, the defined function is called and the arguments are passed to it:

```
reactance = capacitive_reactance(farads, hertz);
```

This results in the variable identifier `reactance` receiving the result of the reactance calculation. This result is then echoed back to the monitor so the program user may observe the result.

Execution of the above program produces the following results (assume the program user has entered the value of a $1\mu f$ capacitor and a frequency of 5 kHz):

```
Input value of the capacitor in farads => 1e-6
Input value of the frequency in hertz => 5e3
The capacitive reactance => 3.18310e+001 ohms.
```

Calling More Than One Function

As shown in the first section a C function may call more than one function. **Calling functions** is illustrated in the following program. What the program does is to solve the capacitive reactance of a capacitor and the inductive reactance of an inductor where the inductive reactance is given by

$$X_L = 2\pi fL$$

Where

X_L = The inductive reactance in ohms.
f = The frequency in hertz.
L = The inductance in henrys.

Program 2-7

```
#include <stdio.h>

/* Function prototype.  */
float capacitive_reactance(float capacitance, float frequency);
/* This function computes the capacitive reactance
   for a given value of capacitor and frequency. */
float inductive_reactance(float inductance, float frequency);
/* This function computes the inductive reactance
   for a given value of inductor and frequency. */


main( )
{
 float farads;       /* Value of the capacitor. */
 float hertz;        /* Value of the frequency. */
 float reactance;    /* Value of the reactance. */

   printf("Input value of the capacitor in farads => ");
   scanf("%f",&farads);
   printf("Input value of the inductor in henrys => ");
   scanf("%f",&henrys);
   printf("Input value of the freqency in hertz => ");
   scanf("%f",&hertz);

   reactance = capacitive_reactance(farads, hertz);

   printf("The reactance of a %e farad capacitor\n", farads);
   printf("at a frequency of %e hertz is %e ohms.\n",hertz, reactance);

   reactance = inductive_reactance(henrys, hertz);

   printf("The reactance of a %e henry inductor\n", henrys);
   printf("at a frequency of %e hertz is %e ohms.\n",hertz, reactance);

   exit(0);
}

float capacitive_reactance(float capacitance, float frequency)
{
  float reactance;   /* The capacitive reactance.  */
```

(continued)

Program 2-7 *(continued)*

```
   float pi = 3.14159;

      reactance =1/(2*pi*frequency*capacitance);

      return(reactance);
   }
 float inductive_reactance(float inductance, float frequency)
   }
   float reactance;   /* The inductive reactance.  */
   float pi = 3.14159;

      reactance = 2 * pi * frequency * inductance;

      return(reactance);

   }
```

Program Analysis

The main difference in the previous program is that there are now two other functions other than **main( )**. These are

float capacitive_reactance(float capacitance, **float** frequency);

and

float inductive_reactance(float inductance, **float** frequency);

Both of these functions have their separate definitions. As before, each function definition starts with the same heading as its prototype, with the required exception of the semicolon:

```
float capacitive_reactance(float capacitance, float frequency)
  {
    float reactance;    /* The capacitive reactance.  */
    float pi = 3.14159;

       reactance = 1/(2 * pi * frequency * capacitance);

       return(reactance);

  }

float inductive_reactance(float inductance, float frequency)
  {
    float reactance;  /* The inductive reactance.  */
    float pi = 3.14159;
```

```
    reactance = 2 * pi * frequency * inductance;

  return(reactance);

}
```

It actually makes no difference in what order these functions appear in the program. Next, the calling to each of these functions is done by **main( )**:

```
reactance = capacitive_reactance(farads, hertz);
```

and

```
reactance = inductive_reactance(henrys, hertz);
```

Notice that the same variable identifier **reactance** is used in both cases. This can be done because after each use its value is displayed:

```
reactance = capacitive_reactance(farads, hertz);

printf("The reactance of a %e farad capacitor\n", farads);
printf("at a frequency of %e hertz is %e ohms.\n",hertz, reactance);

reactance = inductive_reactance(henrys, hertz);

printf("The reactance of a %e henry inductor\n", henrys);
printf("at a frequency of %e hertz is %e ohms.\n",hertz, reactance);
```

Calling Functions from a Called Function

A called function may also call another function. This concept is illustrated in Figure 2–7.

The following program illustrates the process. It solves for the impedance of a series LC circuit. The mathematical relationship is

$$Z = \sqrt{(X_c^2 + X_L^2)}$$

Where

Z = Circuit impedance in ohms.
X_c = Capacitive reactance in ohms.
X_L = Inductive reactance in ohms.

There is a new and necessary feature in this program. That is the **sqrt** function. It is contained in the **library** file and its prototype is in the **math.h** file. Thus this file must be **#included** so that this necessary function may be used. You will see more of these important math functions in later chapters. For now, the important point is to see how a called function in turn calls two other functions. Observe the following program.

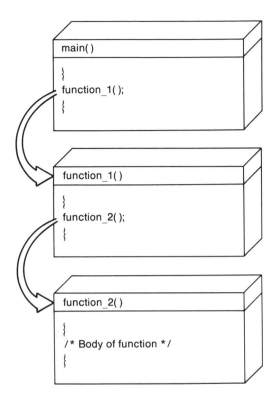

Figure 2-7 A Called Function Calling Another Function

Program 2-8

```
#include <stdio.h>
#include <math.h>

/* Function prototype.  */
float capacitive_reactance(float capacitance, float frequency);
/* This function computes the capacitive reactance
   for a given value of capacitor and frequency. */
float inductive_reactance(float inductance, float frequency);
/* This function computes the inductive reactance
   for a given value of inductor and frequency. */
```

(continued)

Program 2–8 *(continued)*

```c
float series_impedance(float capacitor, float inductor, float frequency);
/* This function computes the series impedance of an inductor and capacitor. */

main( )
{
 float farads;        /* Value of the capacitor. */
 float henrys;        /* Value of the inductor.  */
 float hertz;         /* Value of the frequency. */
 float impedance;     /* Value of the impedance. */

   printf("Input value of the capacitor in farads => ");
   scanf("%f",&farads);
   printf("Input value of the inductor in henrys => ");
   scanf("%f",&henrys);
   printf("Input value of the freqency in hertz => ");
   scanf("%f",&hertz);

    impedance = series_impedance(farads, henrys, hertz);

   printf("The impedance of the circuit is %e ohms.",impedance);

   exit(0);
}

float capacitive_reactance(float capacitance, float frequency)
{
  float reactance;    /* The capacitive reactance.  */
  float pi = 3.14159;

     reactance = 1/(2 * pi * frequency * capacitance);

     return(reactance);
}

  float inductive_reactance(float inductance, float frequency)
  {
  float reactance;  /* The inductive reactance.  */
  float pi = 3.14159;

     reactance = 2 * pi * frequency * inductance;

     return(reactance);

  }

 float series_impedance(float capacitor, float inductor, float frequency)
  {
```

(continued)

Program 2–8 *(continued)*

```
    float cap_react;     /* Resulting capacitive reactance.  */
    float ind_react;     /* Resulting inductive reactance.   */
    float impedance;     /* Resulting impedance.             */

      cap_react = capacitive_reactance(capacitor, frequency);
      ind_react = inductive_reactance(inductor, frequency);

    printf("The capacitive reactance => %e ohms.\n", cap_react);
    printf("The inductive reactance => %e ohms.\n", ind_react);

      impedance = sqrt(cap_react*cap_react + ind_react*ind_react);

      return(impedance);

    }
```

Program Analysis

This program illustrates that a called function may call other functions. Note that the order in which the functions are defined makes no difference in the program. The function called from **main( )** is **series_impedance**. The definition for this function is located at the end of the program. This function then calls two other functions:

```
cap_react = capacitive_reactance(capacitor, frequency);
ind_react = inductive_reactance(inductor, frequency);
```

Note that each of these called functions returns its calculated value back to the calling function (**series_impedance**). These values are then used in the calculation of the impedance. This calculation requires the taking of a square root. In order to do this, the **math.h** file had to be included at the beginning of the program. This is the file that contains the prototype of the built-in function **sqrt( )** used for returning the square root of a number.

The called function **series_impedance** then returns the value of the computation to its calling function **main( )**. Here, the final value is displayed to the program user. Assuming that the program user entered a value of 1 μf for the capacitor and 1 mh for the value of the inductor and a frequency of 5 kHz, then execution of the above program would yield:

```
Input value of the capacitor in farads => 1e-6
Input value of the inductor in henrys => 1e-3
Input value of the frequency in hertz => 5e3
The capacitive reactance => 3.183101e+001 ohms.
The inductive reactance => 3,141590e+001 ohms.
The impedance of the circuit is 4.472329e+001 ohms.
```

How Functions May Be Called

Figure 2–8 illustrates the different ways functions may call other functions. Notice that a function may also call itself. This is called **recursion**. The only restriction is that a function cannot be defined within the body of another function.

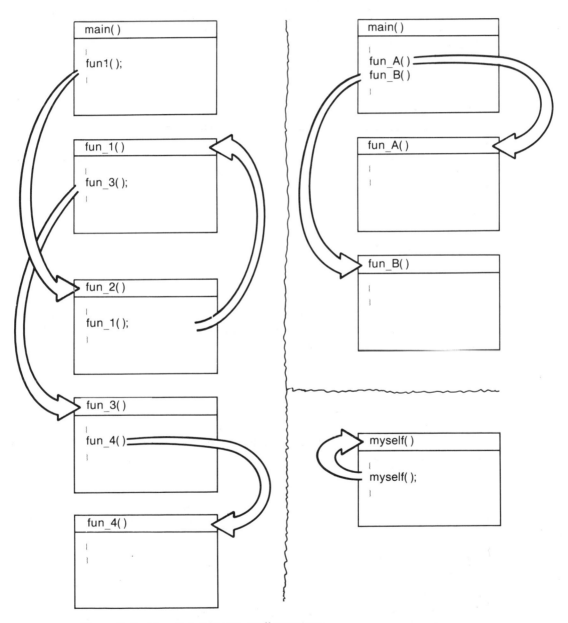

Figure 2–8 How Functions May Call Functions

Conclusion

In this section you learned a lot more about the structure of a C program. You saw that more than one value may be passed to a called function. You saw that a function may call more than one other function. The order of defining functions in the program makes no difference. You also saw that a called function may call another function and that a function may even call itself.

In the next section you will learn how to define items in your C program. As you will see, you may later store this information on your own disk for use by all of your programs. Check your understanding of this section by trying the following section review.

2-4 Section Review

1 May a function pass more than one value to a called function? Explain.
2 Is it possible for a function to call more than one function? Explain.
3 Does it make any difference what order called functions are defined in the body of the C program?
4 Explain what is meant by a called function calling another function.
5 What is recursion?

2-5 Defining Things

Discussion

In the last section you saw more of the power of functions. In this section you will see how you can define your own constants within your programs. Doing this makes your program more portable, easier to change and easier to understand. Here you will see how to use this new and powerful feature of the C system.

Basic Idea

Look at the following program. What it does is take the number 5 and square it. It does this by using the preprocessor command **#define** where the identifier `square` is defined with its parameter (x).

Program 2-9

```
#include <stdio.h>
#define square(x) x*x

main( )
{
  float number;   /* Square of a number. */
```

(continued)

Program 2-9 *(continued)*

```
    number = square(5);

    printf("The square of 5 is %f",number);

}
```

Program Analysis

The above program uses the preprocessor directive **#define**. This defines a preprocessor **macro**. A macro, in this sense, is simply a string of tokens that will be substituted for another string of tokens. As an example, in the preprocessor macro for the above program

#define square(x) x*x

when the statement

square(x)

appears in the program, the compiler will actually substitute

x*x

Thus, in the above program, when the macro is used

number = square(5)

the processor actually substitutes

number = 5*5

As you can see, the blanks on each side of the preprocessor directive serve to separate the tokens to be substituted. This is shown in Figure 2-9.

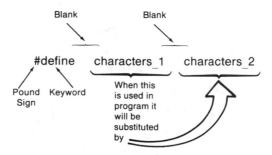

Figure 2-9   Construction of the **#define**

This preprocessor macro may now be used by any function anywhere in the program. Thus, to square any number (constant or variable) simply use

```
answer = square(number);
```

Defining Constants

Program 2-10 illustrates how to define a constant. In this program, the constant PI is defined. It is conventional in C programming to use all uppercase letters for any predefined constant. This helps reading the body of program code because it points out constants that have been defined by the system.

This program computes the area of a circle. It uses the preprocessor directives of defining PI and square(x).

Program 2-10

```c
#include <stdio.h>
#define square(x) x*x
#define PI 3.141592

main( )
{
 float area;    /* Circle area. */
 float radius; /* Circle radius. */

   printf("Give me the radius => ");
   scanf("%f",&radius);

   area = PI * square(radius);

   printf("The area of a circle with a radius of %f\n",radius);
   printf("is %f square units.",area);

}
```

Program Analysis

Observe that the above program now uses a preprocessor macro:

```
#define square(x) x*x
```

and defines a constant (in C, constants are written in uppercase):

```
#define PI 3.141592
```

Both statements start at the left side of the screen with the required ⊦ sign **#** immediately followed by the reserved word (in lowercase letters) **define**. Then come the required space and the tokens you want defined, then another space and the tokens that are to be substituted. From now on, in any function you write in the program using these directives, the token **PI** will be substituted with **3.141592** and the token **square(x)** will be substituted with **x*x**.

Execution of the above program yields

```
Give me a radius => 3
The area of a circle with a radius of 3.000000
is 28.274334 square units.
```

Defining Operations

Mathematical operations themselves that use prior macros may be included in a new macro. This is illustrated in the following program. Note that this time the formula for calculating the area of a circle is now itself a preprocessor macro. Program 2–11 does exactly the same thing as Program 2–10; it just uses more macros.

Program 2–11

```
#include <stdio.h>
#define square(x) x*x
#define PI 3.141592
#define circle_area(r) PI*square(r)

main( )
}
  float area;     /* Area of the circle. */
  float radius;   /* Radius of the circle. */

    printf("Give me the radius => ");
    scanf("%f",&radius);

    area = circle_area(radius);

    printf("The area of a circle with a radius of %f\n",radius);
    printf("is %f square units.",area);

}
```

Again execution of the above program will yield

```
Give me a radius => 3
The area of a circle with a radius of 3.00000
is 28.274334 square units.
```

Forms for #defines

It isn't necessary to use predefined macros for the circle area formula. Consider Program 2–12. Here there is simply one macro that defines circle_area(r). The advantage to this is that there are fewer macro definitions. However, the disadvantage is that some flexibility is lost because the value PI is never given a macro definition. The purpose of this program is to show that such an arrangement can be done for the C precompiler.

Program 2–12

```
#include <stdio.h>
#define circle_area(r) 3.141592*r*r

main( )
{
 float area;        /* Area of the circle. */
 float radius;      /* Radius of the circle. */

   printf("Give me the radius => ");
   scanf("%f",&radius);

   area = circle_area(radius);

   printf("The area of a circle with a radius of %f\n",radius);
   printf("is %f square units.",area);

}
```

Revisiting an Old Program

Recall the program of the last section that computed the reactance of a capacitor. The program could be rewritten so that the formula for capacitive reactance was given as a precompiler directive. Look at Program 2–13. It computes the value of the capacitive reactance. All that is needed is the value of the capacitor and the value of the frequency.

Program 2–13

```
#include <stdio.h>
#define PI 3.141592
#define capacitive_reactance(c,f) 1/(2*PI*f*c)

main( )
{
```

(continued)

Program 2–13 *(continued)*

```
float capacitance;    /* Value of the capacitor in farads. */
float frequency;      /* Value of the frequency in hertz. */
float cap_react;      /* Value of the capacitive reactance in ohms. */

  printf("Give me the capacitance => ");
  scanf("%f",&capacitance);
  printf("Give me the frequency => ");
  scanf("%f",&frequency);

  cap_react = capacitive_reactance(capacitance, frequency);

  printf("The reactance of a %e farad capacitor\n",capacitance);
  printf("with a frequency of %e hertz is %e ohms.",frequency, cap_react);

}
```

Program Analysis

The main feature of the above program is the use of a precompiler macro to define a complex formula:

#define capacitive_reactance(c,f) 1/(2*PI*f*c)

Note again that the tokens to be defined are one space from the **#define** directive. Note also the use of parameters (c,f,) that reserves two spaces for input by the arguments c and f. Following this is another space to indicate that what follows next is what is to be substituted: 1/(2*PI*f*c) A close look at this reveals that it is the formula for the capacitive reactance of a capacitor. This is exactly the same formula that was used in a separate function of a similar program presented in the last section. The difference here is that now the same thing is accomplished by use of a preprocessor macro.

Assuming the program user entered a value of 1 μf for the capacitor and a frequency of 5 kHz, execution of the above program would yield

```
Give me the capacitance => 1e-6
Give me the frequency => 5e3
The reactance of a 1.000000e-006 farad capacitor
with a frequency of 5.000000e+003 hertz is 3.183101e+001 ohms.
```

Expanding the Concept

You can take the idea of defining complex formulas in precompiler macros a step further. Program 2–14 shows that macros with parameters may be used to define other macros with parameters. In this program, the series LC impedance formula is defined by using the predefined formulas for capacitive reactance (now shortened to X_c) and inductive reactance (X_1). There is one addition. Remember that a square root function was needed by the impedance formula. Its prototype is contained in the C file called **math.h**. Thus, as shown in the program, this file has to be **#included**.

Program 2-14

```
#include <stdio.h>
#include <math.h>
#define PI 3.141592
#define X_c(c,f) 1/(2*PI*f*c)
#define X_1(1,f) 2*PI*f*1
#define series_impedance(c,1,f) sqrt(X_c(c,f)*X_c(c,f) + X_1(1,f)*X_1(1,f))

main( )
{
 float capacitance;    /* Value of capacitor in farads.   */
 float inductance;     /* Value of inductor in henrys.  */
 float frequency;      /* Value of frequency in hertz.  */
 float cap_react;      /* Value of capacitive reactance in ohms.   */
 float ind_react;      /* Value of inductive reactance in ohms.   */
 float imped;          /* Value of circuit impedance in ohms.  */

   printf("Give me the capacitance => ");
   scanf("%f",&capacitance);
   printf("Give me the inductor => ");
   scanf("%f",&inductance);
   printf("Give me the frequency => ");
   scanf("%f",&frequency);

    imped = series_impedance(capacitance, inductance, frequency);

   printf("The impedance of a series LC circuit with a \n");
   printf("capacitance of %e farads,",capacitance);
   printf("an inductance of %e henrys\n" inductance);
   printf("is %e ohms.",imped);

 }
```

Program Analysis

This program does exactly the same thing as the one in the last section that computed the value of series LC impedance. The difference with this one is that instead of using three different functions to define the formulas, this was done with the precompiler directives:

```
#define PI 3.141592
#define Xc (c,f) 1/(2*PI*f*c)
#define X_1(1,f) 2*PI*f*1
#define series_impedance(c,1,f) sqrt(X_c(c,f)*X_c(c,f) + X_1(1,f)*X_1(1,f))
```

The order is important—X_c has to be defined before `series_impedance` because it is used there.

All that is now necessary is to use the defined statement

`series_impedance(c,l,f)`

any time you wish to compute the impedance of this type of circuit.

Assuming that the program user entered a value of 2 μf for the value of the capacitor and 1 mh for the inductor at a frequency of 5 kHz, this program will yield an output of

```
Give me the capacitance => 2e-6
Give me the inductor => 1e-3
Give me the frequency => 5e3
The impedance of a series LC circuit with a
capacitance of 1.000000e-006 farads
an inductance of 1.000000e-003 henrys
is 4.472329e+001 ohms.
```

Including the **#define**

As you will see in the next section on program debugging and implementation, you may create a file of your own preprocessor macros and save it on your disk. You can then give the file a name (such as **mystuff.h**). That file could contain all of your necessary electronic formulas or other such formulas in your area of technical interest. Then, any C program you wish to write may use these formulas. All you need to do is to use the **#include** preprocessor directive in much the same way you did to get the **math.h** file so you could use the predefined **sqrt** function.

Conclusion

This was an exciting section. Here you saw a whole new aspect of the C language— a valuable programming tool for creating large and complex technical programs. You will find that these preprocessor directives will become very familiar and useful. For now, check your understanding of this section by trying the following section review.

2-5 Section Review

1 What is a preprocessor directive?
2 Explain what is meant by the **#include** directive.
3 What is a macro?
4 How are constants usually defined in C?
5 May a parameter be used with the **#define**? Give an example.

2-6   Program Debugging and Implementation: Making Your Own Header Files

Discussion

In this section you will discover how to make your own **header files** (`.h`) in C. With these, you can store all of your own preprocessor macros. This means that you can have a library of information for your area of technology to use with any of your programs.

An Example

Look at the following program.

Program 2–15

```
#include <stdio.h>
#define square(x) x*x
#define cube(x) x*x*x

main( )
 {
 float total;                \*  Value of the square.   *\

     total = square(5);

     printf("The square of 5 is %f",total);
 }
```

Program Analysis

The above program is similar to the one presented in the last section. This program uses two **#define** statements:

```
#define square(x) x*x
#define cube(x) x*x*x
```

The first statement defines `square(x)` as `x*x` and the second defines `cube(x)` as `x*x*x`. The **main( )** function uses the `square(x)` definition for demonstration purposes. Execution of the above program yields

```
The square of 5 is 25.000000
```

Saving Your #defines

Figure 2–10 illustrates the concept of taking the **#define** statements of the last program and saving them to a disk file.

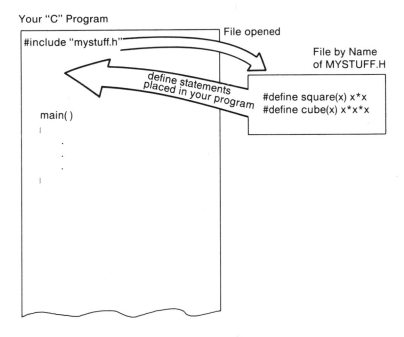

Figure 2-10   Concept of Saving the **#define** Statements

As shown in the figure, you could now use the file by stating

#include "mystuff.h"

Note that the quotation marks are used instead of the < > symbols to enclose the name of the header file. This tells the C compiler to look for this header file on the active drive (on a two drive system it will be drive B:, the one that contains your disk). Once you have done this, the previous program will now appear as follows:

Program 2-16

```
#include <stdio.h>
#include "mystuff.h"

main( )
{
float total;            \*  Value of the square.   *\

    total = square(5);

    printf("The square of 5 is %f",total);
}
```

A Sample Program

Consider the following program. It could be used by a technician at an integrated circuit manufacturing facility. Its **#defines** refer to information about some of the elements used in the manufacturing process of integrated circuits.

Program 2–17

```
#include <stdio.h>
#define head   "Element       Symbol  Atomic Number  Atomic Weight\n"
#define Ge      "Germanium      Ge        32             72.60\n"
#define Si      "Silicon        Si        14             28.09\n"
#define Au      "Gold           Au        79            197.2\n"
#define Ag      "Silver         Ag        47            107.88\n"
#define As      "Arsenic        As        33             74.91\n"
#define Ge_Wt   72.60
#define Si_Wt   28.09
#define Au_Wt   197.2
#define Ag_Wt   107.88
#define As_Wt   74.91

main( )
{
 float total;

 printf(head);
 printf(Ge);
 printf(Ag);
   total = Ge_Wt + Ag_Wt;
 puts("");
 printf("The combined atomic weights of germanium and silver is %f",total);
}
```

Program Analysis

The above program has **#include** statements that define strings:

```
#define head "Element       Symbol  Atomic Number  Atomic Weight\n"
#define Ge     "Germanium      Ge        32             72.60\n"
#define Si     "Silicon        Si        14             28.09\n"
#define Au     "Gold           Au        79            197.2\n"
#define Ag     "Silver         Ag        47            107.88\n"
#define As     "Arsenic        As        33             74.91\n"
```

As an example, the first **#define** defines the token head as the string (note the quotation marks):

```
"Element        Symbol  Atomic Number  Atomic Weight\n"
```

Note that the end of the string has the return to a new line command \n. What this means is when a **printf( )** uses the token **head**, the compiler will substitute this string instead. Thus

printf(head);

will produce

```
Element        Symbol  Atomic Number  Atomic Weight
```

upon program execution. The same is true of all the other tokens that have a string definition. For example,

printf(Ge);

produces

```
Germanium      Ge         32           72.60
```

upon program execution. Thus the source code of

printf(head);
printf(Ge);

produces

```
Element        Symbol  Atomic Number  Atomic Weight
Germanium      Ge         32           72.60
```

upon execution.

Now the technician may easily display information about specific atomic elements anywhere in the program simply by entering the symbol of the element in an output statement.

The remainder of the **#define** statements set numerical values to the tokens. These values represent the atomic weights of each corresponding element symbol.

```
#define Ge_Wt 72.60
#define Si_Wt 28.09
#define Au_Wt 197.2
#define Ag_Wt 107.88
#define As_Wt 74.91
```

Thus, the token Ge_Wt will have the numerical value 72.60 substituted. Assuming that total is a **float** type variable identifier, then

```
total = Ge_Wt + Si_Wt;
```

will have the compiler actually substitute

```
total = 72.60 + 28.09;
```

and produce a value of 100.69 for the sum of the two atomic weights of these elements.

Execution of the above program produces

```
Element      Symbol  Atomic Number  Atomic Weight
Germanium    Ge           32            72.60
Silver       Ag           47           107.88
```

The combined atomic weights of germanium and silver is 180.480000

Making the Header File

To produce the header file that would contain all of the **#define** statements for the above program you should create a program similar to the previous one and test all of the **#define** statements to ensure they produce the expected result(s) in your program. The next step, after saving your program, is to eliminate all of the lines of your program except for the **#define** statements. Your program will now appear as shown:

Program 2-18

```
#define head    "Element    Symbol  Atomic Number  Atomic Weight\n"
#define Ge      "Germanium    Ge           32            72.60\n"
#define Si      "Silicon      Si           14            28.09\n"
#define Au      "Gold         Au           79           197.2\n"
#define Ag      "Silver       Ag           47           107.88\n"
#define As      "Arsenic      As           33            74.91\n"
#define Ge_Wt   72.60
#define Si_Wt   28.09
#define Au_Wt   197.2
#define Ag_Wt   107.88
#define As_Wt   74.91
```

If your B: drive is the active drive, then save the **#defines** of Program 2-18 in a file called B: CHEM.H.

Using Your Header File

The following program illustrates the use of your new header file. Observe that it outputs exactly the same thing as the previous program.

Program 2–19

```
#include <stdio.h>
#include "chem.h"

main( )
 {
 float total;

 printf(head);
 printf(Ge);
 printf(Ag);
   total = Ge_Wt + Ag_Wt;
 puts("");
 printf("The combined atomic weights of germanium and
         silver is %f",total);

 }
```

The following reference will now invoke your header definitions:

#include "chem.h"

When your C system sees this (with the quotation marks instead of the < >), it will go to your disk, look for a file by that name and then actually bring the material in that file into your program.

Token Pasting and String-izing Operators

The ANSI C standard includes the preprocessor operation that allows **token pasting**. The token-pasting operator **##** allows you to join one token to another to create a third token. As an example, if you used the following **#defines** in your program:

```
#define PI_1     3.14159
#define PI_2     6.28318
#define Value(x) PI_##x
```

Then if Value(1) is declared, this will be replaced by PI_1 which has been defined as equal to 3.14159. In the same manner, Value(2) would be replaced by PI_2 which has been defined as equal to 6.28318.

The **string-izing** operator is the single **#** and it allows you to make a string out of an operand with a **#** prefix. This is accomplished by placing the operand in quotes. As an example, if you used the following **#define** at the beginning of your program:

```
#define present_value(x)  printf(#x " = %d",x)
```

then when `present_value(increment)` is declared, the preprocessor will generate the statement

```
printf("increment" " = %d",increment);
```

which reduces to:

```
printf("increment = %d",increment)
```

Conclusion

As you can see, header files are a powerful extension of the C language system. As you develop more programs, your library of header files (developed by you for your particular area of technology) will become one of your most valuable programming tools. The other fact about header files is that they protect your source code. Another person may have a copy of your source code to look at, but without access to your header file code, they can do little to replicate your program.

Check your understanding of this section by trying the following section review.

2–6 Section Review

1 State the advantage of creating your own header files.
2 As presented in this section, what information may be contained in your header files?
3 What is the extension given to a C header file?
4 Explain how you may call your header file from a C program.

2–7   Case Study

Discussion

The case study for this chapter takes you step by step through the development of a C program that utilizes the new information presented in this chapter. Here you will see how decisions are made during the development of a very specialized program. As you increase your understanding of C and its versatility, you will come to see that there are many options available to you concerning how to create your source code. One of the powerful points of this language is this versatility of the source code.

Because of this, what is presented in this case study (as well as the others), is an attempt to increase readability and understanding of what the program will do while still preserving the fundamental characteristics of the C language.

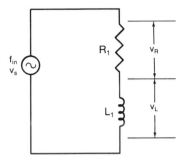

Figure 2-11 Circuit for Case Study

The Problem

Given the circuit of Figure 2-11, develop a C program that will compute the voltage drop across the resistor and the inductor.

First Step—Stating the Problem

Recall from the case study of the last chapter that the first step in designing a computer program is to state the problem in writing. As a guide, the following areas are the minimum requirements:

- Purpose of the Program
- Required Input (source)
- Process on Input
- Required Output (destination)

Stating the case study in writing yields

- Purpose of the Program: Compute the voltage drop in volts across an inductor and a resistor in a series RL circuit.
- Required Input: Values of the following:
  - Inductor in henrys
  - Resistor in ohms
  - Source voltage in volts
  - Applied frequency in hertz
  (Source: keyboard)

- Process on Input: Compute the voltage across the inductor in volts. Compute the voltage across the resistor in volts.

$$X_l = 2\pi fl$$
$$Z = \sqrt{(R^2 + X_l^2)}$$
$$v_l = v_s(X_l/Z)$$
$$v_R = v_s(R/Z)$$

Where

X_l = Inductive reactance in ohms.
f = Applied frequency in hertz.
l = Value of inductor in henrys.
Z = Circuit impedance in ohms.
R = Value of resistor in ohms.
v_l = Voltage across inductor in volts.
v_R = Voltage across resistor in volts.
v_s = Voltage of the source in volts.

- Required Output: The value of the voltage across the resistor: v_R, and the value of the voltage across the inductor: v_L (Displayed on monitor).

From this description, it should be clear to anyone what the program is required to do (and, just as important, what it will not do). The next step in the development of this program is to develop an algorithm. You can think of an algorithm as similar to a cake recipe. An algorithm is a step-by-step explanation of exactly what the program will do, written as concisely as possible (just like the cake recipe would be written).

Developing the Algorithm

If you have reduced the problem to writing including the required four areas (program purpose, required input, process on input and required output), then the algorithm may be developed straight from there. For most technology problems, the program will follow the order shown in the outline below:

The following is all commented (between /* and */, except for the function prototypes) and placed in the programmer's block sometimes referred to as the **prologue**.

    I. Program Information
       A. Name of program
       B. Name of programmer
       C. Date of development
          1. Optional date of last modification
             a. Name of person who did last modification
   II. Program Explanation
       A. What the program will do
       B. What is required for Input
       C. What process will be performed
       D. What the results will be
          1. State all units for variables
  III. Describe All Functions
       A. Use function prototypes
       B. Explain purpose of each prototype.
       C. Define all variables and constants.

The following are the programming steps that are used by most technical programs:

IV. Explain Program to User
   A. Purpose of the program
   B. What the user is required to do
       1. State all units for variables
   C. What the program will do
   D. What the final results will be
       1. State all units
V. Get Information from User
   A. Prompt user for input
   B. Acknowledge the input
VI. Perform the Process
   A. Do the process
VII. Display the Results
   A. Confirm what the user has entered
   B. Display all units
VIII. Ask for Program Repeat
   A. Ask if user wishes to repeat the program
   B. Give option to bypass instructions

The above outline will serve as the guide for this program.

First Coding

The first part of the program coding is to develop the programmer's block, compile it, save it to a disk file and then make a printed copy. There is no output, but the purpose is to ensure that the documentation of the program has already started. This first step is shown in Program 2–20.

Program 2–20

```
/*********************************************************/
/*           RL Circuit Voltage Drops                   */
/*********************************************************/
/*         Developed by: A. Good Programmer             */
/*********************************************************/
/*         Date: October 12, 1992                       */
/*********************************************************/
/*    This program will compute the voltage drop in     */
/* volts across an inductor and a resistor in a series*/
/* RL circuit. The program user must enter the value    */
/* of the inductor in henrys, resistor in ohms,         */
/* applied frequency in hertz and source voltage in     */
/* volts.                                               */
/*********************************************************/
```

(continued)

Program 2–20 *(continued)*

```
    /*-------------------------------------------------------*/
    void explain_program(void);
    /*  This function explains the operation of the      */
    /*  program to the program user.                     */
    /*-------------------------------------------------------*/
    void get_values(void);
    /*  This function gets the value of the resistor,    */
    /*  inductor source voltage and applied frequency,   */
    /*  calculates and displays the results.             */
    /*******************************************************/

    main( )
      {
      }

    void explain_program( )
      {
      }

    void get_values( )
      {
      }
```

Program Analysis

The above program includes a program title, developer's name and date. A brief description of the program follows. Note that all units of measurement are given in the program description. Also note that the programmer decided to use a single programming block that not only gets the values from the program user, but also does the calculations and displays the values. As the program development progresses, this may be too much coding in a single block, and the program may be more readable if this is divided into at least two blocks.

Entering the First Code

For this program, the next step is to do the coding for the user explanation block. After this is done, the program is compiled and executed. The block appears as shown below:

```
main( )
  {
    explain_program( );    /* Explain program to user.  */
    exit(0);
  }
```

```
/*----------------------------------------------------*/

void explain_program( )   /* Explain program to user.  */
{
    printf("\n\n This program will calculate the voltage drop\n");
    printf("across the inductor and the resistor of a series\n");
    printf("RL circuit. You must enter the value of the resistor\n");
    printf("in ohms, the inductor in henrys, the applied frequency\n");
    printf("in hertz and the value of the source voltage in volts.\n\n");

}    /* End of Explain program to user.  */
```

Note that **main( )** now contains a call to the function **explain_program**. This is so the output can be observed. It's important to see the use of the two **\n**'s at the beginning of the first **printf** function. This is good practice because it brings the first sentence down two spaces from the top of the monitor screen, thus making the text a little easier to read. Notice the ending comment **/* End of Explain program to user. */**. This may seem unnecessary for this small a function. However, it is a good practice to develop because it is a great aid in helping you find your way around larger programs.

Formula Decision

The program developer next has to decide if the formulas for the solution of this problem should be put into a separate function or into a header file. It was decided to create a header file for these formulas because they are formulas that will probably be used again in future programs.

The formulas were first developed using **#define** statements as shown below:

```
#include <stdio.h>
#include <math.h>
#define PI              3.141592
#define square(x)       x*x
#define X_L(f,l)        2*PI*f*l
#define Z(f,l,r)        sqrt(square(X_L(f,l)) + square(r))
#define V_L(f,l,r,v)    v*(X_L(f,l)/Z(f,l,r))
#define V_R(f,l,r,v)    v*(r/Z(f,l,r))
```

These are placed at the beginning of the program before the programmer's block. Observe that the **#include** files are used for input/output as well as the math function. Essentially the programmer followed the formulas as they are stated in the written statement of the problem.

Adding the Calculate and Display Block

Once the decision was made to use a header file for the formulas, it was then decided to use a separate program block to do the calculations and display the values.

The header files should not be made until the accuracy of the information to be included in them has been checked. For this program this may be done with a desk-check of the program.

Program 2–21

```c
#include <stdio.h>
#include <math.h>
#define PI          3.141592
#define square(x)   x*x
#define X_L(f,l)    2*PI*f*l
#define Z(f,l,r)    sqrt(square(X_L(f,l)) + square(r))
#define V_L(f,l,r,v)  v*(X_L(f,l)/Z(f,l,r))
#define V_R(f,l,r,v)  v*(r/Z(f,l,r))

/*****************************************************/
/*            RL Circuit Voltage Drops             */
/*****************************************************/
/*       Developed by: A. Good Programmer          */
/*****************************************************/
/*          Date: October 12, 1992                 */
/*****************************************************/
/*   This program will compute the voltage drop in */
/* volts across an inductor and a resistor in a series*/
/* RL circuit. The program user must enter the value */
/* of the inductor in henrys, resistor in ohms,    */
/* applied frequency in hertz and source voltage in */
/* volts.                                          */
/*****************************************************/
/*          Non-Standard Header Files Used:        */
/*-------------------------------------------------*/
/*  serrl.h => Series RL, this file contains the   */
/*             formula for the inductor and resistor */
/*             voltage drops when the above values are*/
/*             given.                              */
/*****************************************************/
/*              Function prototypes                */
/*-------------------------------------------------*/
float calculate_and_display(float f, float l, float r, float v);
/*  f = Frequency in hertz.                        */
/*  l = Inductance in henrys.                      */
/*  r = Resistance in ohms.                        */
/*  v = Source voltage in volts.                   */
/*  This is the function that will calculate and   */
/*  display the voltage drop across the resistor and */
/*  the voltage drop across the inductor.          */
/*-------------------------------------------------*/
void explain_program(void);
/*  This function explains the operation of the    */
/*  program to the program user.                   */
/*-------------------------------------------------*/
```

(continued)

Program 2–21 *(continued)*

```c
void get_values(void);
/*  This function gets the value of the resistor,    */
/*  inductor source voltage and applied frequency.   */
/********************************************************/


main( )
{

    explain_program( );    /* Explain program to user.  */
    get_values( );         /* Get circuit values.  */
    exit(0);
}


/*-------------------------------------------------------*/

void explain_program( ) /* Explain program to user.  */
{

    printf("\n\n This program will calculate the voltage drop\n");
    printf("across the inductor and the resistor of a series\n");
    printf("RL circuit. You must enter the value of the resistor\n");
    printf("in ohms, the inductor in henrys, the applied frequency\n");
    printf("in Hertz and the value of the source voltage in volts.\n\n");

}    /* End of explain program to user.  */
/*-------------------------------------------------------*/

void get_values( )      /* Get circuit values.  */
{
    float resistor;    /* Value of series resistor in ohms. */
    float inductor;    /* Value of the series inductor in henrys. */
    float frequency;   /* Value of the applied frequency in hertz. */
    float voltage_s;    /* Value of the source voltage in volts. */

    printf("Enter the following values as indicated:\n");
    printf("Resistor in ohms => ");
    scanf("%f",&resistor);
    printf("Inductor in henrys => ");
    scanf("%f",&inductor);
    printf("Frequency in hertz => ");
    scanf("%f",&frequency);
    printf("Source voltage in volts => ");
    scanf("%f",&voltage_s);
  calculate_and_display(frequency,inductor,resistor,voltage_s);

}    /* End of get_values  */

/*-------------------------------------------------------*/
```

(continued)

Program 2–21 *(continued)*

```
float calculate_and_display(float f,float l, float r,float v)
{
 float inductor_v;     /* Value of voltage across inductor.  */
 float resistor_v;     /* Value of voltage across resistor.  */

    inductor_v = V_L(f,l,r,v);   /* Calculate inductor voltage.  */

    resistor_v = V_R(f,l,r,v);   /* Calculate resistor voltage.  */

    printf("\n\nFor a series LR circuit consisting of a\n");
    printf("%e henry inductor with a %e ohm resistor\n",l,r);
    printf("with an applied frequency of %e hertz\n",f);
    printf("and a source voltge of %e volts, the\n",v);
    printf("component voltage drops are: \n\n");
    printf("Inductor voltage => %e volts\n",inductor_v);
    printf("Resistor voltage => %e volts\n",resistor_v);

 }   /*  End of calculate_and_display. */

/*-------------------------------------------------------------*/
  /*  End of program.    */
```

Program Analysis

One of the important additions to the above program is the explanation of the variables used as formal arguments in a function prototype. This was done in the function that will be used to calculate and display the resultant answer as shown below:

```
float calculate_and_display(float f, float l, float r, float v);
/*  f = Frequency in hertz.                             */
/*  l = Inductance in henrys.                           */
/*  r = Resistance in ohms.                             */
/*  v = Source voltage in volts.                        */
/*  This is the function that will calculate and        */
/*  display the voltage drop across the resistor and    */
/*  the voltage drop across the inductor.               */
/*----------------------------------------------------- */
```

Observe that the comments define exactly what each identifier uses in the formal function parameters. As always, a brief description of the function purpose follows the prototype. It's important to note that the prototype is not between /* and */ because the prototype is serving as an actual instruction to the C compiler.

The function `get_values` had a call to the `calculate_and_display` function. The actual parameters used by the function in getting information from the program user are

```
calculate_and_display(frequency,inductor,resistor,voltage_s);
```

The `calculate_and_display` function definition uses the same identical header as its prototype (with the exception that the semicolon is omitted). Notice how easy it is to code the calculations for these complex formulas:

```
float calculate_and_display(float f,float l, float r,float v)
{
 float inductor_v;      /* Value of voltage across inductor.  */
 float resistor_v;      /* Value of voltage across resistor.  */

   inductor_v = V_L(f,l,r,v);    /* Calculate inductor voltage.  */

   resistor_v = V_R(f,l,r,v);    /* Calculate resistor voltage.  */

   printf("\n\nFor a series LR circuit consisting of a\n");
   printf("%e henry inductor with a %e ohm resistor\n",l,r);
   printf("with an applied frequency of %e hertz\n",f);
   printf("and a source voltage of %e volts, the\n",v);
   printf("component voltage drops are: \n\n");
   printf("Inductor voltage => %e volts\n",inductor_v);
   printf("Resistor voltage => %e volts\n",resistor_v);

}    /* End of calculate_and_display.  */
```

The output of this program echoes the user input using E notation along with all units of measurement. The solutions are also displayed using E notation as well as the important units of measurement.

Final Program

Since the coding for giving the program user the option of repeating a program over again has not yet been presented, it was omitted from this one. The last step of the coding was to create a header file called `serrl.h` that contains the following information:

```
#define PI              3.141592
#define square(x)       x*x
#define X_L(f,l)        2*PI*f*l
#define Z(f,l,r)        sqrt(square(X_L(f,l)) + square(r))
#define V_L(f,l,r,v)    v*(X_L(f,l)/Z(f,l,r))
#define V_R(f,l,r,v)    v*(r/Z(f,l,r))
```

Thus the top of the final program includes a listing of **#include** statements:

```
#include <stdio.h>
#include <math.h>
#include "serrl.h"
```

Conclusion

You have come a long way in your observation of the development of a C program. It's important that you apply the principles presented here to the programs that you develop. Careful planning and program documentation are important in serious programming. These steps are used by professional programmers who all know the old saying: "The sooner you start to enter program code, the longer it will take to debug the final program." There are usually good reasons for old sayings.

Check your understanding of this section by trying the following section review.

2-7 Section Review

1   State one of the main goals in the development of the case study for this section.
2   What should be included in the programmer's block?
3   Explain the reason for developing **#define** statements.
4   What is the last thing that is usually asked of the program user in a typical technology program?

Interactive Exercises

DIRECTIONS

These exercises require that you have access to a computer and software that supports C. They are provided here to give you valuable experience and most importantly immediate feedback on what the concepts and commands introduced in this chapter will do. They are also fun.

Exercises

1   Predict what the following program will do, then try it. Did you get a call to the next function? You should have.

Program 2-22

```
#include <stdio.h>

main( )
{
  This_One( );
}

This_One( )
{
  puts("Here I am.");
}
```

2   Predict what you think the next program will do, then try it. You should be able to call all of the functions.

Program 2-23

```
#include <stdio.h>

main( )
{
 puts("This is main( )");
 First_One( );
}

 First_One( )
 {
  puts("This is First_One");
   Second_One( );
  }

 Second_One( )
 {
  puts("This is Second_One");
 }
```

3 The next program is similar to the last one in that you will need to press Ctrl-Break to stop it. The difference here is that this function calls itself. This is an example of recursion.

Program 2-24

```
#include <stdio.h>

main( )
{
 Recursive_One( );
}

 Recursive_One( )
 {
  puts("This is an example of recursion.");
  Recursive_One( );
  }
```

4 C is such a versatile language that you can make it look like almost any other language you wish. The following program redefines C commands. These could be put in a header file called **newcode.h**. Try it to prove to yourself that it does work.

Program 2–25

```
#include <stdio.h>
#define START   main( ){
#define END      }
#define WRITE   puts

  START
   WRITE("This is a C program?");
  END
```

Self-Test

DIRECTIONS

Use Program 2–21 developed in the case study section of this chapter to answer the following questions:

Questions

1 How many functions are defined in the program? Name them.
2 How many total functions are used in the program? Name those not named in question 1 above.
3 List the identifiers used for formal parameters in the program.
4 List the identifiers used for actual parameters in the program.
5 Which function passes variables to another function?
6 Explain how values are passed from one function to another using Program 2–21 as an example.
7 How many variable identifiers are used in the program? Name them.
8 What is the minimum change you would make in this program to display the value of the inductive reactance (X_L)?
9 Will the program accept user input in E notation? Explain.

End-of-Chapter Problems

General Concepts

Section 2–1

1 Explain what is meant by block structure.
2 State the completeness theorem.
3 Which type of block has the ability to go back and repeat a part of the program?
4 Which type of block has the ability to execute a different sequence of program code?

Section 2–2

5 In a C program, what gives the compiler specific information about the functions that will be defined in the program?

6   State the difference between the function prototype and the function header used to define the function.

7   How is a function called?

8   What C function causes normal program termination? What value is assigned to it to indicate normal termination?

Section 2-3

9   What type is assigned to a function when no value is to be returned by the function?

10   State what is used in a function to define the number and types of data that will be passed to the function when it is called.

11   What is used to return a value from the called function to the calling function?

Section 2-4

12   What is it called when a function calls itself?

13   May a function call more than one other function?

14   Does is make any difference in what order functions are called?

15   May a called function call another function?

Section 2-5

16   What is a preprocessor directive?

17   Name the preprocessor directive presented in this chapter.

18   Explain how constants are usually defined in C.

19   Give an example of how parameters may be used with the **#define**.

Section 2-6

20   Explain how you can make a file of your own **#define** statements for use in other programs.

21   How do you invoke your own header file?

22   What is the accepted extension given to C header files?

Program Design

In developing the following C programs, use the method described in the case study section of this chapter. For each program you are assigned, document your design process and hand it in with your program. This process should include the design outline stating the purpose of the program, required input, process on the input, and required output as well as the program algorithm. Be sure to include all of the documentation in your final program. This should consist of, but not be limited to, the programmer's block, function prototypes and a description of each function as well as any formal parameters you may use.

Electronics Technology

23   Develop a C program that will compute the power delivered by a voltage source. The relationship is

$$P = I \times E$$

Where

$P = $ The power delivered in watts.

$I = $ The source current in amps.

$E = $ The source voltage in volts.

Do not use any **#define** headers for this program. Use functions for each major part of the program. One function will explain the program to the user and the other will get the values and pass them on to the function that defines the power formula.

24 Modify the above program so that the power formula is now included in a header file instead of being a separate function.

25 Create a C program that will compute the voltage drop across each resistor in a series circuit consisting of three resistors. The user input is to be the value of each resistor and the applied circuit voltage. The mathematical relationships are

$$R_T = R_1 + R_2 + R_3$$
$$I_T = V_S / R_T$$
$$V_N = I_T \times R_N$$

Where

R_T = Total circuit resistance in ohms.
R_1, R_2, R_3 = Value of each resistor in ohms.
I_T = Total circuit current in amps.
V_N = Voltage drop across an individual resistor N in volts.
V_S = Applied circuit voltage.

Use only functions for this program; do not use any **#define** statements.

26 Redo the above program using **#define** statements. Also include the total circuit resistance, total circuit current as well as the voltage drop across each resistor as part of the output.

27 Expand the program in problem 25 to include the power dissipated by each resistor as well as the power dissipated by the voltage source. The relationship for resistor power dissipation is

$$P_{RN} = I_T \times R_N$$

Where

P_{RN} = Power dissipated by individual resistor N in watts.
I_T = Total circuit current in amps.
R_N = Individual resistor value in ohms.

$$P_S = I_T \times V_S$$

Where

P_S = Power of the source.
V_S = Source of voltage.

Manufacturing Technology

28 Create a C program that will compute the volume of a magnetic ring or circular area. The volume is given by

$$V = 2\pi^2 R r^2$$

Where

V = The volume in square units.
R = Radius of the ring.
r = Radius of the cross-sectional area.

See Figure 2–12.

Drafting Technology

29 Create a C program that will compute the area of a segment of a circle. The relationships are shown in Figure 2–13.

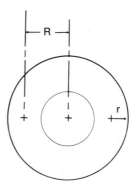

Figure 2-12   Magnetic Ring for Manufacturing Technology

Computer Science

30   Develop a C program that will return to the user the factorial of a number from 1 to 10. The factorial of 3 is expressed as 3! and means $1 \times 2 \times 3 = 6$.

Construction Technology

31   Write a C program that will compute the volume of three different rooms. Assume that each room is a perfect rectangle. User input is the height, width and length of each room. The program is to return the volume of each separate room as well as the total volume of all three rooms.

Agriculture Technology

32   Referring to a standard table of conversions, create a header file that will convert acres to square feet, and convert square feet to square meters as well as hectares. Use this file in a C program that demonstrates the correctness of the conversions.

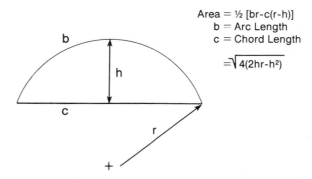

$$\text{Area} = \frac{1}{2} \, [br - c(r-h)]$$
$$b = \text{Arc Length}$$
$$c = \text{Chord Length}$$

$$= \sqrt{4(2hr - h^2)}$$

Figure 2-13   Relationships for Area of Circle Segment

Health Technology

33 Develop a C program that will compute the total bill of a hospital patient. The user inputs are

1. Number of days in hospital
2. Surgery cost
3. Medication cost
4. Miscellaneous cost
5. Cost per day
6. Insurance deductible

The program must compute the following for insurance purposes:

1. Total cost
2. Total cost less insurance deductible
3. Total cost less cost of medication and deductible

Business Applications

34 Design a C program that will balance two different accounts consisting of five items each. The user inputs are

1. Number of widgets ordered
2. Number of widgets shipped
3. Shipping cost per widget
4. Cost of each widget
5. Handling charge per unit

The program output is to display

1. Number of widgets on back order
2. Total cost of widgets shipped
3. Total cost when all widgets are shipped

3 Branching and Logic

Objectives

This chapter gives you the opportunity to learn:

1 The meaning and use of relational operators.
2 The difference between an open branch and a closed branch.
3 Technical applications of opened and closed branches.
4 Logical operations used in C.
5 Applications of logical and relational operations to programming branches.
6 Methods of choosing one of several alternatives.
7 Methods of working with different data types and how to cast a type.
8 The use and methods of conditional compilation.
9 Special decision making used by C.
10 Program development from existing technical flow charts.

Key Terms

Relational Operators
TRUE
FALSE
Assignment
Open Branch
if Statement
Conditional Statement
Compound Statement
Closed Branch
if...else Statement

if...else if...else
Logical AND
Logical OR
Logical NOT
rank
casting
lvalue
Conditional Compilation
Program Stubs

Outline

3-1 Relational Operators
3-2 The Open Branch
3-3 The Closed Branch
3-4 Logical Operation
3-5 Conversion and Type Casting
3-6 The C **switch**

3-7 One More **switch** and the
    Conditional Operator
3-8 Program Debugging and
    Implementation
3-9 Case Study

Introduction

In this chapter you will learn how to develop C programs that make the computer look "smart." You will find out how to create programs that will allow program decisions to be made. Doing this allows the program flow to branch in one direction or another depending upon the input from a user or the results of a computation.

The chapter starts by explaining how quantities may be related. These relationships may then be used in the decision-making process of the program. You will then be introduced to the logic used by C and its symbolism. Here you will discover the simplicity and power contained in the TRUE and FALSE logic of a C program.

The chapter concludes with an interesting case study for troubleshooting a robot. When you finish with the material in this chapter, you will be well on your way to developing programs that solve and analyze technology problems.

3-1   Relational Operators

Discussion

This section introduces **relational operators**. As you will discover, these operators show the relationship between two quantities. You will use this information as a foundation for the material in the next section. There, you will learn how to create C programs that will exhibit decision-making capabilities.

Relational Operators

Relational operators are symbols that indicate a relationship between two quantities. These quantities may be variables, constants or functions. The important point about these relations is that they are either **TRUE** or they are **FALSE**—there is nothing in between!

The relational operators used in C are shown in Table 3-1.

Table 3-1   Relational Operators Used in C

Symbol	Meaning	TRUE Examples	FALSE Examples
>	Greater than.	5 > 3 (3+8) > (5−2)	3 > 5 (12/6) > 18
>=	Greater than or equal to.	10 >= 10 (3*4) >= 8/2	3 >= 5 8+5 >= 10*15
<	Less than.	3 < 5 12/3 < 12*3	5 < 3 9−2 < 3+1
<=	Less than or equal to.	3 <= 15 18/6 <= 9/3	15 <= 3 12+4 <= 12/4
==	Equal to.	5 == 5 2+7 == 18/2	10 == 5 8 − 5 == 2 + 4
!=	Not equal to.	8 != 5 8−5 != 2+4	5 != 5 24/6 != 12/3

For relational operators, the value returned for a TRUE condition is 1 while the value returned for a FALSE condition is a 0. This is illustrated by the following program.

Program 3-1

```
#include <stdio.h>

main( )
{
 float logic_value;    /* Numeric value of relational expression.  */

   printf("Logic values of the following relations:\n\n");

   logic_value = (3 > 5);
   printf("(3 > 5) is %f\n",logic_value);

   logic_value = (5 > 3);
   printf("(5 > 3) is %f\n",logic_value);

   logic_value = (3 >= 5);
   printf("(3 >= 5) is %f\n",logic_value);

   logic_value = (15 >= 3*5);
   printf("(15 >= 3*5) is %f\n",logic_value);

   logic_value = (8 < (10−2));
   printf("(8 < (10−2)) is %f\n",logic_value);
```

(continued)

Program 3–1 *(continued)*

```
    logic_value = (2*3 < 24/3);
    printf("(2*3 < 24/3) is %f\n",logic_value);

    logic_value = (10 < 5);
    printf("(10 < 5) is %f\n", logic_value);

    logic_value = (24 <= 15);
    printf("(24 <= 15) is %f\n",logic_value);

    logic_value = (36/6 <= 2*3);
    printf("(36/6 <= 2*3) is %f\n",logic_value);

    logic_value = (8 == 8);
    printf("(8 == 8) is %f\n",logic_value);

    logic_value = (12+5 == 15);
    printf("(12+5 == 15) is %f\n",logic_value);

    logic_value = (8 != 5);
    printf("(8 != 5) is %f\n",logic_value);

    logic_value = (15 != 3*5);
    printf("(15 != 3*5) is %f\n",logic_value);

}
```

Program Analysis

Execution of the above program yields

```
Logic values of the following relations:
(3 > 5) is 0.000000
(5 > 3) is 1.000000
(3 >= 5) is 0.000000
(15 >= 3*5) is 1.000000
(8 < (10−2)) is 0.000000
(2*3 < 24/3) is 1.000000
(10 < 5) is 0.000000
(24 <= 15) is 0.000000
(36/6 <= 2*3) is 1.000000
(8 == 8) is 1.000000
(12+5 == 15) is 0.000000
(8 != 5) is 1.000000
(15 != 3*5) is 0.000000
```

As you can see from the above output, a relational operation in C returns a value of either 1 or 0. If the operation is TRUE, a value of 1 is returned, if FALSE, a value of 0 is returned.

To illustrate a portion of the program, consider the following program excerpt:

```
logic_value = (3 > 5);
printf("(3 > 5) is %f\n",logic_value);

logic_value = (5 > 3);
printf("(5 > 3) is %f\n",logic_value);
```

The variable `logic_value` has been declared a type **float**. You may wish to try this with other types. It is being set to the value of the relational operation (3 > 5). This statement is FALSE; therefore, the value returned will be a 0. This value is displayed to the monitor as a type **float (%f)** as 0.000000. In the following program line, the variable `logic_value` is being set to the value of the relational operation (5 > 3). This is a TRUE statement and hence the value returned will be 1. This is displayed again as a type **float** producing the display of 1.000000.

The remainder of the program continues with the same type of process.

Equal To

It may at first seem strange that C uses the == (double equals or "equal-equal") to mean equal to. Understandably, many who are new to C may have thought that the use of the = (single equals) meant equal to. The fact is that the = (single equals) does not mean the same in C as it does in ordinary math. What the = means in C is **assignment**. The concept of assignment is illustrated in Figure 3–1.

It's important to make the distinction between the assignment operator (=) and the equals operator (==) in C. The assignment operator takes the value on the right side of the assignment statement and puts it in the memory location of the variable on the left side of the assignment. The equals operator does something different. It simply compares the value of one memory location to the value of another memory location. No transfer of data from one memory location to the other takes place.

Check your understanding of relational operators by trying the following example.

EXAMPLE 3-1

_____

Assume that the following assignments have been made:

$$a = 5, \quad b = 10, \quad c = 5$$

State the final values of each of the following comparisons:

A. a == 5	C. a >= c
B. c == b	D. b != a

Solution

A. a == 5 is TRUE, value is 1.
B. c == b is FALSE, value is 0.
C. a >= c is TRUE, value is 1.
D. b != a is TRUE, value is 1.

_____

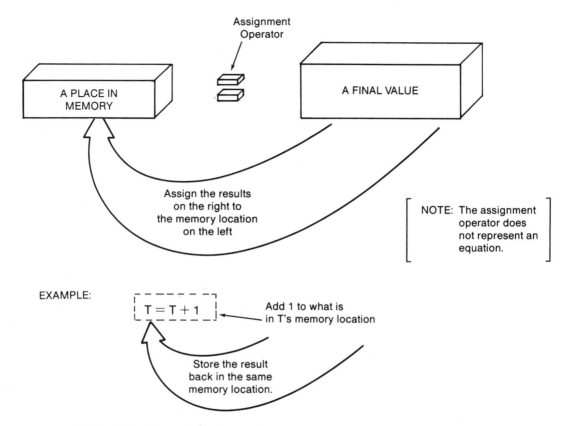

Figure 3-1 Concept of Assignment

Conclusion

In this section you were introduced to relational operators used in C. You saw the symbols used for each of them and saw that there are only two possible conditions—TRUE or FALSE—and nothing in between. You also saw that a C relational operator actually returns a numerical value. The value returned for a TRUE operation is 1, and the value returned for a FALSE operation is 0. This chapter also made the important distinction between the assignment operator (=) and the equals operator (==).

In the next section you will discover how to use what you have learned here to develop programs that will exhibit decision-making capabilities. For now, test your understanding of this section by trying the following section review.

3-1 Section Review

1 What are relational operators?
2 State all of the relational operators used in C.

3   What are the two conditions that relational operators may have?
4   Explain what is meant by the statement that a relational operation in C returns a value.
5   State the difference between the C operation symbols = and ==.

3-2   The Open Branch

Discussion

This section will show you how to use the relational operations presented in the last section. Here you will see how to code decision-making capabilities using C.

The open branch is presented here. This is perhaps the simplest of the computer's decision-making capabilities. Even so, you will find this a powerful addition to your C programming skills.

Basic Idea

The basic idea of an **open branch** is illustrated in Figure 3-2.

There are two important points about the open branch. First, the flow of the program always goes forward to new information. Second, the option may or may not be used—but the remainder of the program is always executed. In the C programming language, the open branch is accomplished by the **if statement**.

The if Statement

The **if** statement in C is referred to as a **conditional statement** because its execution will depend upon a specific condition. The form of the **if** statement is

```
if (expression) statement
```

What this means is if **expression** is TRUE, then **statement** will be executed. If **expression** is FALSE, **statement** will not be executed. This is illustrated in the following program.

Program 3-2

```
#include <stdio.h>

main( )
{
  float number;        /* Value inputted by user.  */

    printf("Give me a number from 1 to 10 => ");
    scanf("%f",&number);
```

(continued)

Program 3-2 *(continued)*

```
        if (number > 5)
        printf("Your number is larger than 5.\n");

   printf("%f was the number you entered.",number);

}
```

Program Analysis

This program asks the user to enter a number from 1 to 10. If the user enters a number from 1 to 5 (assume 3 was entered), the program will display

```
3 was the number you entered.
```

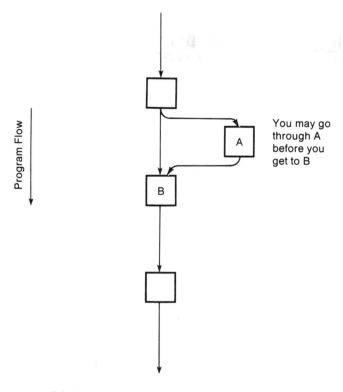

Figure 3-2 Concept of the Open Branch

If the program user enters a number greater than 5 (say 7), then the program displays

```
Your number is larger than 5.
7 was the number you entered.
```

You can see the use of the **if** statement as

```
if (number > 5)
printf("Your number is larger than 5.\n");
```

The expression is a relational one that compares the value of the user input to see if it is greater than 5. If it is, then the expression is TRUE and the statement

```
printf("Your number is larger than 5.\n");
```

will be executed.

If the relational operation inside the expression is FALSE then the statement will not be executed.

Note how the **if** statement is indented from the rest of the program. Also note that there is no semicolon between the expression and the statement (no semicolon following **if** (number > 5)).

The important point to see here is that the program always goes forward and may or may execute the conditional part of the program (depending on whether the value of the user input is greater than 5). However, regardless of the user input, the remainder of the program is always executed (the last **printf** function output is always displayed).

A Compound Statement

The above example illustrates the use of the **if** statement with only one statement following it. This is fine when all you need to output is a simple single statement. But suppose you need to develop a program that has more than one statement as a condition? For example, suppose your task is to develop a C program that takes the value of a voltage measurement across a resistor from the program user. The condition for the program is that if the voltage value is 100 volts or greater, then the program will compute the power dissipation in the resistor. If the voltage value is not equal to or greater than 100 volts, then the resistor power dissipation will not be computed. This idea is illustrated in Figure 3–3.

When more than one statement must be included in the logical space reserved for a statement, then a **compound statement** may be used. This is nothing more then a series of C statements enclosed within braces { }. The following program illustrates.

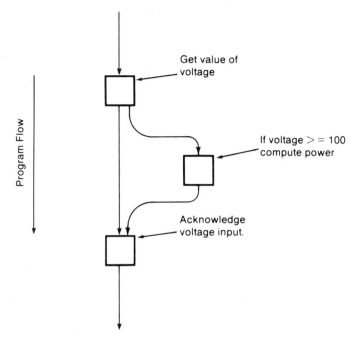

Figure 3-3   Optional Power Dissipation Computation

Program 3-3

```
#include <stdio.h>

main( )
{
  float voltage;          /* Voltage measurement in volts.  */
  float resistor;         /* Resistance value in ohms.      */
  float power;            /* Power calculation in watts.    */

    printf("Enter the voltage reading in volts => ");
    scanf("%f",&voltage);
```

(continued)

Program 3-3 *(continued)*

```
if (voltage >= 100.0)
    {
        printf("Voltage is equal to or greater than 100 V\n");
        printf("Please enter the resistor value => ");
        scanf("%f",&resistor);

        power = voltage*voltage/resistor;

        printf("The power dissipation is %f watts.\n",power);
    }

    printf("Input value of %f volts is acknowledged.",voltage);
    exit(0);
    }
```

Figure 3-4 illustrates the logical construction of the above program.

Program Analysis

The program illustrates the use of a compound C statement:

```
{
        printf("Voltage is equal to or greater than 100 V\n");
        printf("Please enter the resistor value => ");
        scanf("%f",&resistor);

        power = voltage*voltage/resistor;

        printf("The power dissipation is %f watts.\n",power);
}
```

Note that all of this part of the program is enclosed by braces { }. What is between these braces is effectively another program block (just like the { } used in **main** or any other C function). This new program block is viewed by the **if** statement as another statement. This concept is illustrated in Figure 3-5.

So as far as the C program is concerned, the relational operation of

```
if (voltage >= 100.0)
```

will cause the execution of a single statement—which in this case is actually a compound statement.

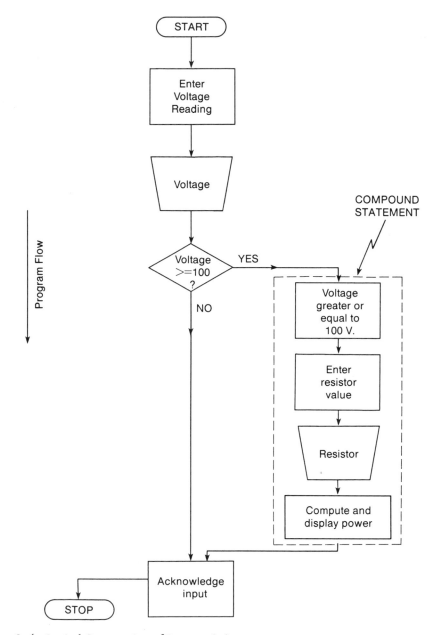

Figure 3-4 Logical Construction of Program 3-3

Thus, in the above program, if the program is executed and the user inputs a value of 25 volts across a 500 Ω resistor the output will be

```
Enter the voltage reading in volts => 25
Value of 25 volts is acknowledged.
```

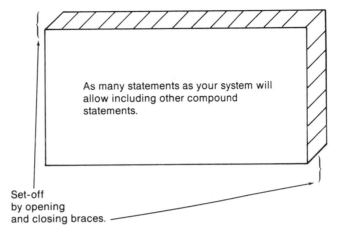

As many statements as your system will allow including other compound statements.

Set-off
by opening
and closing braces.

Figure 3-5   Concept of Viewing a Compound Statement

As you can see from the above output, the **if** statement was not activated because the relation

$$voltage >= 100.0$$

is FALSE.

However, if the program user now enters a voltage reading of 125 volts across a 500 Ω resistor the output is

```
Enter the voltage reading in volts => 125
Voltage is equal to or greater than 100 V
Please enter the resistor value => 500
The power dissipation is 31.250000 watts
Input value of 125 volts is acknowledged.
```

Observe from the above output that the last **printf** function is still executed.

Calling a Function

Since the **if** statement can be used with a compound statement, it seems natural to ask if it can be used to call another C function (which could also contain some more **if** statements). The answer is yes it can. This is illustrated in the following program.

Program 3-4

```
#include <stdio.h>

void power_calculation(float voltage);
```

(continued)

Program 3–4 *(continued)*

```
main( )
{
  float voltage;          /* Voltage measurement in volts.  */

    printf("Enter the voltage reading in volts => ");
    scanf("%f",&voltage);

 if (voltage >= 100.0)
    power_calculation(voltage);

    printf("Input value of %f volts is acknowledged.",voltage);
    exit(0);
 }

  void power_calculation(float voltage)
  {
   float resistor;    /* Resistance value in ohms. */
   float power;       /* Power dissipation in watts. */

    printf("Voltage is equal to or greater than 100 V\n");
    printf("Please enter the resistor value => ");
    scanf("%f",&resistor);

   power = voltage*voltage/resistor;

    printf("The power dissipation is %f watts.\n",power);

 }
```

Program Analysis

This program illustrates good programming practice in that it is divided into two separate parts. This doesn't mean that the more parts a program has the better it is. It means that each distinct task should be distinguished within the program. This program is divided by putting the optional power dissipation calculation in a separate function of its own. Now the **if** statement in the body of **main( )** simply becomes

```
if (voltage >= 100.0)
   power_calculation(voltage);
```

This is much easier to read. It means that if the voltage value entered by the program user is greater than or equal to 100 volts, then do a power calculation using the voltage value as a part of that calculation. Here, you can see that the value of the voltage is passed to the function `power_calculation`. The function prototype at the beginning of the program

```
void power_calculation(float voltage);
```

tells the compiler to expect a **void** function (it will not return any value) that has one argument of type **float**. The function definition header

```
void power_calculation(float voltage)
```

is identical to the prototype (except, of course, there are no ending semicolons). The declarations of the resistor value and power value are now in the function that uses them (**main( )** is no longer cluttered with them):

```
float resistor;    /* Resistance value in ohms. */
float power;       /* Power dissipation in watts. */
```

The output of the program is exactly the same as the one before it. The difference is that this program is easier to read and debug.

Conclusion

This section demonstrated the use of the **if** statement as a conditional C statement in an open branch. Here you saw that the **if** statement consists of an expression and a statement. If the expression is TRUE then the statement will be executed. If the expression is FALSE then the statement will not be executed. In either case, the remainder of the program is always executed.

You were also introduced to the compound statement. In C this consists of using braces { } to set off more than one statement. Doing this allows many different operations to be a part of a conditional statement. The practice of using the **if** statement to call another C function was also presented here.

In the next section, you will be introduced to the closed branch. For now, test your understanding of this section by trying the following section review.

3–2 Section Review

1 Describe an open branch.
2 Explain the operation of the **if** statement.
3 What is a compound statement?
4 May an **if** statement call a function? Explain.

3–3   The Closed Branch

Discussion

This section demonstrates the action of the closed branch in C. The **closed branch** extends the decision-making capabilities of the language. As you will see, the closed branch statement in C is very similar to the open branch statement.

Basic Idea

The basic idea of a *closed branch* is illustrated in Figure 3–6.

There are two important points about the closed branch. First, the flow of the program always goes forward to new information. Second, the program will do one of two options (not both) and then proceed with the rest of the program. In C the closed branch is accomplished by the **if...else statement**.

The if...else Statement

The **if...else** statement in C is another conditional statement. It differs from the **if** statement in that the **if** statement represents an open branch while the **if...else** statement represents a closed branch. The form of this statement is

if (expression) statement₁ else statement₂

What this means is that if **expression** is TRUE, then **statement₁** will be executed, and **statement₂** will not be executed. If, on the other hand, **expression** is FALSE, then **statement₁** will not be executed and **statement₂** will be executed. This is illustrated in the following program.

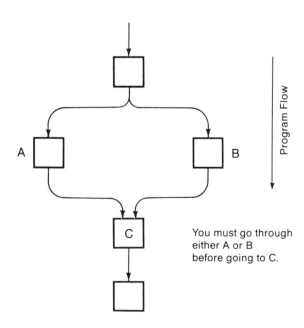

Figure 3–6   Concept of a Closed Branch

Program 3-5

```c
#include <stdio.h>

main( )
{
float number;    /*  User number value.  */

  printf("\n\nGive me a number from 1 to 10 => ");
  scanf("%f",&number);

  if (number > 5.0)
    printf("Your number is greater than 5.\n");
  else
    printf("Your number is less than or equal to 5.\n");

  printf("The value of your number was %f",number);

}
```

Figure 3–7 illustrates the construction of the above program.

Program Analysis

Note the construction of the **if...else** in the above program.

```c
if (number > 5.0)
 printf("Your number is greater than 5.\n");
else
 printf("Your number is less than or equal to 5.\n");
```

The general form of the **if...else** statement presents itself with the expression

```c
if(number > 5.0)
```

and

```c
printf("Your number is greater than 5.\n");
```

as statement$_1$, and

```c
printf("Your number is less than 5.\n");
```

as statement$_2$.

This **if...else** combination forces a selection between one of two **printf** functions. If the number entered by the program user is greater than 5, then the expression (number > 5) will be TRUE and statement$_1$ will be executed yielding

```
Your number is greater than 5.
```

and the program will skip the **printf** function following the **else**.

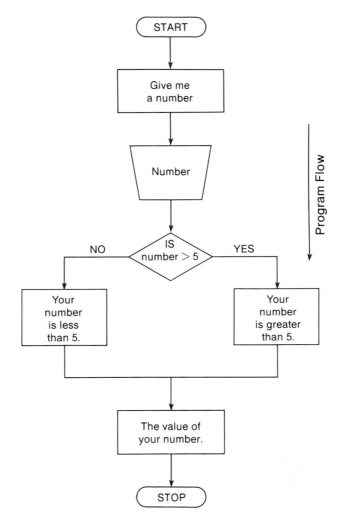

Figure 3-7 Construction of Program 3-5

If the program user inputs a 4, then the expression (`number > 5`) will be FALSE and statement₁ will not be executed. However, now statement₂ following the **else** will be executed.

In either case, program flow always goes forward and the final **printf** function is always executed:

```
printf("The value of your number was %f",number);
```

Compound if...else

You can add even more power to your program's decision making capabilities through the use of compound **if...else** statements. This is illustrated in the following program. The program will compute the area of a square or the area of a circle. The program user selects which computation is to be performed.

Program 3–6

```c
#include <stdio.h>
#define PI 3.141592

main( )
{
float selection;  /* User selection.  */
float length;     /* Length of side or radius.  */
float area;       /* Area in square units.  */

   printf("\n\nThis program will compute the area of\n");
   printf("a square or the area of a circle.\n");

   printf("\nSelect by number:\n");
   printf("1] Area of circle. 2] Area of square.\n");
   printf("Your selection (1 or 2) => ");
   scanf("%f",&selection);

   if (selection == 1)
     {
       printf("Give me the length of the circle radius => ");
       scanf("%f",&length);

     area = PI * length * length;

       printf("A circle of radius %f has an area of ",length);
       printf("%f square units.",area);
     }
   else
   if (selection == 2)
     {
       printf("Give me the length of one side of the square => ");
       scanf("%f",&length);
```

(continued)

Program 3-6 *(continued)*

```
        area = length * length;
        printf("A square of length %f has an area of ",length);
        printf("%f square units.",area);
    }
    else
    {
        printf("That was not one of the selections.\n");
        printf("You must run the program again and\n");
        printf("select either a 1 or a 2.\n");
    }

    printf("\n\nThis concludes the program to calculate\n");
    printf("the area of a circle or a square.");

    exit(0);

}
```

Figure 3-8 illustrates the construction of the above program.

Program Analysis

As you can see from Figure 3-8, this program has three options as indicated by the **if...else if...else** construction. This can be viewed as

if (expression₁) statement₁ **else if** (expression₂) statement₂ **else** statement₃.

What this means is if $expression_1$ is TRUE then $statement_1$ will be executed and none of the other statements. If $expression_2$ is TRUE then $statement_2$ will be executed and none of the others. If neither $statement_1$ or $statement_2$ is TRUE, then $expression_3$ will be executed. This is coded in the above program as

```
if (selection == 1)
{

    [Body of compound statement to compute circle area.]

}
else
if (selection == 2)
{

    [Body of compound statement to compute area of square.]
```

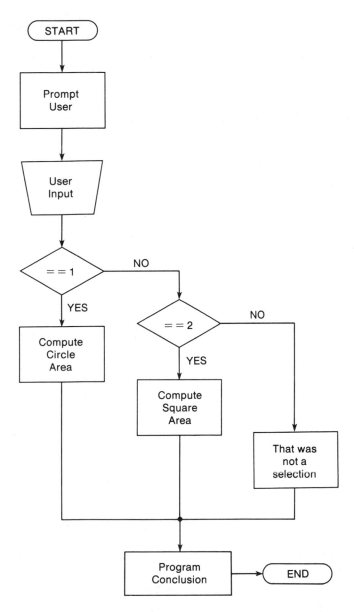

Figure 3-8   Construction of Program 3-6

```
}
else
{

    [Body of compound statement to give message to user.]

}
```

Notice that there are no semicolons between the **else if** or after any of the **if** or **else** keywords.

For this program, the program user enters a 1 or a 2 to select which computation will be performed. If a 1 is entered, then the compound statements for the computation of the area of a circle are executed. If a 2 is entered, then the compound statements for the computation of the area of a square are executed. If a value other than 1 or 2 is entered, then the statement following the last **else** is executed.

Another Way

A more structured way of presenting the same program is shown in Program 3–7. Note that now all of the compound statements are placed in their own functions. Also notice that both formulas have now been **#defined** at the head of the program.

Program 3–7

```
#include <stdio.h>
#define PI 3.141592        /*  The constant pi.   */
#define square(x) x*x      /* Area of a square.   */
#define circle(r) PI*r*r   /* Area of a circle.   */

void user_selection(void);   /* Get selection from user.   */
void circle_data(void);      /* Get circle radius and compute. */
void square_data(void);      /* Get square side and compute. */
void wrong_selection(void);  /* Notify user of wrong selection. */

main( )
{

  printf("\n\nThis program will compute the area of\n");
  printf("a square or the area of a circle.\n");

   user_selection( );  /* Get selection from user.   */

  printf("\n\nThis concludes the program to calculate the\n");
  printf("area of a circle or a square.");

  exit(0);

 }

/*----------------------------------------------------------*/

 void user_selection(void) /* Get selection from user.   */
 {
   float selection;    /*  User selection.   */
```

(continued)

Program 3–7 *(continued)*

```c
    printf("\nSelect by number:\n");
    printf("1] Area of circle. 2] Area of square.\n");
    printf("Your selection (1 or 2) => ");
    scanf("%f",&selection);

      if (selection == 1)
        circle_data( );
      else
      if (selection == 2)
        square_data( );
      else
        wrong_selection( );

    }

/*------------------------------------------------------------*/

  void circle_data(void)  /* Get circle radius and compute.  */
    {
    float radius;          /*  Radius of the circle.  */
    float area;            /*  Circle area in square units.  */

        printf("Give me the length of the circle radius => ");
        scanf("%f",&radius);

    area = circle(radius);

        printf("A circle of radius %f has an area of ",radius);
        printf("%f square units.",area);

    }

/*------------------------------------------------------------*/

  void square_data(void)    /*  Get square side and compute.  */
    {
    float side;    /*  Side of the square.  */
    float area;    /*  Area of the square in square units.  */

        printf("Give me the length of one side of the square => ");
        scanf("%f",&side);

      area = square(side);

        printf("A square of length %f has an area of ",side);
```

(continued)

Program 3-7 *(continued)*

```
      printf("%f square units.",area);

}

/*------------------------------------------------------------*/

void wrong_selection(void)   /*  Notify user of wrong selection.  */
  {
    printf("That was not one of the selections.\n");
    printf("You must run the program again and\n");
    printf("select either a 1 or a 2.\n");
  }

/*------------------------------------------------------------*/
```

Figure 3-9 illustrates the structure of the above program.

As shown in the figure, the structure of the program has now been divided into function blocks.

Program Analysis

The program starts with the **#define** directives:

```
#define PI 3.141592        /*  The constant pi.  */
#define square(x) x*x       /* Area of a square.  */
#define circle(r) PI*r*r    /* Area of a circle.  */
```

Note that each **#define** is commented to exactly describe the purpose of the definition. Next, the function prototypes are presented:

```
void user_selection(void);    /* Get selection from user.  */
void circle_data(void);       /* Get circle radius and compute. */
void square_data(void);       /* Get square side and compute. */
void wrong_selection(void);   /* Notify user of wrong selection. */
```

Each prototype states its type and the type of any arguments. Each one is commented—and this same comment will be used on the header of the function definition in the body of the program. Next follows function **main( )** with its simple coding:

```
main( )
{

  printf("\n\nThis program will compute the area of\n");
  printf("a square or the area of a circle.\n");
```

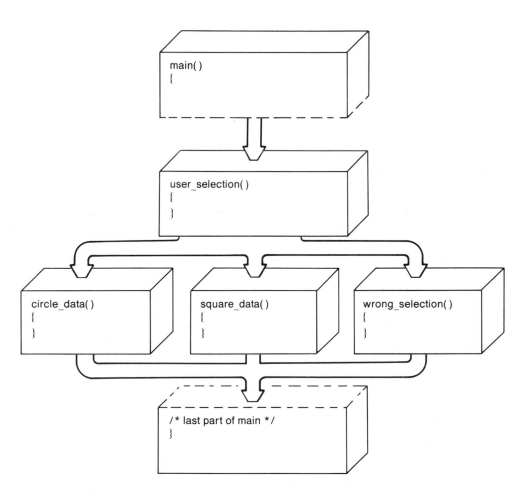

Figure 3-9   Structure of Program 3-7

```
    user_selection( );   /* Get selection from user.   */

    printf("\n\nThis concludes the program to calculate the\n");
    printf("area of a circle or a square.");

    exit(0);

}
```

It is now very easy to read through **main( )** and get the idea of what the program is to do and how it is going to do it. Note that only one function is called **user_selection( )**. This function is now also easy to follow. It contains the necessary **if...else if...else** statements:

```
void user_selection(void) /* Get selection from user. */
{
 float selection;    /*  User selection.  */

 printf("\nSelect by number:\n");
 printf("1] Area of circle. 2] Area of square.\n");
 printf("Your selection (1 or 2) => ");
 scanf("%f",&selection);

   if (selection == 1)
     circle_data( );
   else
   if (selection == 2)
     square_data( );
   else
     wrong_selection( );

}
```

In this function, it is clear that the user will be asked to choose between one of two calculations. The **if...else if...else** statements that follow simply call functions. Thus, it can be seen that **circle_data**, or **square_data** or **wrong_selection** will be called by this function. Also note that the names of each of these functions are descriptive of what they do.

The other functions simply get the information, do the appropriate calculation and display a corresponding message on the monitor. The action of this program, from a user standpoint, will be no different from the action of Program 3–6.

Conclusion

This section presented the powerful decision-making C statement **if...else if....** This feature is used very often in technical programming. You will see many examples of it in this and future chapters. In the next section you will be introduced to another powerful C decision-making structure. For now, test your understanding of this section by trying the following section review.

3–3 Section Review

1 Explain the meaning of a closed branch.
2 What is the difference between the **if** and the **if...else** statement?
3 May compound statements be used with the **if...else**? Explain.
4 May function calls be used with the **if...else**? Explain.
5 Describe the action of the **if...else if...else** statement in C.

3–4   Logical Operation

Discussion

In this section you will learn about the logical operators used in C. Here the information presented in the first part of this chapter will be applied. By using C's logical operators, your programs will have increased decision-making capabilities.

Logical AND

The **logical AND** operation in C is expressed as

`(expression₁)&&(expression₂)`

The above operation will be evaluated TRUE only if `expression₁` is TRUE *and* `expression₂` is TRUE; otherwise the operation will be evaluated as FALSE. Keep in mind that in C a FALSE evaluation is actually a `0` while a TRUE evaluation is actually a non-zero value. Table 3–2 summarizes the AND operation.

Table 3–2 The AND Operation

expression₁	expression₂	Result
FALSE	FALSE	FALSE
FALSE	TRUE	FALSE
TRUE	FALSE	FALSE
TRUE	TRUE	TRUE

Observe that the double **&&** is used to represent this operation. No spaces are allowed between these symbols although spaces are allowed to the left and right of this double symbol.

The following program illustrates the use of the logical AND operation in C.

Program 3–8

```
#include <stdio.h>

main( )
{
float result;    /* Result of logical expression.  */

  result = 0 && 0;
  printf("0 && 0 = %f\n",result);

  result = 0 && 1;
  printf("0 && 1 = %f\n",result);

  result = 1 && 0;
  printf("1 && 0 = %f\n",result);

  result = 1 && 1;
  printf("1 && 1 = %f\n",result);
}
```

Figure 3–10 illustrates the operation of the above program.

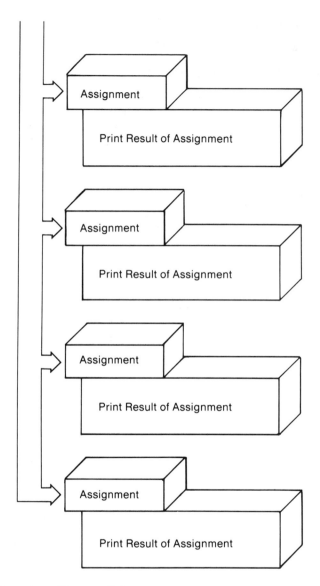

Figure 3–10   Operation of Program 3–8

Program Analysis

Execution of the Program 3–8 yields

```
0 && 0 = 0.000000
0 && 1 = 0.000000
1 && 0 = 0.000000
1 && 1 = 1.000000
```

Notice the the variable `result` is of type **float**. You should try this with other data types as well. It is then used to store the result of each of the AND operations.

The first AND operation

```
result = 0 && 0;
```

causes two FALSE values (represented in C as `0`) to be ANDed. Since the result of this is FALSE, the value of the variable `result` will be `0`.

The next AND operation

```
result = 0 && 1;
```

causes a FALSE and a TRUE (represented in C as a `0` and a `1`) to be evaluated. Since the result of this operation is FALSE, the value of the variable `result` will again be `0`.

The third AND operation

```
result = 1 && 0;
```

causes a TRUE and a FALSE (represented in C as a `1` and a `0`) to be evaluated. This is another operation that is FALSE and the value of `result` is once again `0`.

The last operation is the only one that evaluates to a TRUE:

```
result = 1 && 1;
```

Here, a TRUE is being ANDed with a TRUE (represented as a `1` and `1`). The result of this operation is TRUE and the value of `result` will be evaluated to a `1`.

The OR Operation

The **logical OR** operation in C is expressed as

```
(expression₁) || (expression₂)
```

The above operation will be evaluated FALSE only if `expression₁` is FALSE *and* `expression₂` is FALSE; otherwise the operation will be evaluated as TRUE. As before, in C a FALSE evaluation is actually a `0` while a TRUE evaluation is actually a non-zero value. Table 3–3 summarizes the OR operation.

Observe that the double `||` is used to represent this operation. No spaces are allowed between these symbols although spaces are allowed to the left and right of this double symbol.

Table 3–3   The OR Operation

expression₁	expression₂	Result
FALSE	FALSE	FALSE
FALSE	TRUE	TRUE
TRUE	FALSE	TRUE
TRUE	TRUE	TRUE

The following program illustrates the use of the OR operation in C.

Program 3-9

```
#include <stdio.h>

main( )
{
float result;    /* Result of logical expression.   */

    result = 0 || 0;
    printf("0 || 0 = %f\n",result);

    result = 0 || 1;
    printf("0 || 1 = %f\n",result);

    result = 1 || 0;
    printf("1 || 0 = %f\n",result);

    result = 1 || 1;
    printf("1 || 1 = %f\n",result);

}
```

Program Analysis

Execution of Program 3-9 yields

```
0 || 0 = 0.000000
0 || 1 = 1.000000
1 || 0 = 1.000000
1 || 1 = 1.000000
```

Notice the variable `result` in this case is of type **float**. Again, other types could have been used. It is then used to store the result of each of the OR operations.

The first OR operation

```
result = 0 || 0;
```

causes two FALSE values (represented in C as 0) to be ORed. Since the result of this is FALSE, the value of the variable `result` will be 0.

The next OR operation

```
result = 0 || 1;
```

causes a FALSE or a TRUE (represented in C as a 0 and a 1) to be evaluated. Since the result of this operation is TRUE, the value of the variable `result` will now be 1.

The third OR operation

`result = 1 || 0;`

causes a TRUE and a FALSE (represented in C as a 1 and a 0) to be evaluated. This is another operation that is TRUE, and the value of `result` is once again 1.

The last operation also evaluates to a TRUE.

`result = 1 || 1;`

Here, a TRUE is being ORed with a TRUE (represented as a 1 and 1). The result of this operation is TRUE, and the value of `result` will be evaluated to a 1.

Relational and Logic Operations

Relational operations may be used with the logical operations. This can be done because a relational operation returns a TRUE or FALSE condition. These are the same conditions used by logical operators. Table 3–4 shows the order of *precedence* for the C operators presented up to this point. This means that the ! is evaluated before the * which is evaluated before the <, and so on.

Table 3–4   C Precedence

Operators	Name
!	Not
* /	Multiplication and Division
< <= >= >	Less, Less or Equal, Greater or Equal, Greater
== !=	Equal, Not Equal
&&	Logical AND
||	Logical OR

There is a new logical operator shown in Table 3–4, called the **logical NOT** and represented by the !. You will see an example of this shortly. For now, realize that what the above table shows is which of these operations will be done before others in a program line that contains more than one of them. The use of this table as well as the logical NOT is illustrated in the following example.

EXAMPLE 3-2

Determine the results of each of the following expressions (will they be TRUE or FALSE):

A. $(5 == 5) || (6 == 7)$
B. $(5 == 8) \&\&(6 != 7)$
C. $(8 >= 5) \&\&(!(5 <= 2))$

Solution

A. Analyze the expression by first determining the logical value of the operation in the innermost parentheses. For this one, (5==5) is TRUE. It now makes no difference what the condition of the following operation is because of the OR. The final result of an OR will always be TRUE as long as any one member of the OR expression is TRUE. This analysis method is the same used by the C compiler. If it determines that the first part of an || is TRUE, then it will not bother to check the remainder of the expression. If it finds that the first part is FALSE, then the second part will be evaluated to test its final condition.

B. Using the same analysis method, you see that the expression (5 == 8) is FALSE. Since this is an AND expression, you do not need to evaluate the (6 != 7) (which is TRUE). The reason for this is that any AND that has one member FALSE will always be FALSE. Again, this is how C evaluates an **&&** expression. If the first part is FALSE, it does not go on to evaluate the second part. The only time the second part will be evaluated is when the first part is TRUE. Thus, the result of this expression is FALSE.

C. This expression uses the NOT (!) logical operator. Therefore !TRUE means FALSE (NOT TRUE is FALSE), and !FALSE means TRUE (NOT FALSE is TRUE). Evaluation of the first part: (8>=5) is TRUE. Since this is an **&&** operation, it's now necessary to evaluate the next part: !(5 < 2). Since (5 < 2) is FALSE then NOT(5 < 2) is TRUE. Since both sides of the AND are TRUE, the final result is also TRUE.

Program 3–10 shows an example of using relational and logical operations together. The program gets a number from the program user and then checks to see if its value is between 1 and 10. If it is, a message is printed to the screen.

Program 3–10

```
#include <stdio.h>

main( )
{
float number;    /* User input number. */

  printf("\n\nGive me a number from 1 to 100 => ");
  scanf("%f",&number);

  if ((number >= 1.0) && (number <= 10.0))
      printf("You gave me a number between 1 and 10.");

}
```

Program Analysis

Note the program line that combines a relational and logical expression:

`if ((number >= 1.0) && (number <= 10.0))`

The above statement can be thought of as shown in Figure 3-11.

The parentheses around the relational operations may be omitted because the => has a higher precedence than the **&&**. However, there is no harm in putting them there, and they should be used if they improve clarity.

Compounding the Logic

Any combination of relational and logical operations may be used. For example, Program 3-11 uses a complex logical operation consisting of OR as well as AND with relational operations. The program checks the user input of a number between 1 and 100 to see if the number entered is in the top or bottom 10%.

Program 3-11

```
#include <stdio.h>

main( )
{
float number;    /* User input number. */

 printf("\n\nGive me a number from 1 to 100 => ");
 scanf("%f",&number);

  if (((number>=1.0)&&(number<=10.0))||((number>=90.0)&&(number<=100.0)))
     printf("You gave me a number in the top or bottom 10%%.");

}
```

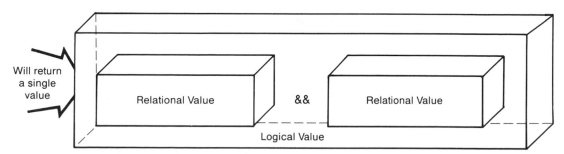

Figure 3-11 Relational and Logical Operation

Program Analysis

The key to the above program is the statement

```
if (((number>=1.0)&&(number<=10.0))||((number>=90.0)&&(number<=100.0)))
```

An analysis of this statement is shown in Figure 3–12.

Again, parentheses are used around the relational operations to help read the logic of what the program is to do.

Application Program

Program 3–12 illustrates a technical application using relational and logic operations. The program reads the temperature of a process. In this program, this is entered by the program user. However, it could also be done directly from a temperature probe interfaced with the computer. The result of the program would then depend upon the temperature reading of the probe. The logic of this program is illustrated in Figure 3–13.

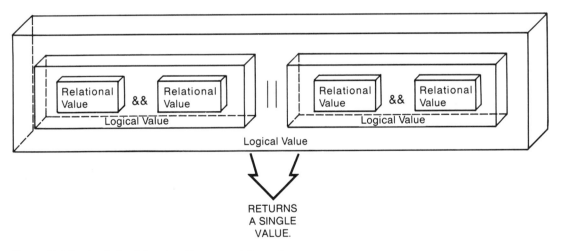

Figure 3–12 Analysis of Another Compound Operation

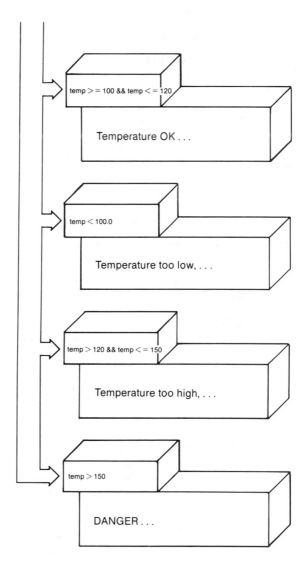

Figure 3-13   Temperature Probe Program

Program 3-12

```c
#include <\include\stdio.h>

main( )
{
float temp;    /* Temperature reading. */

 printf("\n\nGive me the temperature reading => ");
 scanf("%f",&temp);

  if ((temp>=100.0)&&(temp<=120.0))
      printf("Temperature OK continue process.");

  if (temp < 100.0)
      printf("Temperature too low, increase energy.");

  if ((temp > 120.0)&&(temp <= 150.0))
      printf("Temperature too high, decrease energy.");

  if (temp > 150.0)
      printf("Danger! Shut down systems.");

}
```

Conclusion

This section demonstrated the power of combining relational and logical operations. You saw some very simple examples as well as a more complex technical application. In the next section, you will learn about another powerful decision-making tool available in the C language. First check your understanding of this section by trying the following section review.

3-4 Section Review

1 Describe the operation of the logical AND. What symbol is used to represent this in C?
2 Describe the operation of the logical OR. What symbol is used to represent this in C?
3 Describe the operation of the logical NOT.
4 Give an example of combining a relational and a logical operation.
5 State how C evaluates the AND operation and the OR operation.

3-5 Conversion and Type Casting

This section presents information regarding what may seem to be a small detail. You may not have had to pay attention to it up to this point, but this "detail" could become very important in more complex C programs.

Data Types

You will recall that every piece of data in C has a type. It may be an **int**, **char** or other type. The C language allows you to add a type **int** to a type **float**. Doing this is called mixing types.

When data types are mixed in an expression, the compiler will convert all of the variables to a single compatible type. Then the operation will be carried out. This is done according to the **rank** of the data type. Variables of a lower rank are converted to the type of variables with a higher rank. The ranking of variables is

Low Rank $<=$ **char**, **int**, **long**, **float**, **double** $=>$ High Rank

This means if an operation of a **char** and **int** takes place, the resulting value will be of type **int**. If an operation with a **float** and an **int** takes place, the resulting output will be of type **float**. A good rule to follow is to not mix data types. If it is necessary to do so, use the method of type **casting**.

Type Casting

A type cast provides a method for converting a variable to a particular type. To do this, simply precede the variable by the desired type. As an example, if **value** were originally declared as a type **int**, then to convert it to a type **float**

```
result = (float)value;
```

The variable **result** will now be of type **float**.

You should be cautious when demoting a variable from a high rank to a lower rank. In doing this, you are asking a larger value to fit into a smaller value, and this usually results in corrupted data.

You cannot use the cast on a type **void**. You can cast any type to **void**, but you cannot cast a type **void** to any other type.

Lvalue

There are times in programming that you will see the term **lvalue**. It literally means left-handed value because an assignment operation assigns the value of the right-hand operand to the memory location indicated by the left-hand operand. As an example: **value = 3 + 5;** is a legal statement in C; however, **3 + 5 = value;** is not a legal statement in C because the left-hand operand is not an lvalue (does not represent a unique memory location). What this means is that the left-hand operand of an assignment operation must be an expression that refers to a modifiable memory location, and expressions that refer to memory locations in this manner are called lvalue expressions.

Conclusion

This section presented some important details about C. Here you learned about mixed data types and how they are converted in C. You also saw how to cast to a different data type, and you discovered the meaning of lvalue. Check your understanding of this section by trying the following section review.

3-5 Section Review

1 Explain what is meant by mixing data types.
2 What happens if a type **int** is added to a type **float**?
3 State the rule for mixed data types in C.
4 What is meant by a cast in C?
5 What is an lvalue expression?

3-6 The C **switch**

Discussion

This section presents a method used by C to make one selection when there are several choices to be made. Here you will see one of the most powerful decision-making commands in the C language. First, the need for this command will be demonstrated. Then the command will be defined and applied to a technology problem. You will also see the need for using **char** or **int** types as presented in the last section.

Decision Revisited

Program 3–13 is an example of having one selection from several different alternatives. What the program does is to display one of three forms of Ohm's law depending upon the choice made by the program user. In the program, the program user selects one of three letters, and the corresponding formula is displayed on the monitor.

Program 3–13

```
#include <stdio.h>

main( )
{
  char selection;    /*  Item to be selected by program user.  */

  printf("\n\nSelect the form of Ohm's law needed by letter:\n");
  printf("A] Voltage B] Current C] Resistance\n");
  printf("Your selection (A, B or C) => ");
  scanf("%c",&selection);
```

(continued)

Program 3-13 *(continued)*

```
if (selection == 'A')
    printf("V = I X R");
  else
  if (selection == 'B')
    printf("I = V/R");
  else
  if (selection == 'C')
    printf("R = V/I");
  else
    printf("That was not one of the proper selections.");
}
```

Program Analysis

Figure 3-14 illustrates the logic of Program 3-13.

As shown in the figure, the program allows the program user three choices. This is achieved in the program by the use of the **if...else if** in C. Note the use of the **char** variable as the selection variable. The program user must enter one of the three uppercase letters A, B, or C.

The selection is made by the relational statements. As an example,

```
if (selection == 'A')
  printf("V = I X R");
```

If **selection** equals the character A, then the **printf( )** function will be executed. If none of the correct selections are made, then the program will default to the last else:

```
else
    printf("That was not one of the proper selections.");
```

This type of program, where there is a selection from a list of different possibilities, is so common that a special C statement is used for it.

How to Switch

In C the **switch** statement is an easier way to code multiple **if...else if** statements. The **switch** statement has the form shown below:

```
switch (expression₁)
  {
    case : ( constant-expression )
    default : ( expression₂)
  }
```

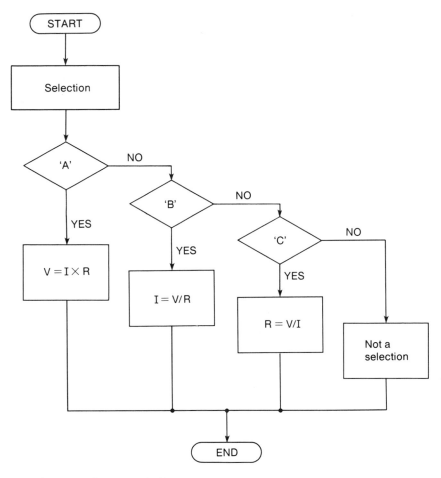

Figure 3–14   Logic of Program 3–13

Where
          switch = Reserved word indicating that a switch statement is about to take place.

      expression₁ = Any legal C expression.

              {= Defines the beginning of the **switch** body.

          case = Reserved word indicating that what follows is the constant-expression required for a match.

  constant-expression = Identifies what is required for a match. This must be of the same data type as expression₁ (sometimes called the case label).

default = C reserved word indicating the option to
be exercised if no match is made.

}= Defines the ending of the **switch** body.

The construction of the **switch** statement requires the introduction of two other C key words: **case** and **break**. Essentially the switch statement identifies a variable whose value will determine which **case** will be activated. The **break** lets C know when the selected program code is to end. Program 3-14 does exactly the same thing as Program 3-13 did; the program user gets to select one of three forms of Ohm's law. The difference is that now the **switch** statement is used in place of the **if...else if** statements.

Program 3-14

```
#include <stdio.h>

main( )
{
 char selection;     /*  Item to be selected by program user.  */

  printf("\n\nSelect the form of Ohm's law needed by letter:\n");
  printf("A] Voltage  B] Current   C] Resistance\n");
  printf("Your selection (A, B or C) => ");
  scanf("%c",&selection);

  switch (selection)
    {
     case 'A' : printf("V = I X R");
                break;
     case 'B' : printf("I = V/R");
                break;
     case 'C' : printf("R = V/I");
                break;
      default : printf("That was not one of the proper selections.");
    } /* end of switch */

}
```

Program Analysis

It should be clear from the above program that it is now easier to read program code that uses the **switch** instead of multiple **if...else if** statements. From the above program:

```
switch (selection)
  {
   case 'A' : printf("V = I X R");
              break;
```

```
    case 'B' : printf("I = V/R");
              break;
    case 'C' : printf("R = V/I");
              break;
    default  : printf("That was not one of the proper selections.");
    } /* end of switch */
```

Observe that the **switch** statement is followed by the **char** variable selection. This means that each of the **case** labels must be of type **char** (and they are: 'A', 'B' and 'C'). Note also the use of the opening { and closing }. These are both brought out to the left so that it is clear exactly where the body of the **switch** statement begins and ends. One other important item is the comment following the closing }. It is there to remind you that this is the end of the body of a **switch** statement. This is a good habit to get into—especially when you have programs with long selection lists. Note that the **break** statement is aligned so that it clearly identifies where the execution of the selection option will terminate.

Again, if none of the correct inputs is selected, the program will **default** to

```
default  : printf("That was not one of the proper selections.");
```

Compound statements may also be used with the **switch**. This is presented below.

Compounding the **switch**

You may use compound statements with the C **switch**. This is illustrated in Program 3–15. This program is an expansion of the previous ones in this section. Now the program user may actually perform a calculation with the selected form of Ohm's law.

Program 3–15

```
#include <stdio.h>

main( )
{
  char selection;      /*  Item to be selected by program user.  */

  float voltage;       /*  Circuit voltage in volts.  */
  float current;       /*  Circuit current in amps.  */
  float resistance;    /*  Circuit resistance in ohms.  */

  printf("\n\nSelect the form of Ohm's law needed by letter:\n");
  printf("A] Voltage B] Current  C] Resistance\n");
  printf("Your selection (A, B or C) => ");
  scanf("%c",&selection);
```

(continued)

Program 3–15 *(continued)*

```
switch (selection)
 {
   case 'A' : {   /*  Solve for voltage.  */
                printf("Input the current in amps => ");
                scanf("%f",&current);
                printf("Value of the resistance in ohms => ");
                scanf("%f",&resistance);
                   voltage = current * resistance;
                printf("The voltage is %f volts.",voltage);
              }
            break;
   case 'B' : {   /*  Solve for current.  */
                printf("Input the voltage in volts => ");
                scanf("%f",&voltage);
                printf("Value of the resistance in ohms => ");
                scanf("%f",&resistance);
                   current = voltage/resistance;
                printf("The current is %f amps.",current);
              }
            break;
   case 'C' : {   /*  Solve for the resistance.  */
                printf("Input the voltage in volts => ");
                scanf("%f",&voltage);
                printf("Value of the current in amps => ");
                scanf("%f",&current);
                   resistance = voltage/current;
                printf("The resistance is %f ohms.",resistance);
              }
            break;
   default  : {
                printf("That was not a correct selection\n");
                printf("Please go back and select A, B or C");
              }
 } /* end of switch */

}
```

The general structure of Program 3–15 is shown in Figure 3–15.

Observe the block structure used in the C **switch** statement. This structure makes it easier to read and understand the program code. Notice the indentation of the compound statement and how its beginning and closing are clearly defined with the { and }. Also note the location of the **break**. It lets you clearly see where the body of the option ends. Again, structure makes very little difference to the program user. Program structure is for you and others who will be modifying or trying to understand your program.

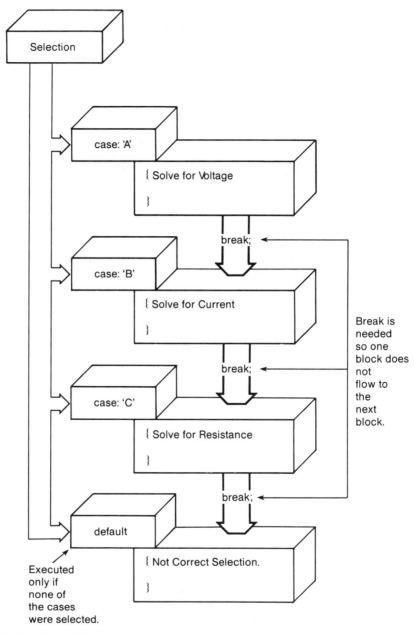

Figure 3-15 General Structure of Program 3-15

Program Analysis

The important characteristic of this program is the use of a compound C statement
with the **switch** statement.

```
switch (selection)
  {
  case 'A' : {   /* Solve for voltage.  */
             printf("Input the current in amps => ");
             scanf("%f",&current);
             printf("Value of the resistance in ohms => ");
             scanf("%f",&resistance);
                voltage = current * resistance;
             printf("The voltage is %f volts.",voltage);
             }
             break;
```

If the above selection is made (**selection == 'A'**), then the whole body
of statements defined by the compound statement between the opening { and the
closing } will be executed. The concept of the **switch** statement here is the same
as in the previous program. The main difference is the use of a compound statement.
As you might suspect, the **switch** statement also allows you to call other functions.

Functional Switching

Program 3–16 illustrates another program structure using the **switch** statement. This
program does exactly the same thing as the previous one. It allows the program
user to select any one form of Ohm's law and then permits a calculation. The difference
in this program structure is that now each of the compound statements is replaced
by a function call. In turn, each of these function definitions contains the statements
needed to implement the user selection.

Program 3–16

```
#include <stdio.h>

void resistance_solution(void);  /*  Solve for resistance in ohms. */
void current_solution(void);     /*  Solve for current in amps. */
void voltage_solution(void);     /*  Solve for voltage in volts. */

main( )
{
 char selection;    /*  Item to be selected by program user.  */
```

(continued)

Program 3–16 *(continued)*

```c
printf("\n\nSelect the form of Ohm's law needed by letter:\n");
printf("A] Voltage  B] Current   C] Resistance\n");
printf("Your selection (A, B or C) => ");
scanf("%c",&selection);

 switch (selection)
  {
   case 'A' : voltage_solution( );
             break;
   case 'B' : current_solution( );
             break;
   case 'C' : resistance_solution( );
             break;
  } /*  end of switch  */

}

/*-------------------------------------------------------------*/

void voltage_solution(void)    /* Solve for voltage in volts.  */
{
float resistance;   /* Value of circuit resistance in ohms. */
float voltage;      /* Value of circuit voltage in volts.   */
float current;      /* Value of circuit current in amps.    */

    printf("Input the current in amps => ");
    scanf("%f",&current);
    printf("Value of the resistance in ohms => ");
    scanf("%f",&resistance);
       voltage = current * resistance;
    printf("The voltage is %f volts.",voltage);

}

/*-------------------------------------------------------------*/
 void current_solution(void)   /* Solve for current in amps. */
 {
 float resistance;   /* Value of circuit resistance in ohms. */
 float voltage;      /* Value of circuit voltage in volts. */
 float current;      /* Value of circuit current in amps. */
```

(continued)

Program 3-16 *(continued)*

```
    printf("Input the voltage in volts => ");
    scanf("%f",&voltage);
    printf("Value of the resistance in ohms => ");
    scanf("%f",&resistance);
        current = voltage/resistance;
    printf("The current is %f amps.",current);
}


/*-----------------------------------------------------------*/

    void resistance_solution(void)  /* Solve for resistance in ohms. */
    {
      float resistance;    /* Value of circuit resistance in ohms.  */
      float voltage;       /* Value of circuit voltage in volts.    */
      float current;       /* Value of circuit current in amps.     */

    printf("Input the voltage in volts => ");
    scanf("%f",&voltage);
    printf("Value of the current in amps => ");
    scanf("%f",&current);
        resistance = voltage/current;
    printf("The resistance is %f ohms.",resistance);

    }
```

Program Analysis

Note how easy it now is to read the **switch** statement:

```
switch (selection)
  {
    case 'A' : voltage_solution( );
               break;
    case 'B' : current_solution( );
               break;
    case 'C' : resistance_solution( );
               break;
  } /*  end of switch  */
```

This is much cleaner than the compound statements used in Program 3-15; as such this is the preferred method. If the actual function definition is needed for any one of the called functions, it is only necessary to scan down the program and find the proper one.

What May Be Switched

In C the **switch**, expression$_1$ may be of type **int** as well as **char**. A type **float** is not permitted as an expression$_1$ type in a **switch** statement.

Switching within Switches

You may have multiple **switch** statements within other switch statements. This concept is illustrated in Figure 3–16.

Conclusion

This section presented the powerful C **switch** statement. You will find that the correct writing of technical programs will require the use of the **switch** statement. Here you were introduced to the **switch** structure and several different ways of applying this general structure. You were also shown a suggested program structure both for single and compound statements. The last program in this section presented the concept of function calls from the **switch** statement.

There is one more C decision-making statement. This is presented in the next section. For now, test your understanding of this section by trying the following section reveiw.

3–6 Section Review

1 State the purpose of the **switch** statement in C.
2 What other keywords must be used with the **switch** statement?
3 State the purpose of the keyword **default** in the **switch** statement.
4 May function calls be made from a **switch** statement? Explain.

3-7   One More **switch** and the Conditional Operator

Discussion

This section will present one more application of the C **switch**. Here you will also be introduced to the final decision-making process used in C. This section will conclude the discussion of the important decision-making capabilities of the C language.

One More **switch**

Recall from the last section that the C **switch** has the form

```
switch (expression)
   {
   case label : statement;
               break;
   }
```

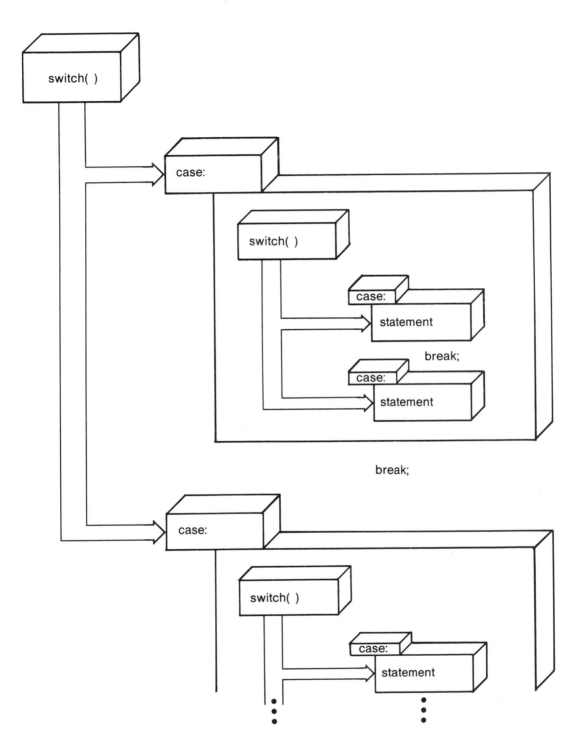

Figure 3-16   Concept of Switches within a **switch**

As you will see, the **break** is important in determining the logic of the **switch**. This is illustrated by the following program. This program displays formulas required to compute the power delivered by the power source in a series circuit consisting of three resistors. The program user only needs to enter what conditions are known about the circuit. Note where the **break** has been omitted from the C **switch**.

Program 3–17

```
#include <stdio.h>

main( )
{
char selection;   /* User input selection.  */
 printf("This program will show the formulas necessary\n");
 printf("to compute the power delivered by a voltage source\n");
 printf("to a series circuit consisting of three resistors\n");
 printf("when the value of the source voltage is known.\n");
 printf("\n\nSelect by letter:\n");
 printf("A] Resistor values known.  B] Total resistance known.\n");
 printf("C] Total current known.\n");
 printf("Your selection => ");
 scanf("%c",&selection);

   switch (selection)
   {
   case 'A' : printf("Rt = R1 + R2 + R3\n");
   case 'B' : printf("It = Vt/Rt\n");
   case 'C' : printf("Pt = Vt * It\n");
             break;
   default  : printf("That was not a correct selection!");
   } /* end of switch.  */
 }
```

Program Analysis

What follows is the monitor display for four different responses by the program user:

1. Program user enters the letter A:

```
Your selection => A
Rt = R1 + R2 + R3
It = Vt/Rt
Pt = Vt * It
```

2. Program user enters the letter B:

```
Your selection => B
It = Vt/Rt
Pt = Vt * It
```

3. Program user enters the letter C:

```
Your selection => C
Pt = Vt * It
```

4. Program user enters the letter T:

```
That was not a correct selection!
```

What has happened here is that the **break** between the labels in the C **switch** has been omitted:

```
switch (selection)
{
case 'A' : printf("Rt = R1 + R2 + R3\n");
case 'B' : printf("It = Vt/Rt\n");
case 'C' : printf("Pt = Vt * It\n");
           break;
default  : printf("That was not a correct selection!");
}  /* end of switch. */
```

Observe from the indicated output that if there is a match with **case 'A'**, then all remaining statements are executed up to the first **break**. If the **break** were not present, then the **default** statement would be executed as well. As you can see from the above discussion, wherever a match is made every statement from that point on is executed up to the first **break** or the end of the **switch** statement. This concept is illustrated in Figure 3–17.

The Conditional Operator

The last decision maker in C is the conditional operator. This has the form of an **if...else** statement. It is

expression$_1$? expression$_2$: expression$_3$

What happens is that when **expression$_1$** is TRUE (any value other than zero), then the whole operation becomes the value of **expression$_2$**. If, on the other hand, **expression$_1$** is FALSE (equal to zero), then the whole operation becomes the value of **expression$_3$**. A simple illustration follows.

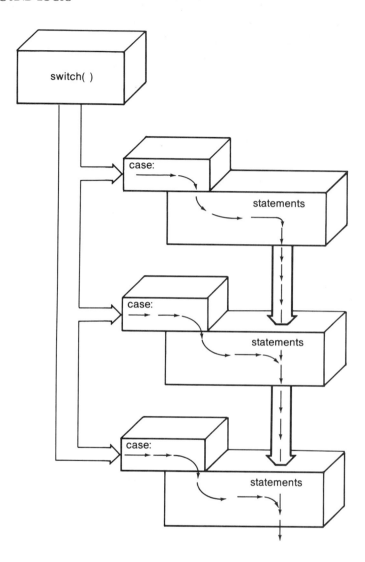

Figure 3–17 Switching without a Break

Program 3-18

```
#include <stdio.h>

main( )
{
  int selection;    /*  User input selection.   */

  printf("Enter a 1 or a 0 => ");
  scanf("%d",&selection);

    selection? printf("A one.") : printf("A zero.");

}
```

Program Analysis

In the above program, if the program user inputs a 0, then the second **printf** will be evaluated, and the monitor will display

A zero.

If the program user inputs a 1 (or any non-zero value), then the first **printf** will be evaluated, and the monitor will display

A one.

What has happened here is that the condition of **selection** determines which of the following two expressions will be evaluated. Actually, any value entered by the user that is not equal to zero will cause evaluation of the first expression. This means that if the program user entered the value of 12, then the first expression would be evaluated and the monitor would display

A one.

Conditional Operator Application

For an application of the conditional operator, consider the electrical circuit of Figure 3-18.

In this particular circuit, the device called the LED (light emitting diode) will not conduct current until the voltage applied across it is greater than 2.3 volts. Once this happens, further increasing the source voltage will not significantly change the voltage across the LED, and the remaining voltage will be dropped across the resistor.

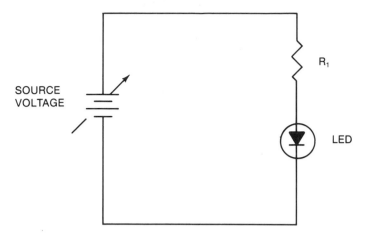

Figure 3-18   Application Circuit for Conditional Operator

To calculate the circuit current, the voltage across the resistor is divided by the value
of the resistor. This type of problem is easily solved by the C conditional operator
as illustrated in the following program.

Program 3-19

```c
#include <stdio.h>

main( )
{
   float led_voltage;        /* Voltage across LED in volts.   */
   float resistor_voltage;   /* Voltage across resistor in volts.  */
   float source_voltage;     /* Voltage of the source in volts.  */
   float circuit_current;    /* Current in the LED in amps.  /*
   float resistor_value;     /* Value of resistor in ohms.  */

    printf("\n\nEnter the source voltage in volts => ");
    scanf("%f",&source_voltage);
    printf("Enter value of resistor in ohms => ");
    scanf("%f",&resistor_value);

    led_voltage = (source_voltage < 2.3)? source_voltage : 2.3;
    resistor_voltage = source_voltage - led_voltage;
    circuit_current = resistor_voltage/resistor_value;

    printf("Total circuit current is %f amps.",circuit_current);

}
```

Program Analysis

Assume that the program user enters a value of 2 volts. This means that the expression

(source_voltage < 2.3)

will be TRUE. Thus led_voltage will be evaluated to source_voltage (which, as entered by the program user, is 2 volts).

Next, the evaluation

resistor_voltage = source_voltage − led_voltage;

becomes

resistor_voltage = 2 − 2

and the resistor voltage is zero (as it should be under these conditions). Now, when the next expression is evaluated

circuit_current = resistor_voltage/resistor_value

the evaluation becomes

circuit_current = 0/resistor_value

which evaluates to 0. Again this is as it should be because if the voltage across the LED is not more than 2.3 volts, there will not be any current flow in the circuit.

Next, assume the program user enters a value greater than 2.3 volts—say 5 volts—for the source voltage. Note the calculations will show a circuit current (as they should).

(source_voltage < 2.3)

will be FALSE. Thus led_voltage will now be evaluated to 2.3 volts (the second expression on the right of the : in the operator)

(source_voltage < 2.3)? source_voltage : 2.3

Now the following calculation becomes

resistor_voltage = 5 − 2.3 = 2.7

and the circuit current will now have a value different from zero. The exact value will be determined by the value of the resistor entered by the program user.

Conclusion

This section demonstrated how to use the C **switch** with variations of the **break**. You also saw an application of this concept. The conditional operator was also presented here along with an important application. Check your understanding of this section by trying the following section review.

1   State the effect of omitting the **break** in a C **switch**.
2   What do you need to do in order to make sure that **default** is always executed in a C **switch**?
3   Explain the operation of the conditional operator.
4   What values makes the conditional operator TRUE? What values make it FALSE?

3-8   Program Debugging and Implementation

Discussion

This section illustrates a method of selecting parts of your program to be compiled while other parts are ignored. This is called **conditional compilation**. This is often used in large programs where many different conditions must be considered.

Conditional Directives

The conditional compilation directives are listed in Table 3–5.

Table 3–5   Conditional Compilation Directives

Directive	Meaning
#ifdef	Permits code following it to be compiled under some circumstances but not others.
#endif	Ends selected compilation.
#else	Presents a compilation option.

Use of these directives is illustrated in Program 3–20. The program illustrates how these directives could be used for a program that has built-in debugging commands. Actual debugging commands are replaced by **printf( )** statements to give you the idea of how these directives function.

Program 3–20

```
#include <stdio.h>

#define DEBUGGING     /* Debugging is now in effect.  */

main( )
 {

#ifdef DEBUGGING  /* If in debugging, then do the following: */
```

(continued)

Program 3–20 *(continued)*

```
        printf("This is the debugging section of the program.\n");
        printf("This will only appear if DEBUGGING is defined.\n");

#endif DEBUGGING    /* End the debugging session. */

        printf("This is the main part of the program.\n");
        printf("It has nothing to do with the DEBUGGING.\n");

#ifdef DEBUGGING
        printf("This ends the debugging."); /* If in debugging.*/
#else DEBUGGING
        printf("No debugging used in the program."); /* If debug off.*/
#endif DEBUGGING

    }
```

For the above program, the output will be

```
This is the debugging section of the program.
This will only appear if DEBUGGING is defined.
This is the main part of the program.
It has nothing to do with the DEBUGGING.
This ends the debugging.
```

To cause the debugging part of the program to be ignored by the compiler, you only need to *not* define DEBUGGING. This can be done by commenting it out:

```
/* #define DEBUGGING */ /* Debugging is not in effect. */
```

When this is done, the resulting output of the program will be

```
This is the main part of the program.
It has nothing to do with the DEBUGGING.
No debugging used in the program.
```

Conclusion

Here you saw how you could use condition compilation directives to selectively compile parts of your C program. This is a useful feature commonly found in larger C programs. Check your understanding of this section by trying the following section review.

3–8 Section Review

1 Explain what is meant by conditional compilation.
2 What is a compilation directive?
3 State the purpose of the **#ifdef** directive.
4 When are conditional compilation directives normally used?

3-9   Case Study: A Robot Troubleshooter

Discussion

The case study for this chapter demonstrates an application of program decision making in technology. Troubleshooting charts are quite popular today. These charts are essentially flow diagrams that direct the technician during the troubleshooting of a specific system. This case study demonstrates how one of these troubleshooting charts for a robot was duplicated by the C language. This is the first step toward computer aided troubleshooting (CAT) where the computer makes the actual measurements (through appropriate interface circuits) and then diagnoses the actual problem.

The Hypothetical Robot

In order to keep the case study program down to a reasonable size, only a part of the complete robot troubleshooting flow chart will be investigated. The flow chart that will be used for this study is illustrated in Figure 3-19.

The purpose here is to develop a C program that will duplicate exactly what the chart is doing. For the purpose of this case study, the program will be developed so that the program user may enter the required values. This simulates the action a technician may take if actual measurements were being taken.

First Step

The first step, as always, is to write out the following:

- Purpose of the program
- Required input (source)
- Process on the input
- Required output (destination)

Here is the written purpose of the case study for the hypothetical robot:

- Purpose of the Program: Replicate the troubleshooting logic of Figure 3-19 so it can interact with the program user.
- Required Input: Numerical values that represent voltage readings at given test points. Other inputs require a multiple choice answer.
- Process on Input: Input values entered by program user will determine the next set of instructions to the program user or termination of the program.
- Required Output: Instructions to the program user for information to be entered into the program.

The above makes it clear exactly what is to be done (and not done) by the C program. The next step is to develop the algorithm.

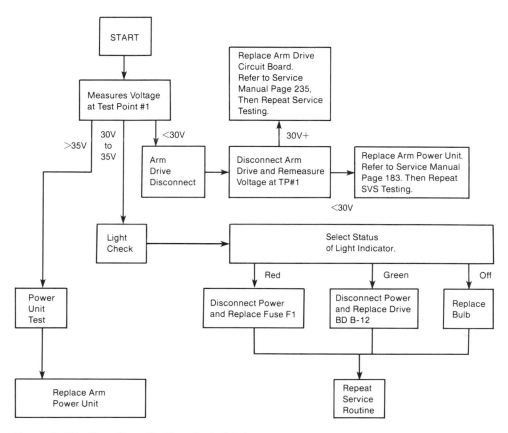

Figure 3-19   Flow Chart for Hypothetical Robot

Developing the Algorithm

The algorithm development for this program is essentially done. The troubleshooting chart is a graphical algorithm. All that you need to do is follow the logic of the chart. The other required information will be the standard prologue that includes the program information, program explanation and the function descriptions. This information is contained in the first pass of the program.

First Development Stage

Program 3-21 shows the first step in the development of the hypothetical robot problem.

Program 3-21

```c
#include <stdio.h>

/***************************************************/
/*              Robot Troubleshooter              */
/***************************************************/
/*           Developed by: Robert Shooter         */
/***************************************************/
/*              Date: October 24, 1992            */
/***************************************************/
/* This program will take the user step by step   */
/* through the analysis and troubleshooting of a */
/* hypothetical robot. The program also            */
/* demonstrates the various kinds of decision      */
/* making available with the C language.           */
/***************************************************/
/*          Non-Standard Header Files Used:        */
/*-----------------------------------------------*/
/*                   None                          */
/***************************************************/
/*              Function Prototypes                */
/*-----------------------------------------------*/
void explain_program(void);
/* This function explains the operation of the    */
/* program to the program user.                    */
/***************************************************/

main( )
 {

    explain_program( );     /* Explain program to user.  */

    exit(0);

 }

/*-------------------------------------------------*/

  void explain_program(void)   /* Explain program to user. */
  {

    printf("\n\nThis program represents the troubleshooting\n");
    printf("of a hypothetical robot system.\n");
    printf("\nThe information you enter will simulate actual\n");
    printf("measurements you would make on this robot.\n");
    printf("\nThe program will instruct you as to what measurement\n");
    printf("you are to enter next.\n");

  }
```

Program Analysis

The programmer's block clearly gives the name of the program, date of development, name of developer, and the purpose of the program. The only function prototype is the one for the function that will be used to explain the purpose of the program to the program user. This is enough to compile and execute. You will see nothing more than the instructions explaining what the program will do. However, it is now documented as a part of the program exactly what it is to do. There is no longer any reason to refer back to written documentation for this information. As the program is developed further, all the programmer now needs to do is look at the programmer's block to be reminded of what is needed.

Adding Some Structure

Program 3–22 illustrates the next step in the development of this program.

Program 3-22

```
#include <stdio.h>

/**************************************************/
/*              Robot Troubleshooter              */
/**************************************************/
/*         Developed by: Robert Shooter           */
/**************************************************/
/*            Date: October 24, 1992              */
/**************************************************/
/* This program will take the user step by step  */
/* through the analysis and troubleshooting of a */
/* hypothetical robot. The program also           */
/* demonstrates the various kinds of decision     */
/* making available with the C language.          */
/**************************************************/
/*          Non-Standard Header Files Used:       */
/*----------------------------------------------*/
/*                  None                          */
/**************************************************/
/*              Function Prototypes               */
/*----------------------------------------------*/
void explain_program(void);
/*   This function explains the operation of the */
/* program to the program user.                   */
/*----------------------------------------------*/
float arm(void);
/*   This function contains the arm service       */
/* routine for the robot.                         */
/*----------------------------------------------*/
```

(continued)

Program 3–22 *(continued)*

```c
void power_unit(void);
/*   This function is the power unit service   */
/* routine. It is a final testing function.    */
/*----------------------------------------------*/
void light_check(void);
/*   This function presents instructions on how */
/* to check the status light indicator.         */
/*----------------------------------------------*/
void arm_drive_disconnect(void);
/*   This function presents instructions on the */
/* testing of the arm drive disconnect circuit. */
/*----------------------------------------------*/

main( )
 {
  float measurement;    /* Measurement conducted by program user. */

    explain_program( );      /* Explain program to user. */
    measurement = arm( );    /* Arm measurement value.    */

      if (measurement > 35) power_unit( );
      else
      if ((measurement <= 35)&&(measurement >= 30)) light_check( );
      else
      if (measurement < 30) arm_drive_disconnect( );

    exit(0);

  }


/*----------------------------------------------*/

  void explain_program(void)    /* Explain program to user. */
  {

    printf("\n\nThis program represents the troubleshooting\n");
    printf("of a hypothetical robot system.\n");
    printf("\nThe information you enter will simulate actual\n");
    printf("measurements you would make on this robot.\n");
    printf("\nThe program will instruct you as to what measurement\n");
    printf("you are to enter next.\n");

  }
```

(continued)

Program 3–22 *(continued)*

```
/*--------------------------------------------------*/

    float arm(void)      /* Do arm service routine.  */
    {
    float measurement;   /* User measurement in volts. */

      printf("\nMeasure the voltage at test point #1\n");
      printf("Your measurement in volts => ");
      scanf("%f",&measurement);

      return(measurement);

    }
/*--------------------------------------------------*/

  void power_unit( )     /*  Power unit service instructions.  */
  {

     printf("\nRefer to power unit test in service manual\n");
     printf("and replace arm power unit.");

  }

/*---------------------------------------------------*/

  void light_check( )    /* Checking the status light indicator.  */
  {

     printf("\nThis is the status light indicator check.");

  }

/*--------------------------------------------------*/

  void arm_drive_disconnect( )   /* Arm drive instructions.  */
  {

     printf("\nThis is the arm drive disconnect instructions.");

  }

/*--------------------------------------------------*/
```

Program Analysis

Observe the new structure of **main( )**. Essentially its logic reads the same as the first part of the troubleshooting flow chart. In the flow chart, the first measurement to be made is the voltage at TP#1. From this measurement, the program can branch in one of three directions depending on the value of the voltage. After calling **explain_program**, **main( )** sets the variable **measurement** equal to the value of the function **arm( )**. This function has already been identified as a type **float** prototype. This means it will **return( )** a value. This value will be passed on to **measurement**, and the rest of the program will be determined by the value of **measurement**. This will be done by a series of **if...else if** statements that follows.

```
main( )
{
 float measurement;    /* Measurement conducted by program user. */

   explain_program( ); /* Explain program to user.  */
   measurement = arm( );    /* Arm measurement value.  */

      if (measurement > 35) power_unit( );
      else
      if ((measurement <= 35)&&(measurement >= 30)) light_check( );
      else
      if (measurement < 30) arm_drive_disconnect( );
      exit(0);

}
```

Thus, reading the code of the above function gives you the overall structure of this program. The point is, you don't have to read through a lot of program code in order to get the idea of what the structure of the program is or how the program is doing what it is supposed to do. The details of the code are there in the function definitions if and when you need them.

All of the new functions have their function prototypes in the programmer's block along with the explanation of what they are to do.

```
/*              Function Prototypes          */
/*-----------------------------------------------*/
void explain_program(void);
/*   This function explains the operation of the */
/* program to the program user.                  */
/*-----------------------------------------------*/
float arm(void);
/*    This function contains the arm service     */
/* routine for the robot.                        */
/*-----------------------------------------------*/
void power_unit(void);
/*   This function is the power unit service     */
/* routine.  It is a final testing function.     */
/*-----------------------------------------------*/
```

```
void light_check (void);
/*   This function presents instructions on how  */
/* to check the status light indicator.          */
/*---------------------------------------------*/
void arm_drive_disconnect(void);
/*   This function presents instructions on the  */
/* testing of the arm drive disconnect circuit.  */
/*---------------------------------------------*/
```

This type of documentation makes it easy to come back to the program at any time and see exactly how many functions are used in the program, what type of functions each is, and exactly what each is supposed to do. This makes program modification, debugging, and maintenance a much easier and less frustrating task.

Note that not all of the functions are completed. There is just enough information there to let you know that the program is working up to this point.

```
/*---------------------------------------------*/

   float arm(void)      /*  Do arm service routine.  */

   float measurement;   /* User measurement in volts. */

     printf("\nMeasure the voltage at test point #1\n");
     printf("Your measurement in volts => ");
     scanf("%f",&measurement);

     return(measurement);

   }

/*---------------------------------------------*/

   void power_unit( )    /*  Power unit service instructions.  */
   {

     printf("\nRefer to power unit test in service manual\n");
     printf("and replace arm power unit.");

   }

/*---------------------------------------------*/

   void light_check( )   /*  Checking the status light indicator.  */
   {

       printf("\nThis is the status light indicator check.");

   }

/*---------------------------------------------*/
```

```
void arm_drive_disconnect( )    /* Arm drive instructions.  */
{

    printf("\nThese are the arm drive disconnect instructions.");

}
```

```
/*------------------------------------------------------------*/
```

Programmers frequently put only part of the information in a program—just enough to test the overall flow of the program—to catch bugs during the development of the program. Such undeveloped functions are called **program stubs**.

Final Program

Program 3–23 represents the final development of the program for the hypothetical robot troubleshooter. Note the decision-making statements used include most of the techniques presented in this chapter.

Program 3–23

```
#include <stdio.h>
/*----------------------------------------------------*/
/*               Robot Troubleshooter              */
/*----------------------------------------------------*/
/*            Developed by: Robert Shooter           */
/*----------------------------------------------------*/
/*              Date:  October 24, 1992              */
/*----------------------------------------------------*/
/*    This program will take the user step by step  */
/* through the analysis and troubleshooting of a    */
/* hypothetical robot. The program also             */
/* demonstrates the various kinds of decision       */
/* making available with the C language.            */
/*----------------------------------------------------*/
/*           Non-Standard Header Files Used:         */
/*----------------------------------------------------*/
/*                      None                         */
/*----------------------------------------------------*/
/*                Function Prototypes                */
void explain_program(void);
/*    This function explains the operation of the   */
/* program to the program user.                     */
/*----------------------------------------------------*/
float arm(void);
/*    This function contains the arm service        */
/* routine for the robot.                           */
/*----------------------------------------------------*/
```

(continued)

Program 3–23 *(continued)*

```c
void power_unit(void);
/*   This function is the power unit service    */
/*   routine. It is a final testing function.   */
/*-----------------------------------------------*/
void light_check (void);
/*   This function presents instructions on how  */
/*   to check the status light indicator.        */
/*-----------------------------------------------*/
void arm_drive_disconnect(void);
/*   This function presents instructions on the  */
/* testing of the arm drive disconnect circuit.  */
/*-----------------------------------------------*/

main( )
 {
 float measurement;      /* Measurement conducted by program user. */

   explain_program( );      /* Explain program to user.  */
   measurement =  arm( );   /* Arm measurement value.    */

     if (measurement > 35) power_unit( );
     else
     if ((measurement <= 35 && measurement >= 30)) light_check( );
     else
     if (measurement < 30) arm_drive_disconnect( );

   exit(0);

 }    /* end of main.  */

/*-----------------------------------------------*/

  void explain_program(void)    /* Explain program to user. */
  {

    printf("\n\nThis program represents the troubleshooting\n");
    printf("of a hypothetical robot system.\n");
    printf("\nThe information you enter will simulate actual\n");
    printf("measurements you would make on this robot.\n");
    printf("\nThe program will instruct you as to what measurement\n");
    printf("you are to enter next.\n");

  }    /* end of explain_program.  */

/*-----------------------------------------------*/
```

(continued)

Program 3-23 *(continued)*

```c
    float arm(void)      /*  Do arm service routine.  */
    {
    float measurement;   /* User measurement in volts. */

      printf("\nMeasure the voltage at test point #1\n");
      printf("Your measurement in volts => ");
      scanf("%f",&measurement);

      return(measurement);

    }    /*   end of arm.   */

/*-----------------------------------------------------*/

  void power_unit( )     /*  Power unit service instructions.  */
  {

    printf("\nRefer to power unit test in service manual\n");
    printf("and replace arm power unit.");

  }   /*  end of power_unit.  */

/*-----------------------------------------------------*/

    void light_check( )   /*  Checking the status light indicator.  */
    {
    int light_status;    /*  User input of light status.  */

      printf("\nInput the condition of the status light:\n");
      printf("1] Red 2] Green 3] Off \n");
      printf("Enter number => ");
      scanf("%d",&light_status);

  switch (light_status)
    {
    case  1  : printf("Disconnect power and replace fuse F1\n");
             break;
    case  2  : printf("Disconnect power and replace arm drive board");
             break;
    case  3  : printf("Replace bulb.");
             break;
    default  : printf("That was not a correct selection!");
    }  /* end of switch.  */

      printf("\n Repeat service routine. \n");
```

(continued)

Program 3–23 *(continued)*

```
    }   /*  end of light_check.  */

/*-------------------------------------------------------*/

    void arm_drive_disconnect( )   /* Arm drive instructions.  */
    {
    float measurement;       /*  Measurement made by user.  */

        printf("Disconnect arm drive and remeasure voltage at TP#1\n");
        printf("Your measurement in volts => ");
        scanf("%f",&measurement);

        measurement = (measurement > 30)? 30 : measurement;

        if (measurement < 30)
    {
      printf("Replace arm power unit.  Refer to \n");
      printf("service manual page 183.  Then repeat \n");
      printf("SVS testing.");
    }
        else
        if (measurement == 30)
    {
      printf("Replace arm drive circuit board.  Refer to \n");
      printf("service manual page 235, then repeat service \n");
      printf("testing.");
    }

    }  /* end of arm_drive_disconnect.  */

/*-------------------------------------------------------*/
```

Program Analysis

A complete analysis of the final program is left for the Self-Test of this chapter. This will help you test your understanding of the material presented in this chapter as well as your ability to apply what you have learned.

Conclusion

This section presented a case study that used most all of the decision-making capabilities of the C language. Here you were again taken step by step through the development of a major program for an area of technology. Check your understanding of this section by trying the following section review.

3-9 Section Review

1 State the first step in the development of any program.
2 Explain the purpose of a troubleshooting flow chart.
3 Define a program stub.
4 For the program developed in this case study, what did the user input represent?

Interactive Exercises

DIRECTIONS

These exercises require that you have access to a computer and software that supports C. They are provided here to give you valuable experience and most importantly immediate feedback on what the concepts and commands introduced in this chapter will do. They are also fun.

Exercises

1 Keep in mind that relational statements in C all return a value. Look at the following program. It works. Try it and see what is displayed.

Program 3–24

```
#include <stdio.h>

main( )
{
  printf("What is this => %d",(5 > 3));
}
```

2 The next program uses a logic statement. Remember, these too return a value. Predict what the following output will be, and then try it.

Program 3–25

```
#include <stdio.h>

main( )
{
  printf("What is this => %d",(1 && 1));
}
```

3   In C any value other than 0 is considered to be TRUE. To demonstrate this, try the following program.

Program 3–26

```
#include <stdio.h>

main( )
{
   printf("What is this => %d",(1 && 23));
}
```

4   The next program presents a demonstration with the logical OR. Predict what the program will do and then try it.

Program 3–27

```
#include <stdio.h>

main( )
{
   printf("What is this => %d",(1 || 1));
}
```

5   The next program again emphasizes that any value other than 0 is considered TRUE with C. The next program illustrates this fact with a logical OR statement.

Program 3–28

```
#include <stdio.h>

main( )
{
   printf("What is this => %d",(0 || 38));

}
```

6   The next program gives you an opportunity to test your ability to predict the outcome of a complex logical statement. Try to predict the outcome, and then try the program.

Program 3-29

```
#include <stdio.h>

main( )
{
 int logic_variable;

  logic_variable = (5 < 7)&&((8 > 5)||(7 > 3));

  printf("What is this => %d",logic_variable);

}
```

7   What do you think of the way the next program is written? Do you find it easier to read and predict what the outcome will be?

Program 3-30

```
#include <stdio.h>
#define TRUE 1
#define FALSE 0
#define AND &&
#define OR ||
#define NOT !

main( )
{
 int logic_value;

   logic_value = TRUE AND TRUE;

  if (logic_value) printf("TRUE");
  else printf("FALSE");

}
```

8   Read through the next program and see if you find it easier to read the C **switch** statements. Predict the outcome, and then try it.

Program 3–31

```
#include <stdio.h>
#define TRUE 1
#define FALSE 0
#define AND &&
#define OR ||
#define NOT !

main( )
{
 int logic_value;

    logic_value = TRUE OR NOT TRUE;

    switch (logic_value)
 {
 case FALSE : printf("FALSE");
               break;
 case TRUE  : printf("TRUE");
 }

 }
```

9 Read through the next program. Then try it.

Program 3–32

```
#include <stdio.h>
#define TRUE 1
#define FALSE 0
#define AND &&
#define OR ||
#define NOT !
#define Check_This_Out switch
#define Stop break;
#define Is_It case

main( )
{
 int logic_value;

    logic_value = NOT(TRUE OR NOT TRUE) AND TRUE;
```

(continued)

Program 3-32 *(continued)*

```
     Check_This_Out (logic_value)
   {
   Is_It FALSE : printf("FALSE");
                 Stop
   Is_It TRUE  : printf("TRUE");
   }

   }
```

Self-Test

DIRECTIONS

Use Program 3-23 developed in the case study section of this chapter to answer the following questions:

Questions

1 How many functions are defined in the program? Name them.
2 Are there any functions that contain an open branch? If so, which one(s)?
3 Are there any functions that contain a closed branch? If so, which one(s)?
4 Which function(s) uses the C **switch**?
5 State the meaning of

   `measurement = (measurement > 30)? 30 : measurement;`

6 How many variable identifiers are used in the program?
7 Which functions return a value?
8 Of the functions that return a value, what value(s) is/are returned?
9 State why the variable **light_status** is not of type **float**.

End-of-Chapter Problems

General Concepts

Section 3-1

1 Write the six relational operators presented in this chapter. State what each means.
2 What is the numerical value used by C to represent a FALSE? To represent a TRUE?
3 If the == symbol in C means equal to, what does the = sign mean in C?
4 Indicate the numerical value of the following relational operations, given that A = 0, B = 3 and C = 8:

   A. !A
   B. B == B
   C. C > A

Section 3-2

5 Explain the concept of an open branch.

6 State the C statement that is used for the open branch.

7 What is meant by a compound statement in C?

8 Can a function be called as a branch option? Explain.

Section 3-3

9 Explain the concept of a closed branch.

10 State the C statement that is used for the closed branch.

11 What is a compound **if...else** in C?

12 May compound statements and function calls be made from the **if...else if** statement?

Section 3-4

13 What is meant by a logical operation?

14 State what is meant by the logical AND operation. What symbol is used to represent this in C?

15 State what is meant by the logical OR operation. What symbol is used to represent this in C?

16 State what is meant by the logical NOT operation. What symbol is used to represent this in C?

Section 3-5

17 Is it legal in C to add a type **int** to a type **char**? What is this called?

18 What resulting type would the addition given in problem 17 produce?

19 State the purpose of a cast in C.

20 What is an lvalue expression?

Section 3-6

21 State when the C **switch** is used.

22 Identify the purpose of the keyword **case** in a C **switch**.

23 Identify the purpose of the keyword **break** in a C **switch**.

24 Can compound statements and function calls be used as part of a C **switch**?

Section 3-7

25 State the purpose of the keyword **default** in a C **switch**.

26 Explain the operation of the conditional operator in C.

Section 3-8

27 What is a command to the compiler called?

28 State what is meant by conditional compilation.

29 Name one conditional compiler directive.

Section 3-9

30 State what a troubleshooting flow chart does for a technician.

31 What property of the C language allows the information in a troubleshooting flow chart to be developed into an interactive computer program?

Program Design

In developing the following C programs, use the method described in the case study section of Chapter 2 and used again in the case study section of Chapter 3. For each program you are assigned, document your design process and hand it in with your program. This documentation should include the design outline that states the purpose of the program, required input, process on the input, and required output as well as the program algorithm. Be sure to include all of the documentation in your final program. This should consist of, but not be limited to, the programmer's block, function prototypes, and a description of each function as well as any formal arguments you may use.

Electronics Technology

32 Develop a C program that will compute the power dissipation of a resistor where the user input is the value of the resistor and the current in the resistor. The program is to warn the user when the power dissipation is above 1 watt. The power dissipation of a resistor is:

$$P = I^2R$$

Where

P = Power dissipation in watts.
I = Current in amps.
R = Resistance in ohms.

33 Create a C program that will convert an input number into the resistor color code. The relationship is as follows:

0 = Black	3 = Orange	6 = Blue	9 = White
1 = Brown	4 = Yellow	7 = Violet	
2 = Red	5 = Green	8 = Gray	

The program user inputs the number, and the corresponding color is displayed.

34 Write a C program that determines standard capacitor values between which the value of a given capacitor is. The program user inputs a capacitor value and the program will indicate between what two standard values it belongs. For the sake of simplicity, assume the standard capacitor values to be used by the program are as follows:

$0.001\mu F$	$0.1\mu F$	$10\mu F$	$1000\mu F$
$0.0015\mu F$	$0.15\mu F$	$15\mu F$	$1500\mu F$
$0.0022\mu F$	$0.22\mu F$	$22\mu F$	$2200\mu F$
$0.0033\mu F$	$0.33\mu F$	$33\mu F$	$3300\mu F$

Business Applications

35 Create a C program where the prices of the following items are already entered:

Soap = $12.50   Asphalt = $27.59   Glue = $2.33   Gum = $0.57

The program user only needs to select the item by number and indicate the quantity. The program computes the total cost.

Computer Science

36 Develop a C program that demonstrates to the program user all forms of branching used by C.

Drafting Technology

37 Create a C program where the program user selects the name of a figure and the program presents the formula that is used for calculating its area. Use a rectangle, triangle, circle, and parallelogram as the figure choices.

Agriculture Technology

38 Develop a C program that will simulate computer aided troubleshooting of a hypothetical diesel engine. The troubleshooting chart is shown in Figure 3–20.

Health Technology

39 Create a C program that will allow the program user to convert from or to metric. The user can then choose length, volume, or weight for the conversion.

Manufacturing Technology

40 Develop a C program that will compute the weight of a machined part. The program user can select from three forms—cylinder, rectangle, or cone—and can select from three materials—copper, aluminum, or steel. The program must automatically compute the volume and use the density of the selected material to compute the total weight.

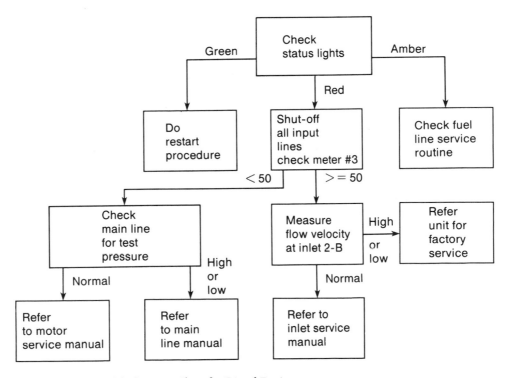

Figure 3–20 Troubleshooting Chart for Diesel Engine

4 Loops

Objectives

This chapter gives you the opportunity to learn:

1 The purpose of program loops.
2 Different types of loops.
3 Loops used by C.
4 The structure and coding of C loops.
5 Practical applications of loops.
6 The nesting of loops.
7 Debugging techniques used in C.
8 The development of a complex technical program that requires the use of loops.

Key Terms

for Loop
Compound Statement
Post-Incrementing
Pre-Incrementing
Post-Decrementing
Pre-Decrementing
Comma Operator
Sequential-Evaluation Operator

while Loop
Conditional Loop
Sentinel
do while Loop
Nested Loops
Run Time Error
Tracing
Auto Debug Function

Outline

4-1 The **for** Loop
4-2 The **while** Loop
4-3 The **do while** Loop
4-4 Nested Loops

4-5 Program Debugging
   and Implementation
4-6 Case Study

Introduction

Recall that three different types of program blocks are all you need to solve for any programming logic, no matter how complex. These are the action, branch, and loop blocks. You have already studied branch blocks in the previous chapter and have worked with action blocks throughout the previous chapters. Once you have completed this chapter you will have used all three kinds of blocks.

You will discover that one of the powers of using the computer for the analysis and solution of technical problems comes from the use of the loop block. Using the loop block allows you to quickly test conditions for a range of values. By doing this, you can easily spot minimum or maximum values. You can also see the direction of change for many different conditions. This is an interesting and important chapter.

4-1   The **for** Loop

Discussion

This section shows you how to develop a C program that will allow you to do something an exact number of times. This has many useful applications in the world of technology. Here you are no longer limited to solving a problem only once for a single answer; you will now be able to solve a problem any number of times with different values each time.

What the **for** Loop Looks Like

The C **for loop** contains three major parts:

1. The value at which the loop starts.
2. The condition under which the loop is to continue.
3. What changes are to take place for each loop.

The above is accomplished in the C **for** loop as follows:

```
for(initial-expression; conditional-expression; loop-expression);
```

The use of the C **for** loop is shown in Program 4-1. The program increments the value of a variable called **time** by one each time through the loop. The variable starts with a value of 1, and the loop continues to repeat as long as the variable time is less than or equal to 5. As soon as time is larger than 5, the program breaks out of the loop.

Program 4-1

```
#include <stdio.h>

main( )
{
int time;      /* Counter variable. */

   /* The loop starts here. */
   for(time = 1; time <= 5; time = time + 1)
      printf("Value of time = %d \n",time);
   /* The loop ends here. */

   printf("This is the end of the loop.");

}
```

Execution of the above program produces:

```
Value of time = 1
Value of time = 2
Value of time = 3
Value of time = 4
Value of time = 5
This is the end of the loop.
```

Program Analysis

The C **for** loop starts with the keyword **for** followed by the loop expression. The loop expression is divided into three parts:

- The value at which the loop starts (the initial-expression)
- The condition under which the loop is to continue (the conditional-expression)
- What changes are to take place for each loop (the loop-expression)

In the case of this program, the C **for** loop is

```
for(time = 1; time <= 5; time = time + 1)
```

This means the value at which the loop starts is `time = 1`, the conditions under which the loop is to continue are `time <= 5`, and the the changes that are to take place are `time = time + 1`. The structure of the C **for** loop is shown in Figure 4-1.

for Loop Facts

Note that the **for** loop expression is not terminated with a semicolon. That is because the whole combination of the keyword **for**, the loop expression, and the statement making up the body of the loop are one single C statement:

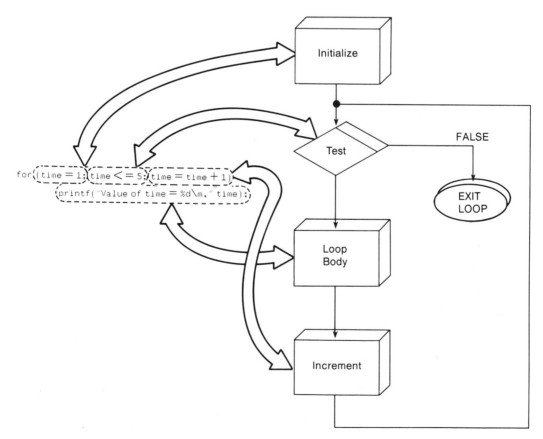

Figure 4-1   Structure of the C **for** Loop

```
for(time = 1; time <= 5; time = time + 1)
    printf("Value of time = %d \n",time);
```

More Than One Statement

You can have more than one statement in a C **for** loop. To do this, you need to indicate the beginning of the loop statements with an opening { and the end of these statements with a closing }. This is often called a **compound statement** and is illustrated in Program 4-2. The program computes the distance a body falls in feet per second, for the first 5 seconds of free fall as given by the equation

$$S = 1/2 \, at^2$$

Where

S = The distance in feet.
a = Acceleration due to gravity (32 ft^2).
t = Time in seconds.

Program 4-2

```
#include <stdio.h>
#define a 32.0

main( )
{
int time;      /* Counter variable. */
int distance;  /* Distance covered by falling body. */

for(time = 1; time <= 5; time = time + 1)
{
distance = 0.5 * a * time * time;
printf ("Distance at the end of %d seconds is %d feet. \n",time,distance);

}  /* End of for loop */

  printf("This is the end of the loop.");

}
```

Execution of the above program produces

```
Distance at the end of 1 seconds is 16 feet.
Distance at the end of 2 seconds is 64 feet.
Distance at the end of 3 seconds is 144 feet.
Distance at the end of 4 seconds is 256 feet.
Distance at the end of 5 seconds is 400 feet.
This is the end of the loop.
```

Note that each of the statements within the { } ends with a semicolon. This is because each is a complete statement. Also note that the ending brace of the compound statement does not require a semicolon. This is no different from other compound statements.

The Increment and Decrement Operators

C offers a shorthand notation for a common programming operation. This is the ability to increment or to decrement a value. The C **for** loop of the previous program uses time = time + 1 to increment the variable time. In C, this could have been condensed to time++. The meanings of the increment and decrement operators are given in Table 4-1.

Program 4-3 illustrates the increment and decrement operators in action.

Program 4-3

```
#include <stdio.h>

main( )
{
int count;       /* Program counter. */
int number1 = 0;         /* Number to post-increment. */
int number2 = 0;         /* Number to pre-increment. */

  for(count = 1; count <= 5; count++)
   {                                   /* A compound statement. */
    printf("Post-incrementing number = %d",number1++);
    printf(" Pre-incrementing number = %d\n",++number2);
   }
}
```

Table 4-1   Increment and Decrement Operators

Operation	Meaning
X++	Increment X after any operation with it (called **post-incrementing**).
++X	Increment X before any operation with it (called **pre-incrementing**).
X--	Decrement X after any operation with it (called **post-decrementing**).
--X	Decrement X before any operation with it (called **pre-decrementing**).

Execution of the above program produces

```
Post-increment number = 0 Pre-increment number = 1
Post-increment number = 1 Pre-increment number = 2
Post-increment number = 2 Pre-increment number = 3
Post-increment number = 3 Pre-increment number = 4
Post-increment number = 4 Pre-increment number = 5
```

Note that the number++ does not get incremented until after it is used. However, the ++number is incremented before it is used. Observe that both numbers are initially set to 0 in the declaration part of the program.

EXAMPLE 4-1

For the given values, determine the results of the following operations:

$$i = 3 \quad c = 10$$

A. $x = i++$ C. $x = c++$ E. $x = i-- + ++c$
B. $x = ++i$ D. $x = --c + 2$

Solution
A. $x = i++ = 3++ =>$ increment after process thus $x = 3$
B. $x = ++i = ++3 =>$ increment before process thus $x = 4$
C. $x = c++ = 10++ =>$ increment after process thus $x = 10$
D. $x = --c + 2 = --10 + 2 = 9 + 2 = 11$
E. $x = i-- + ++c = 3-- + ++10 = 3 + 11 = 14$

The results of the above operations are cumulative as shown in the following program.

Program 4-4

```c
#include <stdio.h>

main( )
{
int i = 3;
int c = 10;

    printf("i = %d\n",i);
    printf("c = %d\n",c);
    printf("x = i++ => %d\n",i++);
    printf("x = ++i => %d\n",++i);
    printf("x = c++ => %d\n",c++);
    printf("x = --c + 2 => %d\n",(--c + 2));
    printf("x = i-- + ++c => %d\n",(i-- + ++c));

}
```

Execution of the above program yields

```
i = 3
c = 10
x = i++ => 3
x = ++i => 5
x = c++ => 10
x = --c + 2 => 12
x = i-- + ++c => 16
```

Program Analysis

For Program 4–4, the result of each action is cumulative. This means that when x = i++ => 3 is completed, the variable i now equals 4 (it gets incremented after the operation). Thus, when the next operation happens (x = ++i => 5), the incrementing takes place before the operation, and since i now has a value of 4, the increment before operation makes it a 5.

In the --c + 2 operation, the variable c starts with the value of 11 (it was incremented after the previous operation x = c++ => 10). Since this is a decrement before operation, --c causes the value of c to become 10; and thus 10 + 2 = 12.

For the last operation, x = i-- + ++c => 16, i starts with its previous value from its last operation (5), and c has a value of 10 from its previous operation. Since ++c causes an increment before operation, this makes it an 11 and 5 + 11 = 16.

The Comma Operator

As a general rule, two C expressions may be separated by a **comma operator**:

expression$_1$, expression$_2$

This means that in a C **for** loop, you may have more than one expression within the **for** loop expression. This means that you could initialize two or more variables at the same time:

for(count = 0, value = 2; count <= 3, count++)

Note that two variables, separated by a comma, are being initialized: count and value. You could have also used multiple statements in the increment part of the **for** statement. However, only one expression is allowed in the test part of the expression. Items separated by the comma operator are evaluated from left to right. The comma operator is sometimes referred to as the **sequential-evaluation operator**.

Conclusion

This section presented the operation of the C **for** loop. Here you saw the main elements of the C **for** along with some examples of how it may be used. You were also introduced to the C increment and decrement operations as well as the comma operator. Check your understanding of this section by trying the following section review.

4-1 Section Review

1   Name the three major parts of the C **for** loop.
2   Can you have more than one statement in a C **for** loop? Explain.
3   Explain the meaning of ++Y.
4   What is the comma operator?

4-2   The **while** Loop

Discussion

This section will introduce you to another kind of loop structure offered by C. It is called the **while** loop. As you will see, this loop has the same elements as the C **for** loop. The difference is that its elements are distributed throughout the program. You will find that it's best to use the **while** loop in situations where you don't know ahead of time how many times the loop will be repeated (such as creating a loop in your program that lets the program user automatically repeat the program).

Structure of the **while** Loop

The structure of the C **while** loop is

```
while(expression)
    statement;
```

The statement may be a single statement or a compound statement (enclosed in braces { }). The statement is executed zero or more times until the expression becomes FALSE.

The way the C **while** loop works is, **expression** is first evaluated. If this evaluation is FALSE (0), then **statement** is never executed, and control passes from the **while** statement to the rest of the program. If the evaluation is TRUE (not zero), then **statement** is executed and the process is repeated again.

Program 4–5 illustrates the C **while** loop. The program does the same thing that Program 4–1 did. It evaluates a variable called **time** for five different values.

Program 4-5

```c
#include <stdio.h>

main( )
{
int time;   /* Counter variable. */

  time = 1;

  while(time <= 5)
  {
   printf("Value of time = %d\n",time);
   time++;
  } /* end of while */

  printf("End of the loop.");

}
```

Execution of the above program produces

```
Value of time = 1
Value of time = 2
Value of time = 3
Value of time = 4
Value of time = 5
```

Program Analysis

The operation of the **while** loop does the test before execution. The logic diagram of the above program is illustrated in Figure 4-2.

As long as the **while** condition is TRUE (meaning not zero), the statement following will be executed. In the program

```
time = 1;
  while(time <= 5)
```

it was first necessary to initialize the counting variable **time** to a value of 1. This was done to ensure its starting value. The **while** will be TRUE as long as the counter **time** is less than or equal to 5.

It is also necessary to change the value of the counter while in the loop. If this is not done, the loop will repeat itself forever. The counter **time** is incremented within the body of the loop:

```
{
 printf("Value of time = %d\n",time);
```

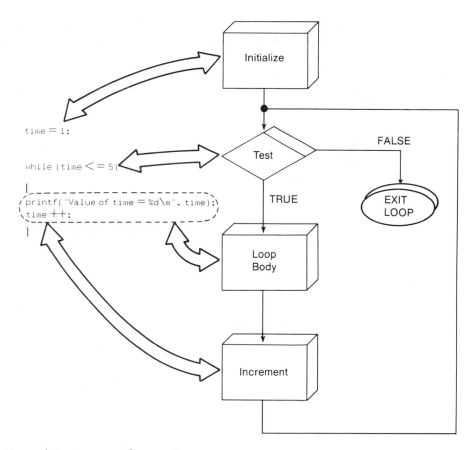

Figure 4-2   Structure of **while** Loop

```
time++;
} /* end of while */
```

Observe that this is a compound statement requiring two statements enclosed by the opening { and closing }.

while Application

An application of the C **while** is illustrated in Program 4-6. C **while** loops are more appropriate than C **for** loops when you do not know when the condition that terminates the loop will occur.

A new C function is used here called **getche**( ). This function gets a single character from the keyboard and echoes it to the screen. It is a useful function whenever a single character is needed for the program.

Program 4-6 simply checks the user input to see if the body of the program is to be repeated. Note that to terminate the loop, the program user must input a capital N.

Program 4-6

```
#include <stdio.h>

main( )
{
   char answer = 'Y';    /* Response of program user. */

    while(answer != 'N')
    {
      printf("This is the body of the program.\n");
      printf("Do you want to repeat this program (Y/N) => ");
      answer = getche( );
      printf("\n");
   }  /* End of while. */

      printf("Thanks for using the program!");

}
```

A sample output of the program is

```
This is the body of the program.
Do you want to repeat this program (Y/N) => Y
This is the body of the program.
Do you want to repeat this program (Y/N) => n
This is the body of the program.
Do you want to repeat this program (Y/N) => N
Thanks for using the program!
```

Program Analysis

First the character variable is declared and initialized to Y. This was done to ensure that it does not come up with the value of N all by itself (a remote chance, but possible).

```
char answer = 'Y';    /* Response of program user. */
```

Next the condition for continuing the loop is given:

```
while(answer != 'N')
```

This means that as long as the character variable `answer` does not equal N, the body of the loop will be activated:

```
{
  printf("This is the body of the program.\n");
  printf("Do you want to repeat this program (Y/N) => ");
  answer = getche( );
  printf("\n");
}  /* End of while.  */
```

Once the condition for the C **while** is FALSE (meaning answer != 'N' is FALSE), the program continues on to the next statement:

```
printf("Thanks for using the program!");
```

Functions Inside the **while**

You can place a C function inside a C **while** loop expression. This practice is often done to reduce the amount of source code. As an example,

```
while(getche( ) != '\r')
```

This will allow the program user to input characters and have them echoed to the screen until the carriage return is pressed. What happens here is that the `getche`( ) function is activated first, and then the evaluation is made. You should be aware of this practice.

Conclusion

The unique characteristic of the C **while** loop is that the loop condition is tested *before* the loop is executed. This type of loop should be used in situations when you do not know ahead of time how many times the loop is to be repeated (sometimes called a **conditional** or a **sentinel loop**).

Check your understanding of this section by trying the following section review.

4–2 Section Review

1   State the construction of the C **while** loop.
2   Under what conditions will a **while** loop be repeated?
3   When is the loop condition tested in a C **while** loop?
4   What is a good use of a **while** loop?

4–3   The **do while** Loop

The last of the three C loop types is the C **do while** loop. As you will see, this loop structure is similar to the C **while** loop, the difference being that the test condition is evaluated *after* the loop is executed. Recall that the C **while** loop tests the condition *before* the loop is executed.

What the do while Loop Looks Like

The C **do while** statement has the form

```
do
   statement
while(expression);
```

where **statement** may be a single or a compound statement. **statement** is executed one or more times until **expression** becomes FALSE (a value of **0**). Execution is done by first executing **statement**, then testing **expression**. If **expression** is FALSE the **do** statement terminates, and control passes to the next statement in the program. Otherwise, if **expression** is TRUE, **statement** is repeated, and the process starts over again.

Using the C do while

Program 4-7 illustrates the action of the C **do** loop. Note that it is a counting loop. The counting variable is **time**. The program simply increments the counter from 1 to 5 in steps of 1.

Program 4-7

```
#include <stdio.h>

main( )
{
int time;   /* Counter variable */

  time = 1;

  do
  {
   printf("Value of time = %d\n",time);
   time++;
  }
   while(time  <= 5);

   printf("End of the loop.");

  }
```

Program output is

```
Value of time = 1
Value of time = 2
Value of time = 3
Value of time = 4
Value of time = 5
```

Note that the output is no different from the similar program using the C **while** loop. The program logic is different in that the action part of the loop is done before the test.

Program Analysis

The counter variable is first initialized to a 1 as before:

```
time = 1;
```

Now the program gets into the C **do**:

```
do
{
printf("Value of time = %d\n",time);
time++;
}
```

Observe that a compound statement follows the keyword **do**. This is the action part of the loop. What must immediately follow it is the condition for doing the action part again. This is achieved by the **while** statement:

```
while(time <= 5);
```

This says that **while** the counter time is less than or equal to 5, the loop will continue. The construction of the C **do while** loop is illustrated in Figure 4–3.

Loop Details

To emphasize that a C **do** loop is always executed at least once, look at Program 4–8. Here the **while** is made false (by setting its argument to zero).

Program 4–8

```
#include <stdio.h>

main( )
{

  do
    printf("This always happens at least once...");

  while(0);

    printf("End of the loop.");

}
```

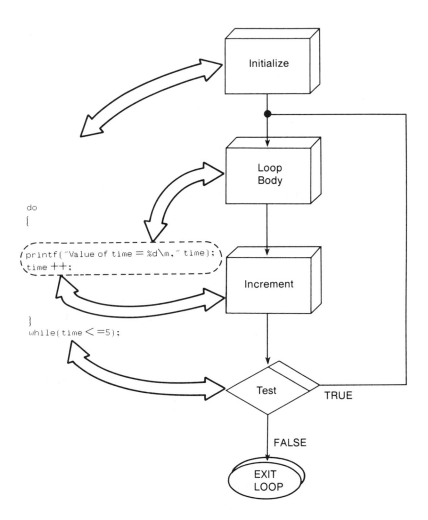

Figure 4–3   Structure of the **do while** Loop

The output of the above program is:

`This always happens at least once...`

Now look at Program 4–9. This is a C **while** loop. Again, however, the **while** argument is set to FALSE. Since for a C **while** the test is always made first, the action part of the loop is never executed!

Program 4–9

```
#include <stdio.h>

main( )
{

 while(0)
   printf("This will not happen if test is FALSE...");

   printf("End of the loop.");

}
```

Sentinel Loops

Both the C **do** and the C **while** loops may be used as sentinel loops. Program 4–10 illustrates the C **do** used as a sentinel loop. Here the program user is asked to enter a value that is less than 5. The program tests to see if this happened. If not, it warns the program user and repeats the loop so that a value within the requested range may be entered.

Program 4–10

```
#include <stdio.h>

main( )
{
 float input;   /* User input number */

   do
   {
   printf("Input a number less than 5 => ");
   scanf("%f",&input);
    if(input >= 5)
      printf("That value is too large, try again:\n");
   }
   while(input >= 5);

   printf("Thank you...");

}
```

A sample execution of the above program could produce

```
Input a number less than 5 => 7
That value is too large, try again:
Input a number less than 5 => 25
That value is too large, try again:
Input a number less than 5 => 3
Thank you...
```

Program Analysis

There is no counter in the above program because it is a sentinel loop. The program executes the C **do** first:

```
do
{
printf("Input a number less than 5 => ");
scanf("%f",&input);
```

This part of the program waits for a user input. The next part of the program starts with a C **if**:

```
 if(input >= 5)
   printf("That value is too large, try again:\n");
}
```

This part of the program is necessary in order to prompt the program user to the fact that the input is not acceptable (if it is 5 or larger) and that another attempt to input a value of within the correct range is needed. If a value within the correct range is entered, then the **printf** function within the loop is never evaluated.

The C **do while** loop ends with the required **while** test:

```
while(input >= 5)
```

Loop Comparisons

Both the C **while** and the C **do while** loops can be used when you do not know ahead of time how many times a loop is to be repeated. When choosing between the two, it is generally considered good programming practice to use the C **while** over the C **do while**. The reason for this is that the C **while** will first test the condition before executing the loop. This is analogous to testing the temperature of the water before you dive in.

Conclusion

Here you were introduced to the last of the three different loop structures available in C. Here you saw the C **do while** loop and how it compares to the C **while** loop. You saw that with the C **do while**, the loop is tested after the execution of the loop statement. Check your understanding of this section by trying the following section review.

4-3 Section Review

1 State the construction of the C **do while** loop.
2 Under what conditions will the **do while** loop be repeated?
3 When is the loop condition tested in a C **do while** loop?
4 Which is preferred, the **do while** or the **while**? Explain.

4-4 Nested Loops

Discussion

In this section you will be introduced to the concept of having one loop inside another. Being able to do this gives you great flexibility when developing C programs to solve technical problems. Here you will see methods to help you structure nested loops to make them easier to read and less prone to error.

Basic Idea

Program 4-11 contains a loop within a loop. This is known as a **nested loop**. The program has two counters and displays the value of each of these counters. Both loops are C **for** loops, and each uses a different counter.

Program 4-11

```c
#include <stdio.h>

main( )
{
int outside_counter;      /* Counter for outside loop.  */
int inside_counter;       /* Counter for inside loop.   */

   for(outside_counter=0;outside_counter<=3;outside_counter++)
     {
      printf("Start of outside loop.\n");
      printf("Outside loop counter => %d\n\n",outside_counter);

       for(inside_counter=0;inside_counter<=3;inside_oounter |+)
         {
          printf("Inside loop counter => %d\n",inside_counter);
         } /* End of inside loop. */

     printf("\nEnd of outside loop.\n\n");
     } /* End of outside loop.  */

}
```

Execution of the above program produces

```
Start of outside loop
Outside loop counter => 0

Inside loop counter => 0
Inside loop counter => 1
Inside loop counter => 2
Inside loop counter => 3

End of outside loop.

Start of outside loop
Outside loop counter => 1

Inside loop counter => 0
Inside loop counter => 1
Inside loop counter => 2
Inside loop counter => 3

End of outside loop.

Start of outside loop
Outside loop counter => 2

Inside loop counter => 0
Inside loop counter => 1
Inside loop counter => 2
Inside loop counter => 3

End of outside loop.

Start of outside loop
Outside loop counter => 3

Inside loop counter => 0
Inside loop counter => 1
Inside loop counter => 2
Inside loop counter => 3

End of outside loop.
```

Program Analysis

The outside loop is the first C **for** in the program:

```
for(outside_counter=0;outside_counter<=3;outside_counter++)
  {
  printf("Start of outside loop.\n");
  printf("Outside loop counter => %d\n\n",outside_counter);
```

The loop counter starts at 0 and on its first pass activates the nested loop:

```
for(inside_counter=0;inside_counter<=3;inside_counter++)
 {
   printf("Inside loop counter => %d\n",inside_counter);
 } /* End of inside loop. */
```

The nested loop is also a C **for** loop, and its counter starts at 0. Once this loop is initialized, it will repeat itself until the condition of its loop is met. This will happen when `inside_counter` is no longer less than or equal to 3. When this happens, the loop quits and program flow goes back to the remainder of the outer loop:

```
printf("\nEnd of outside loop.\n\n");
 } /* End of outside loop.  */
```

Here the outer loop terminates and goes back to its beginning, incrementing its own counter by one. Thus, the inner loop increments itself from 0 to 3 for each single increment of the outer loop.

Nested Loop Structure

Figure 4–4 shows the recommended structure of nested loops.

As you can see from the figure, loops should be nested in such a manner as to encourage understanding of where the loops begin and end. This structure should include comments to make it clear to which loop the closing } of each loop belongs. Doing this will help minimize programming errors.

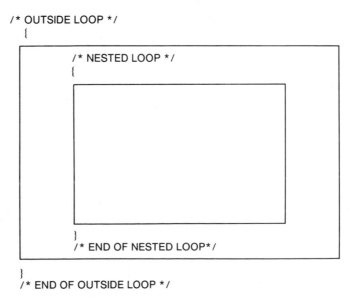

Figure 4–4   Recommended Structure of Nested Loops

Nesting with Different Loops

You can nest as many levels of loops as your hardware will allow. However, when you nest several levels and different types of loops you should be careful because the results may not be exactly what you may expect. Consider, for example, the nested loops of Program 4-12. This program has three nested loops. The outer loop is a C **for**, the next inner loop is a C **while**, and the innermost loop is a C **do**. Look through the program and see if you can predict what the results will be. Then look at the actual output presented in the text at the end of the program for what may be some surprising results.

Program 4-12

```
#include <stdio.h>

main( )
{
int counter_1 = 0;      /* Counter for first loop.  */
int counter_2 = 0;      /* Counter for second loop. */
int counter_3 = 0;      /* Counter for third loop.  */

   for(counter_1 = 0; counter_1 <= 2; counter_1++)
   {

      while(counter_2++ <= 3)
      {

         do
         {
         printf("counter_1 = %d\n",counter_1);
         printf("counter_2 = %d\n",counter_2);
         printf("counter_3 = %d\n",counter_3);
         }
         while(counter_3++ <= 3);   /* end of third loop. */
      } /* end of second loop. */
   } /* end of first loop.   */

}
```

When executed, the above program produces

```
Counter_1 = 0
Counter_2 = 1
Counter_3 = 0
Counter_1 = 0
Counter_2 = 1
Counter_3 = 1
Counter_1 = 0
```

```
Counter_2 = 1
Counter_3 = 2
Counter_1 = 0
Counter_2 = 1
Counter_3 = 3
Counter_1 = 0
Counter_2 = 1
Counter_3 = 4
Counter_1 = 0
Counter_2 = 2
Counter_3 = 5
Counter_1 = 0
Counter_2 = 3
Counter_3 = 6
Counter_1 = 0
Counter_2 = 4
Counter_3 = 7
```

Program Analysis

The results of the above program may not be what you expected. A close analysis of the program follows.

First, all three counter variables are initialized to 0:

```
int counter_1 = 0;    /* Counter for first loop.  */
int counter_2 = 0;    /* Counter for second loop. */
int counter_3 = 0;    /* Counter for third loop.  */
```

Next, the outer C **for** loop is activated:

```
for(counter_1 = 0; counter_1 <= 2; counter_1++)
  {
```

The condition for this loop is that it will continue as long as its counter is equal to or less than 2. For now, its counter has been set to 0 for the first pass.

Next, the inner loop is activated (it is the first instruction of the active outer loop):

```
while(counter_2++ <= 3)
{
```

This is a C **while** loop. Its counter is being tested and then incremented to a 1. Since its test condition is TRUE, it will be active and activate its first instruction which is the beginning of a C **do**:

```
do
{
  printf("counter_1 = %d\n",counter_1);
  printf("counter_2 = %d\n",counter_2);
  printf("counter_3 = %d\n",counter_3);
}
while(counter_3++ <= 3); /* end of third loop.  */
```

Recall that a C **do** loop does its loop before any test for a condition. Thus, no matter what the condition of the counter for the C **do**, all of the **printf** functions will be activated, and the program will display the values for each counter which are

```
counter_1 = 0
counter_2 = 1
counter_3 = 0
```

The reason why the counter for the third loop is still a 0 is because it hasn't yet been incremented. It won't be until the ending **while** where the counter is incremented and tested.

```
while(counter_3++ <= 3); /* end of third loop. */
```

Since the test is still TRUE, this inner C **do** loop will repeat itself until its counter has reached a count of 4. Then it will fall out of the loop and go to the next level which is the C **while**.

Here, the second counter gets incremented. (Note that the first counter is not incremented because the program has only fallen out to the next highest level of loops—which is from the innermost to the next level. This next level has not yet completed its counting condition, so it is still active.)

Since this C **while** loop is still active, it activates its first instruction which is the C **do**. Even though the C **do** counter has met its count, it is still activated because a C **do** always gets done every time it is active (no matter what its condition is). Thus, its counter is now forced to be incremented again! This is what eventually brings counter_3 up to the surprising value of 7! This is another reason why the C **while** loop is preferred over the C **do**.

Once the C **while** loop meets its condition for no longer looping, it falls back out to the C **for** loop. This loop now increments its counter and goes to the first instruction of its loop which is the C **while**. However, the C **while** has already completed its requirement, and thus the C **do** is not activated—and neither are its **printf** functions. Thus, you never see the results of the outer loop counting.

Conclusion

This section presented the concept of nested program loops in C. However, unlike many other programming languages, nesting loops in C can be a bit tricky. It's important to understand how the loops behave and the action of the c ++ when used within the loop instruction.

Check your understanding of this section by trying the following section review.

4–4 Section Review

1 What is a nested loop?
2 Explain the structure that should be used when nesting loops.
3 Which of the three C loop types may be nested?
4 State a potential problem when using the C **do** loop.

4-5   Program Debugging and Implementation

Run Time Errors

You have probably experienced the frustration of developing programs that compiled correctly yet when you went to execute them, you didn't get the answers you expected. This kind of error is called a **run time error**. This can happen because of an incorrect formula or a mistake made entering the formula. It can also happen because you forgot to initialize certain variables or you didn't structure loops or branches correctly. (If you have the LEARN C option, refer to Appendix D for information about the built-in debugger.)

Many times, run time errors can be the most difficult bugs to analyze. The compiler is of no help because everything is correct as far as the C coding goes. The problem is embedded somewhere in the program because you are not getting the required results or what you are getting is incorrect.

There are several techniques for finding these types of errors. One way is to carefully read through your program (or have someone else read through it with you). This is a fine technique for smaller programs. However, in larger programs consisting of many functions with complex loops and branches, taking the time to read through every one may not be the most efficient method. One technique called **tracing** is used by many professional programmers. It's actually an easy addition to any program, and you may want to consider this method for your own use.

A Sample Problem

First, a program containing a problem will be presented, and the problem will be explained. Then a debug routine will be embedded in the program. You will then see how the the debug function can be activated to give you an insight into what is taking place in the program. The sample program follows:

Program 4-13

```c
#include <stdio.h>

main( )
{
int counter;    /* Loop counter. */

   counter = 0;

   while(counter != 9)
   {
    counter += 2;

    /*  Body of loop...  */

   } /* end of while.  */

}
```

The above program will loop forever because the condition for stopping the loop is never met. The entire program will compile successfully because there are no syntax errors. Thus, you will experience a run time error.

What you need to do is to see what is actually going on while this part of the program is executing. A debug block usually consists of two parts, a visual section and a start/stop section.

The visual part causes the values of one or more variables to be displayed while the stop/start part allows you to single step through the program. A debug block for the following loop is shown in Program 4–14.

Program 4–14

```
#include <stdio.h>

main( )
{
int counter;     /* Loop counter. */
char c;          /* Input for return. */

  counter = 0;

  while(counter != 9)
  {
   counter+ =2;

   /* Debug block  */  printf("Counter = %d\n",counter);
                       scanf("%c",c);

   /* Body of loop...  */

  } /* end of while.  */

}
```

When the above program is executed, the output will be

```
Counter = 0
 [Return]
Counter = 2
 [Return]
Counter = 4
 [Return]
Counter = 6
 [Return]
Counter = 8
 [Return]
Counter = 10
 [Return]
```

Notice that now you can single step though the program and observe the action of the loop counter. Its continued increase confirms what you should have suspected. However, the debug routine quickly shows you that the program never achieves the loop exit requirements.

Auto Debug

If you are in the process of developing a large program and you are testing various partially completed functions within the program, you may consider using an auto debug function. This is similar to the above method with the added advantage that you can easily turn it on or off with one simple command. This is especially helpful if you have several functions with debug routines in them. Program 4–15 demonstrates this method.

Program 4–15

```c
#include <stdio.h>
#define TRUE 1
#define FALSE 0

main( )
{
int counter;    /* Loop counter. */
char c;         /* Input for return. */
int debug;      /* Debug flag. */

  debug = TRUE;

  counter = 0;

   while(counter != 9)
   {
    counter += 2;

   /* Debug block  */ if(debug)
                      {
                       printf("Counter = %d\n",counter);
                       scanf("%c",c);
                      }

   /* Body of loop...  */

  } /* end of while.  */

}
```

In the above program, note that TRUE and FALSE have been **#defined**. If you set `debug = FALSE`, then the debug block will never be executed.

Auto Debug Function

The **auto debug function** is simply a C function that is included in your program. This is not feasible with all types of programs, but for certain programs where the variable types are the same, it is particularly useful. Program 4–16 illustrates a debug function.

Program 4-16

```c
#include <stdio.h>
#define TRUE 1
#define FALSE 0

void debug(int active, int display, int startstop);

main( )
{
int counter;    /* Loop counter.  */

  counter = 0;

    while(counter != 9)
    {
     counter += 2;

     /* Debug routing */ debug(TRUE, counter, TRUE);

     /* Body of loop...  */

  } /* end of while.  */

   exit(0);

 {

void debug(int active, int display, int startstop)
{
char c;    /* Input for return.  */

  if (active == TRUE)
  {
   printf("%d",display);
    if (startstop == TRUE)
    scanf("%c",&c);
  } /* end of if */

}
```

Note that now this function can be called by any part of the program. It will output the value of a given variable and offers a stop/start option as well as an on/off option. This can be a very useful method for debugging large programs quickly. If you use such auto debug functions, I suggest that you highlight them so they can be removed from the final program easily. Many programmers will keep two source copies of their programs, one with the auto debug and the other without it. The procedure for doing this is to modify the version with the debug functions in it, make a back-up copy, then remove the debug commands and you will have a new copy of the modified version without the debug functions. Another option is to use the conditional compilation feature presented in the debugging section of the last chapter. Turbo C and Microsoft C both have built-in debuggers. Refer to the Instructor's Guide for the Turbo C debugger and to Appendix D for the Microsoft C debugger.

Conclusion

This section presented some of the debugging methods for large C programs. Here you saw how to include a basic debug command and then how to make it so you could easily deactivate it. You also saw how to create an automatic debugging feature that can be used for certain types of programs.

Try some of these methods in the next program you develop. For now, test your understanding of this section by trying the following section review.

4-5 Section Review

1 Explain what is meant by a run time error.
2 Does the compiler catch run time errors? Explain.
3 State what is usually contained in a debug function.
4 What is meant by an auto debug function?

4-6   Case Study

Discussion

The case study for this chapter shows an application using programming loops. In the previous sections of this chapter, the counting loops that you encountered had their counts determined at programming time. The case study for this chapter follows the development of a C program where the values for a counting loop are determined at run time. Neither the programmer or the program user will know these exact values. As you will see, this type of counting loop has a wide range of applications.

Background Information

Figure 4-5 shows a series resonant RLC circuit and a corresponding plot of the impedance of the circuit over a range of frequencies. A series RLC circuit is an electrical

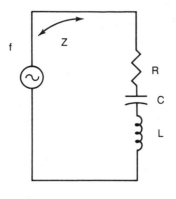

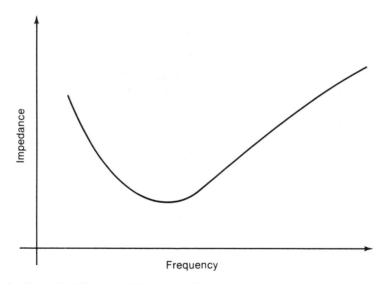

Figure 4-5 Series RLC Circuit and Frequency Plot

circuit that consists of an inductor, a capacitor and a resistor connected to a voltage source in such a way that the current from the source must trace a path through all three circuit components.

Impedance is defined as total opposition to current flow. For a series RLC circuit it is

$$Z = \sqrt{((X_L - X_C)^2 + R^2)}$$

Where

Z = Circuit impedance in ohms.
X_L = Inductive reactance in ohms.
X_C = Capacitive reactance in ohms.
R = Circuit resistance in ohms.

Inductive reactance (X_L) is given by

$$X_L = 2\pi fL$$

And capacitive reactance (X_C) is given by

$$X_C = 1/(2\pi fC)$$

Where

f = Applied frequency in hertz.
L = Inductance of the inductor in henrys.
C = Capacitance of the capacitor in farads.

As you can see from the impedance graph of the circuit in Figure 4–5, as the applied frequency of the source changes, so does the value of the circuit impedance. At one frequency the impedance of the circuit is a minimum and equal to the resistive value of the circuit. This is the point where

$$X_L = X_C$$

and is called resonance. The frequency where this happens is called the resonant frequency and is defined as

$$f_r = 1/(2\pi \sqrt{LC})$$

Where

f_r = The resonant frequency measured in hertz.
L = Inductance of the inductor measured in henrys.
C = Capacitance of the capacitor measured in farads.

This series RLC circuit is sometimes referred to as a series resonant circuit.

The Problem

Create a C program that will allow the program user to input the value of an inductor, capacitor and resistor in a series RLC circuit. The program will compute the resonant frequency of the circuit and then display the value of the circuit impedance for a range of frequencies that are 10% below resonance to 10% above resonance. There will be ten of these calculations below resonance and ten calculations above resonance.

The First Step—Stating the Problem

Stating the case study in writing yields

- Purpose of the Program: Compute and display the impedance of a series RLC circuit for a range of frequencies 10% below to 10% above resonance of the circuit. Calculate and display 10 values of circuit impedance in the range below resonance and 10 values of circuit impedance in the range above resonance.
- Required Input Value of the following:

  - Inductor in henrys.
  - Capacitor in farads.
  - Circuit resistance in ohms.

- Process on Input: Compute resonant frequency. Compute frequencies 10% below and 10% above resonance. Computer incremental value of frequency change. Compute circuit impedance for each of these frequency values:

$$f_r = 1/(2\pi\sqrt{LC})$$
$$X = 2\pi f L_L$$
$$X_C = 1/(2\pi f C)$$
$$Z = \sqrt{((X_L - X_C)^2 + R^2)}$$
$$10\% \text{ below } f_r = f_r - 0.1f_r$$
$$10\% \text{ above } f_r = f_r + 0.1fr$$
$$\text{Incremental change in } f_r = (10\% \text{ above} - 10\% \text{ below})/20$$

Where

f_r = Resonant frequency in hertz.
X_L = Inductive reactance in ohms.
X_C = Capacitive reactance in ohms.
L = Inductance of inductor in henrys.
C = Capacitance of capacitor in farads.
R = Resistance of circuit in ohms.
Z = Circuit impedance in ohms.

- Required Output: Value of the frequency and circuit impedance for 10 readings below resonance starting at 10% below f_r and 10 readings above resonance ending at 10% above f_r.

From the above description, it should be clear exactly what the program should do. Understanding what the program should do must not depend upon any knowledge of programming.

Developing the Algorithm

The steps in this program are the same as for most technology problems:

- Explain program to user.
- Get values from user.
- Do calculations.
- Display results.

For this program, a loop is required that will calculate and display the value of the circuit impedance for a range of frequencies. Since this is a counting loop that will have a definite beginning and ending condition, it was decided to use a C **for** loop.

The required programmer's block was developed as shown in Program 4–17.

Program 4–17

```c
#include <stdio.h>
#include <math.h>

#define PI              3.141592
#define square(x)       x*x
#define X_L(f,l)        2*PI*f*l
#define X_C(f,c)        1/(2*PI*f*c)
#define F_R(l,c)        1/(2*PI*sqrt(l*c))
#define Z(f,l,r)        sqrt(square((X_L(f,l)-X_C(f,l))) + square(r))
/*****************************************************************/
/*              Series Resonant Circuit Analyzer               */
/*****************************************************************/
/*              Developed by: A. Technology Student            */
/*****************************************************************/
/*              Date: October 3, 1992                          */
/*****************************************************************/
/*   This program will display the impedance of a series       */
/* resonant circuit for a range of frequencies that are 10%    */
/* above and 10% below the resonant frequency of the circuit.  */
/* There are ten frequency calculations below and above        */
/* resonance for this circuit.                                 */
/*****************************************************************/
/*              Function prototypes                            */
/*-------------------------------------------------------------*/
void explain_program(void);
/*  This function explains the operation of the program to the*/
/* program user.                                               */
/*-------------------------------------------------------------*/
void get_values(void);
/*   This function gets the circuit values from the program    */
/* user.                                                       */
/*****************************************************************/

main( )
 {
    explain_program( );        /* Explain program to user.  */
    get_values( );             /* Get circuit values.  */
    exit(0);
 }

/*-------------------------------------------------------------*/
```

(continued)

Program 4-17 *(continued)*

```
void explain_program( )          /* Explain program to user.  */
  {
      printf("\n\n This program will display the impedance of a\n");
      printf("series resonant circuit for a range of frequencies\n");
      printf("that are 10%% above and 10%% below the resonant\n");
      printf("frequency of the circuit.\n");

  }    /* End of Explain program to user.  */


/*--------------------------------------------------------------*/

void get_values( )                /* Get circuit values.  */
  {
    float resistor;        /* Value of series resistor in ohms.    */
    float inductor;        /* Value of series inductor in henrys. */
    float capacitor;       /* Value of series capacitor in farads.*/
  }
```

Program Analysis

Note from the above program that all of the formulas are defined using the **#define** directive:

```
#define PI                  3.141592
#define square(x)           x*x
#define X_L(f,l)            2*PI*f*l
#define X_C(f,c)            1/(2*PI*f*c)
#define F_R(l,c)            1/(2*PI*sqrt(l*c))
#define Z(f,l,r)            sqrt(square((X_L(f,l)-X_C(f,l))) + square(r))
```

In order to use the **sqrt** function, the math header file had to be included:

#include <math.h>

The programmer's block then presents the necessary information (the name of the program, name of the program developer, date of development, description of the program and preliminary function prototypes). The **main( )** function is then developed to present the structure of the program:

```
main( )
  {

      explain_program( );          /* Explain program to user.  */
      get_values( );               /* Get circuit values.  */
      exit(0);

  }
/*--------------------------------------------------------------*/
```

Here you can see the use of top-down design where the most general idea of the program is dealt with first, and the details are left for later. You can also see the structure of the program as shown in **main( )**. What will happen is that the program will be explained to the program user and values will be received. No attempt has been made at this initial stage to go any further in the development of other program blocks.

Next, the major portion of each of the prototyped functions is described.

```
void explain_program( )              /* Explain program to user. */
  {
      printf("\n\n This program will display the impedance of a\n");
      printf("series resonant circuit for a range of frequencies\n");
      printf("that are 10%% above and 10%% below the resonant\n");
      printf("frequency of the circuit.\n");

  }    /* End of Explain program to user.  */

/*-------------------------------------------------------------------*/

void get_values( )                   /* Get circuit values.  */
  {
  float resistor;          /* Value of series resistor in ohms.   */
  float inductor;          /* Value of series inductor in henrys. */
  float capacitor;         /* Value of series capacitor in farads.*/
  }
```

Note that all of the explanation to the program user is developed at this first stage, but only the values for the **get_values** function are declared. This is simply a matter of individual programmer choice. But it is suggested that the first coding be as simple and straightforward as possible. This is necessary in order to keep any program debugging time low and to make it so that the general structure of the program will compile and execute.

Designing the get_values Function

The **get_values( )** function design is straightforward. It simply consists of a **printf** user prompt followed by a **scanf** to get the user input. As with most technology problems, the type **float** is used for the input values. This is done so that E notation can be used with the **%f** input. The developed function is shown below. Good program design would call for the addition of calling another function that would calculate and display the required values.

```
void get_values( )                   /* Get circuit values.  */
  {
  float resistor;          /* Value of series resistor in ohms.   */
  float inductor;          /* Value of series inductor in henrys. */
  float capacitor;         /* Value of series capacitor in farads.*/
```

```
printf("\n\n Enter the following values as indicated:\n");
printf(" Resistor in ohms => ");
scanf("%f",&resistor);
printf(" Inductor in henrys => ");
scanf("%f",&inductor);
printf(" Capacitor value in farads => ");
scanf("%f",&capacitor);

}    /* End of get_values. */
```

Designing the calculate_and_display Function

The function for calculating and displaying the required information requires several variables. It was decided that this function would be called from `get_values( )`. Because of this, `calculate_and_display` requires three arguments in its parameter list so that the value of the inductor, capacitor and resistor can be passed to it.

```
float calculate_and_display(float l, float c, float r)
```

Since this function will also contain the C **for** loop, the values of the beginning and ending frequency range as well as the frequency change are needed for the loop counter.

```
{
  float below_fr;      /* 10% below resonant frequency. */
  float above_fr;      /* 10% above resonant frequency. */
  float freq_change;   /* Change in frequency.   */
  float counter;       /* Loop counter.  */
  float impedance;     /* Circuit impedance in ohms. */
```

It was decided that the values for the C **for** loop will be computed by two other functions not yet defined. The first of these, called `calculate_below`, will calculate the frequency that is 10% below resonance and the second, called `calculate_above`, will calculate the frequency that is 10% above resonance. These values will then be used to set the counting range of the C **for** loop.

```
below_fr = calculate_below(l,c);
above_fr = calculate_above(l,c);
```

The change in frequency can then be calculated using the values of the frequency range.

```
freq_change = (above_fr - below_fr)/20.0;
```

For the convenience of the programmer, the values of the C **for** loop ranges are displayed during program execution:

```
printf("below_fr => %f Hertz. \n",below_fr);
printf("above_fr => %f Hertz. \n",above_fr);
printf("freq_change => %f Hertz. \n\n",freq_change);
```

Next, the C **for** statement was developed:

```
for(counter=below_fr; counter<=above_fr; counter += freq_change)
```

The body of this loop consists of displaying the value of the frequency (done by the value of `counter`), calculating the circuit impedance (by using the **#define** for Z) and then using an **if...else**. This **if...else** is used to point out to the user what frequency is the resonant frequency of the circuit. By definition, this is the frequency when the circuit impedance is equal to the value of the circuit resistance.

```
{
    printf("Frequency => %f Hertz",counter);
    impedance = Z(counter,l,c,r);
     if (impedance == r)
      printf(" Impedance = %f Ohms. <== Resonance!\n",impedance);
     else
      printf(" Impedance = %f Ohms.\n",impedance);

}   /* End of for */
```

The next step was the creation of the two functions that will calculate the frequency values 10% below and 10% above resonance. This could have been accomplished by **#defines** at the beginning of the program, but the functions are used in this program to illustrate how the C **return( )** function can be used.

Designing the calculate_below and calculate_above Functions

Since these two functions will be called, they need values passed to them. These values are the ones needed to calculate the resonant frequency of the RLC circuit. Thus, the inductor and capacitor values entered by the program user need to be formal arguments.

```
float calculate_below(float l, float c)
 {
     return(F_R(l,c) - 0.1*F_R(l,c));
 }

/*-------------------------------------------------------------------*/

float calculate_above(float l, float c)
 {
     return(F_R(l,c) + 0.1*F_R(l,c));
 }
```

Note how the C **return** function is used in both of these. The calculation is done within the argument of this function. The resulting value of this calculation is then returned back to the calling function. Doing this saves programming space. However, keep in mind that your primary rule in developing this type of program is to keep the source code easy to read and understand.

Final Program Phase

The final completed program is shown as Program 4–18. Note that all of the functions in the program have a function prototype with comments that explain the purpose of the function as well as the meanings of all variables in the function parameter list.

Program 4-18

```
#include <stdio.h>
#include <math.h>

#define PI              3.141592
#define square(x)       x*x
#define X_L(f,l)        2*PI*f*l
#define X_C(f,c)        1/(2*PI*f*c)
#define F_R(l,c)        1/(2*PI*sqrt(l*c))
#define Z(f,l,c,r)      sqrt(square((X_L(f,l)-X_C(f,c))) + square(r))

/*************************************************************/
/*              Series Resonant Circuit Analyzer           */
/*************************************************************/
/*              Developed by: A. Technology Student        */
/*************************************************************/
/*                  Date: October 3, 1992                  */
/*************************************************************/
/*    This program will display the impedance of a series  */
/* resonant circuit for a range of frequencies that are 10% */
/* above and 10% below the resonant frequency of the circuit. */
/* There are ten frequency calculations below and above     */
/* resonance for this circuit.                              */
/*************************************************************/
/*                  Function prototypes                    */
/*---------------------------------------------------------*/
void explain_program(void);
/*  This function explains the operation of the program to the*/
/*  program user.                                          */
/*---------------------------------------------------------*/
void get_values(void);
/*    This function gets the circuit values from the program */
/*  user.                                                  */
/*---------------------------------------------------------*/
float calculate_and_display(float l, float c, float r);
/*    l = Inductance in henrys. c = Capacitance in farads.  */
/*    r = Resistance in ohms.                              */
/*    This is the function that will calculate and display the */
/*    value of the circuit impedance for the given range of  */
/*    frequencies.                                         */
/*---------------------------------------------------------*/
float calculate_below(float l, float c);
/*    l = Inductance in henrys. c = Capacitance in farads.  */
```

(continued)

Program 4-18 *(continued)*

```
/*    This function calculates the frequency that is 10% below */
/*    resonance.                                                */
/*------------------------------------------------------------*/
float calculate_above(float l, float c);
/*    l = Inductance in henrys. c = Capacitance in farads.    */
/*    This function calculates the frequency that is 10% above*/
/*    resonance.                                               */
/*************************************************************/

main( )
  {
    explain_program( );        /* Explain program to user.  */
    get_values( );             /* Get circuit values.  */
    exit(0);
  }
/*------------------------------------------------------------*/

void explain_program( )        /* Explain program to user.  */
  {
    printf("\n\n This program will display the impedance of a\n");
    printf("series resonant circuit for a range of frequencies\n");
    printf("that are 10%% above and 10%% below the resonant\n");
    printf("frequency of the circuit.\n");

  }  /* End of Explain program to user.  */

/*------------------------------------------------------------*/

void get_values( )                  /* Get circuit values.  */
  {
  float resistor;        /* Value of series resistor in ohms.    */
  float inductor;        /* Value of series inductor in henrys. */
  float capacitor;       /* Value of series capacitor in farads.*/

    printf("\n\n Enter the following values as indicated:\n");
    printf(" Resistor in ohms => ");
    scanf("%f",&resistor);
    printf(" Inductor in henrys => ");
    scanf("%f",&inductor);
    printf(" Capacitor value in farads => ");
    scanf("%f",&capacitor);

    calculate_and_display(inductor, capacitor, resistor);

  }    /* End of get_values.  */

/*------------------------------------------------------------*/

float calculate_and_display(float l, float c, float r)
  {
```

Program 4–18 *(continued)*

```
    float below_fr;        /* 10% below resonant frequency.  */
    float above_fr;        /* 10% above resonant frequency.  */
    float freq_change;     /* Change in frequency.  */
    float counter;         /* Loop counter.  */
    float impedance;       /* Circuit impedance in ohms.  */

        below_fr = calculate_below(l,c);
        above_fr = calculate_above(l,c);
        freq_change = (above_fr - below_fr)/20.0;

        printf("below_fr => %f Hertz. \n",below_fr);
        printf("above_fr => %f Hertz. \n",above_fr);
        printf("freq_change => %f Hertz. \n\n",freq_change);

        for(counter=below_fr; counter<=above_fr; counter += freq_change)
          {
            printf("Frequency => %f Hertz",counter);
            impedance = Z(counter,l,c,r);
             if (impedance == r)
              printf(" Impedance = %f Ohms. <== Resonance!\n",impedance);
             else
              printf(" Impedance = %f Ohms.\n",impedance);

          }  /* End of for */
    }  /*  End of calculate and display.  */
/*------------------------------------------------------------------*/

float calculate_below(float l, float c)
  {
      return(F_R(l,c) - 0.1*F_R(l,c));
  }

/*------------------------------------------------------------------*/

float calculate_above(float l, float c)
  {
      return(F_R(l,c) + 0.1*F_R(l,c));
  }

/*------------------------------------------------------------------*/
```

Program Execution

Execution of the above program produces

```
 This program will display the impedance of a
series resonant circuit for a range of frequencies
```

```
that are 10% above and 10% below the resonant
frequency of the circuit.

 Enter the following values as indicated:
  Resistor in ohms => 10
  Inductor in henrys => 1e-3
  Capacitor value in farads => 1e-6
below_fr => 4529.629883 Hertz.
above_fr => 5536.214355 Hertz.
freq_change => 50.329224 Hertz.
Frequency => 4529.629883 Hertz Impedance = 12.023640 Ohms.
Frequency => 4579.958984 Hertz Impedance = 11.648333 Ohms.
Frequency => 4630.288086 Hertz Impedance = 11.308162 Ohms.
Frequency => 4680.617188 Hertz Impedance = 11.004684 Ohms.
Frequency => 4730.946289 Hertz Impedance = 10.739361 Ohms.
Frequency => 4781.275391 Hertz Impedance = 10.513482 Ohms.
Frequency => 4831.604492 Hertz Impedance = 10.328093 Ohms.
Frequency => 4881.933594 Hertz Impedance = 10.183921 Ohms.
Frequency => 4932.262695 Hertz Impedance = 10.081312 Ohms.
Frequency => 4982.591797 Hertz Impedance = 10.020183 Ohms.
Frequency => 5032.920898 Hertz Impedance = 10.000000 Ohms. <==Resonance!
Frequency => 5083.250000 Hertz Impedance = 10.019782 Ohms.
Frequency => 5133.579102 Hertz Impedance = 10.078132 Ohms.
Frequency => 5183.908203 Hertz Impedance = 10.173290 Ohms.
Frequency => 5234.237305 Hertz Impedance = 10.303208 Ohms.
Frequency => 5284.566406 Hertz Impedance = 10.465627 Ohms.
Frequency => 5334.895508 Hertz Impedance = 10.658155 Ohms.
Frequency => 5385.224609 Hertz Impedance = 10.878351 Ohms.
Frequency => 5435.553711 Hertz Impedance = 11.123784 Ohms.
Frequency => 5485.882813 Hertz Impedance = 11.392090 Ohms.
Frequency => 5536.211914 Hertz Impedance = 11.681007 Ohms.
```

Conclusion

This case study took you step by step through the design and development of a technology program that required the use of a program loop. More specifically, this loop has its values for counting determined at program execution time. In this case, neither the programmer nor the program user needs to know the beginning, ending, or incremental loop values. All of this information is supplied by the program. Check your understanding of this section by trying the following section review.

4-6 Section Review

1   What was the first step in the design of this case study program?
2   State why a C **for** loop was used to meet the programming requirements of this program.
3   How were the beginning and ending values of the program loop determined?
4   Explain how the C **return** is used in this program.

Interactive Exercises

DIRECTIONS

These exercises require that you have access to a computer and software that supports C. They are provided here to give you valuable experience and immediate feedback on what the concepts and commands introduced in this chapter will do. They are also fun.

Exercises

1   Try this one with a character. Guess what the output will be and then give it a try for what may be a surprising result!

Program 4-19

```
#include <stdio.h>

main( )
{
char c;    /* A character.  */

    for(c = 'a'; c <= 'z'; c++)
     printf(" c = %c",c);

}
```

2   Note what is different about this loop. Here a number is being changed; however, it is being displayed as a character (%c) not as a decimal (%d). What do you think this loop produces? Give it a try and see!

Program 4-20

```
#include <stdio.h>

main( )
{
int i;    /* A number.  */

    for(i = 97; i <= 122; i++)
     printf(" i = %c",i);

}
```

3   Your computer system probably has an extended ASCII character code set. If you tried the previous problem, you probably know what the next program will do. Don't pass up trying it!

Program 4–21

```
#include <stdio.h>

main( )
{
int i; /* A number. */

    for(i = 128; i <= 255; i++)
     printf(" i = %c",i);

}
```

4   The following program shows all! It displays everything old ASCII is capable of doing on the monitor. Listen as well as watch on this one.

Program 4–22

```
#include <stdio.h>

main( )
{
int i;    /* A number.   */
int j;    /* Another number. */

    j = 0;

    for(i = 0; i <= 255; i++)
    {
     printf(" %d = %c",j,i);
     j++;
    } /* End of for */

}
```

5   In the next C **for** loop, the loop counter has a ++i instead of an i++. What difference do you suppose this will make in the outcome of the loop counter values? Try it and see. (Another good test question!)

Program 4-23

```
#include <stdio.h>

main( )
{
int i;    /* A number.  */

    for(i = 0; i <= 5; ++i)
    printf(" i = %d",i);

    printf(" last i = %d",i);

}
```

6 What follows is a "null **while**." This means there is nothing in the argument of the C **while** statement. What do you think will happen when you try to compile this program? Check your conclusion.

Program 4-24

```
#include <stdio.h>

main( )
{

    while( )
     printf("Is this ever executed?");

}
```

7 Predict the output of the following program. Then try it.

Program 4-25

```
#include <stdio.h>

main( )
{
int i;    /* A number.  */
```

(continued)

Program 4-25 *(continued)*

```
    for(i = 97; i <= 122; i++)
      printf(" i = %c",i);

}
```

8   How many times does the following program loop? Try it. What causes it to stop?

Program 4-26

```
#include <stdio.h>

main( )
{
  int i; /* A number. */

    i = 10;

    while(i)
      {
      printf("How many times does this loop?");
       printf(" i = %d\n",i--);

      }
}
```

Self-Test

DIRECTIONS

Use Program 4-18 developed in the case study section of this chapter to answer the following questions.

Questions

1   How many function prototypes are used in the program?
2   What type of a loop is used in the program?
3   In the loop used, what are the initial loop conditions?
4   What are the final loop conditions?
5   State what is changed each time through the loop.

End-of-Chapter Problems

General Concepts

Section 4–1

1 What is a compound statement? How is it set off in a C program?
2 Give the C notation for pre-incrementing the variable `counter`.
3 In C, what allows you to have two sequential C statements?
4 What are the major parts of a C **for** loop?

Section 4–2

5 State the construction of the C **while** loop.
6 When is the loop condition tested in a C **while** loop?
7 Give the condition for repeating a C **while** loop.
8 What is a good use of the C **while** loop?

Section 4–3

9 Give the construction of the C **do while** loop.
10 When is the loop condition of a C **do while** loop tested?
11 What are the conditions for repeating a C **do while** loop?
12 Which loop is preferred, the C **while** or the C **do while**? Explain.

Section 4–4

13 What is meant by a nested loop?
14 Which of the C loops may be nested?
15 Is there a problem with nesting a C **do** loop? Explain.

Section 4–5

16 Explain what is meant by a run time error.
17 What is a debug function? How is it used?
18 How is the auto debug function used?

Program Design

In developing the following C programs, use the method described in the case studies. For each program you are assigned, document your design process and hand it in with your program. This process should include the design outline, process on the input, and required output as well as the program algorithm. Be sure to include all of the documentation in your final program. This should consist of, but not be limited to, the programmer's block, function prototypes and a description of each function as well as any formal arguments you may use.

Electronics Technology

19 Create a C program that will compute the voltage drop across a resistor for a range of current values selected by the program user. The program user is to input the value of the resistor, the beginning and ending current values, and the incremental value of the current. The relationship between the resistor voltage and current is given by

$$V = IR$$

Where

V = Voltage across the resistor in volts.
I = Current in resistor in amps.
R = Value of resistor in ohms.

20   Develop a C program that will calculate the power dissipated in a resistor for a range of voltage values across the resistor. The program user is to select the value of the resistor and the beginning as well as the ending resistor voltages. The relationship is given by

$$P = V^2/R$$

Where

P = Power dissipation of the resistor in watts.
V = Voltage across the resistor in volts.
R = Value of the resistor in ohms.

21   Modify the above program so that the program user will be warned when the power dissipation of the resistor exceeds a maximum power dissipation determined by the program user. An example warning could be a monitor display of

WARNING! Power exceeds 2 watts!!!

22   Modify the program in problem 20 by adding an inner loop so that the program user may observe the power dissipation for a range of resistors as well as a range of voltage values. The program user would now select the voltage range and resistance range as well as the incremental values for each.

Business Applications

√23   Develop a C program that will display the interest compounded annually from one to thirty years. The program user is to input the principal and the rate of interest. The mathematical relationship is

$$Y = A(1 + N)^T$$

Where

Y = The amount.
A = The principal invested.
N = Interest rate.
T = Number of years.

Computer Science

24   Create a C program that will take any number to any power. The program user is to select both the base and the power.

Drafting Technology

25   Develop a C program that will show the change in volume of a sphere for a given change in radius. The program user selects the initial, incremental, and ending values of the radius. The mathematical relationship is

$$V = 4/3\pi r^3$$

Where

V = Volume of the sphere in cubic units.
π = The constant pi.
r = Radius of the sphere in linear units.

Agriculture Technology

26   Construct a program in C that will demonstrate the change in the height of water in a cylindrical water tank as the volume of water in the tank changes (increases or decreases). The program user is to input the radius and height of the water tank as well as the incremental increase or decrease of the amount of water in the tank, expressed in gallons.

Health Technology
27   Construct a program in C that will compute and display the values of a range of temperatures in degrees Fahrenheit and degrees centigrade. The program user is to select the lowest, highest, and incremental temperature change. The mathematical relationship is

$$F = (9/5)C + 32$$

Where

F = Temperature in degrees Fahrenheit.
C = Temperature in degrees centigrade.

Manufacturing Technology
28   Develop a C program that will show the change in the volume of a metal cone as it is machined. Assume that the machining reduces the height of the metal cone. The program user enters the dimensions of the cone as well as the incremental changes in the cone height. The mathematical relationship is

$$V = (h/3)(A_1 + A_2 + \sqrt{A_1 A_2})$$

Where

V = Volume of the cone in cubic units.
h = Height of the cone in linear units.
A_1 = Area of the lower base in square units.
A_2 = Area of the upper base in square units.

Business Applications
29   Modify the program for problem 23 so that the program contains an inner loop that will display the interest for different amounts of money. The program user now selects the range for the principal and the incremental difference as well as entering the rate of interest.

Agriculture Technology
30   Modify the program for problem 26 so that the program user is warned when the water tank is empty or overflowing.

Drafting Technology
31   Create a C program that will display a range of the same lengths in feet, meters, and inches. The program user can select the beginning, ending, and incremental measurements in feet.

Agriculture Technology
32   The relationship between the number of units of fertilizer and the expected crop yield is

$$Y = F/(2^F + 1)$$

Where

Y = The yield improvement factor.
F = Arbitrary units of fertilizer per acre.

Develop a C program to find what value of F produces a maximum value of Y.

Health Technology
33   Create a C program that will display the heart beat of a patient in beats per minute, beats per hour, and beats per day. The program user can enter the minimum and maximum number of beats per minute and the value of the increment.

Manufacturing Technology

34 Develop a C program that will show the range of weights in pounds and kilograms. The program user can enter the minimum and maximum weights along with the increment value either in pounds or kilograms.

Business Applications

35 Write a C program that will compute the sales tax for a range of values and round the result off to the nearest cent. The program user can input the minimum and maximum dollar amounts and the percentage of the sales tax. The increment is to be determined by the program so that a printout is given for every one-cent change in the total amount.

5 Pointers, Scope, and Class

Objectives

This chapter gives you the opportunity to learn:

1 The method used by your computer to store information in its memory.
2 Different ways memory is used in your computer by the C language.
3 Methods used to represent values in your computer's memory.
4 The way negative numbers are represented in memory.
5 What pointers are in C and how they are used.
6 Methods of passing variables by using pointers.
7 Which constants and variables are known to what parts of your C program.
8 Various methods of classifying your C variables.
9 Pitfalls common to new C programmers.
10 The meaning of bit manipulation and how to use it in C.

Key Terms

Address
Fetch/Execute
Immediate Addressing
Direct Addressing
Byte
Word
Ones Complement
Twos Complement

Nibble
Pointer
Address Operator
Pointer Declaration
Local Variable
Variable Scope
Variable Life
Global Variable

Automatic Variable
External Variable
Static Variable
Register Variable
Initialization
Bit Manipulation

Boolean Operators
Bitwise Complement
Bitwise AND
Bitwise OR
Bitwise XOR
Bit Shifting

Outline

5-1 Internal Memory Organization
5-2 How Memory Is Used
5-3 Pointers
5-4 Passing Variables
5-5 Scope of Variables

5-6 Variable Class
5-7 Program Debugging and
     Implementation
5-8 Case Study

Introduction

This chapter introduces the important concept of *pointers.* As you will see, this is simply another method of getting data in and out of your computer's memory. But, unlike other methods you have been using, this is a very powerful one that will give you great programming flexibility.

You should have a background in binary, octal, and hexadecimal number systems.

In this chapter, you will first be introduced to some of the internal structure of your computer. Here you will discover how the computer's memory is organized and how it is used. Having this information as a background, you will be ready to explore the wonders of the C pointer.

5-1   Internal Memory Organization

Discussion

This section presents important concepts about how information is stored inside your computer. Understanding memory storage will open a whole new world of programming opportunities for you. This section lays the foundation for the rest of the material presented in this chapter.

Basic Idea

Figure 5-1 presents a basic model of how a program is stored in a computer's memory.

As shown in Figure 5-1, you may visualize a computer's memory as a pile of storage locations. In order to distinguish one storage location from the other, each

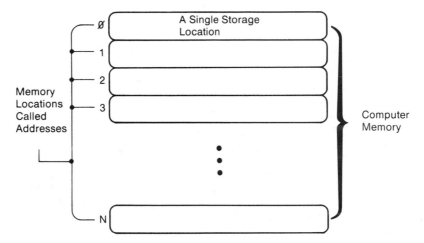

Figure 5-1 Visualization of Computer Memory

location is assigned a unique number. This number is called an **address**. Thus, each memory location inside your computer has a unique address.

Storing a Program

Consider a program where you want to add two numbers and store the answer back in memory. Such a program, stored in a computer, may be visualized as shown in Figure 5-2.

Look at Figure 5-2. There are seven memory locations, each identified by an address. Observe that a memory location may contain an instruction (an action to be taken by the computer) or it may contain data (an item that is acted upon by the action of the instruction). This program is executed by having the CPU (central processing unit) start getting information from memory at memory location 1. As shown in Figure 5-2, the instruction located here tells the CPU to get the number in the next memory location (location 2). After it does that, the CPU goes to the following memory location (location 3) and gets the next instruction. This process of getting an instruction, doing the instruction, and getting another instruction is referred to as the fetch/execute cycle of a computer.

The CPU will add the two numbers (3 and 2) and store the answer (5) in memory location 6. Addresses are used to distinguish one memory location from the other.

Another Way

It isn't common practice to have the data in a program in the very next memory location following an instruction. For example, consider the program shown in Figure 5-3. It does the same thing as the program shown in Figure 5-2. It adds the same two numbers (3 and 2). The difference is how this is done.

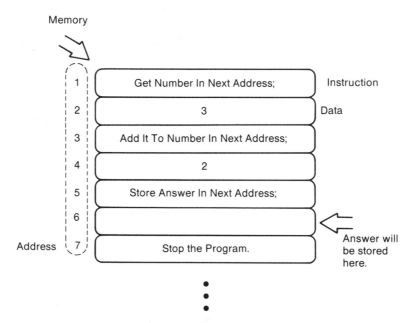

Figure 5-2   Concept of a Program in Memory

The program of Figure 5-3 executes the following process.

1. Get number located at following address: This instruction requires the CPU to go to the next memory location to get the address (new location) where the number it is to work with will be located. It gets the number 8 from the next memory location. Now the first instruction is completed and it means:
2. Get number located at address 8: So now the CPU skips down to address 8 and gets the value stored there (which is the number 3). What happens here is the first two memory locations are actually instructions to the CPU. The first piece of data for the CPU is the number 3 located at address 8.

   The CPU now returns to the address following its last instruction (its last instruction was at address 2) and goes to address 3 and gets its next instruction:
3. Add to it the number located at the following address: Again, this is only part of a complete instruction. As before, the CPU has to go down to the next address (location 4) to find the address where the next piece of data is located. Once it does this, the completed instruction is now interpreted as:
4. Add it to the number located at address 9: And once again, the CPU skips down to address 9 where it finds the next piece of data (the number 2). This is the number to be added to 3. Now the CPU returns back to the address following its last instruction (location 5).
5. Store the answer at the following address: The CPU goes to the next memory location in order to find out where it is to store the answer. Thus it goes

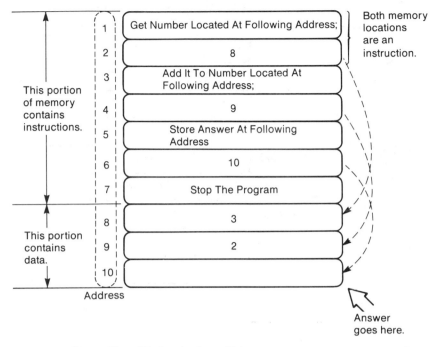

Figure 5-3   A Different Way of Doing the Same Thing

from address 5 to address 6 to complete this instruction. There it finds the number 10. This instruction is now completed.

6. Store the answer at address 10: Thus, the CPU stores the answer (2 + 3 = 5) at memory location 10. Having done this, it goes back to the address following its last instruction (memory location 7) to get its next instruction.

7. Stop the program: And the CPU stops all further processing.

This may seem like a strange way of processing instructions and data, but both processes are used by modern-day computers. In order to use many of the capabilities of the C language, it's important to understand both processes.

The first method (like the program in Figure 5-2), where the <u>data immediately follows the instruction, is called</u> **immediate addressing**. The second method (like the program in Figure 5-3), where the two memory locations are needed for the instruction—one to direct the program to where the data is located in memory—is called **direct addressing**.

These processes are going on every time you use your computer. The difference is the example in Figure 5-3 uses ten memory locations. Your computer has thousands of memory locations.

EXAMPLE 5-1

What does the following program do? Assume that the CPU reads the first instruction from address 3438. See Figure 5-4.

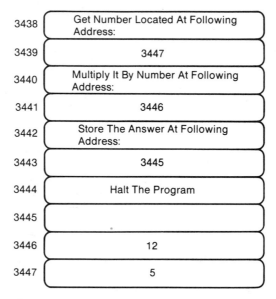

3438	Get Number Located At Following Address:
3439	3447
3440	Multiply It By Number At Following Address:
3441	3446
3442	Store The Answer At Following Address:
3443	3445
3444	Halt The Program
3445	
3446	12
3447	5

Figure 5-4   Program for Example 5-1

Solution

The program shown in Figure 5-4 will multiply 5 × 12 = 60 and store the answer in memory location 3445.

Conclusion

This section presented an important conceptual framework for understanding how instructions and data may be stored in your computer. Here you saw the important difference between immediate and direct addressing. Check your understanding of this section by trying the following section review.

5-1 Section Review

1   State how a computer's memory may be visualized.
2   Explain the difference between an instruction and data.
3   What is an address?
4   State what is meant by immediate addressing.
5   State what is meant by direct addressing.

5-2   How Memory Is Used

Discussion

In order to benefit from the full programming power of C, you must understand how data is organized in memory. The information presented here will use material on binary, octal, and hexadecimal number systems.

Storage Size

Figure 5-5 shows how information may be classified in computer memory. Keep in mind that all instructions and data are represented by groups of electrical ON's and OFF's.

Computer memory is usually organized in groups of **bytes** called **words**. Some of the most common word sizes for computers are 8 bits (1 byte), 16 bits (2 bytes), and 32 or 64 bits. As an example, memory using an 8-bit word is illustrated in Figure 5-6.

char Type

In C, the type **char** occupies a single byte (8 bits) of memory. This means that the range of values that may be stored in a type **char** range from $0000\ 0000_2$ to $1111\ 1111_2$. Expressed in hex this is 00_{16} to FF_{16}.

This is equal to 0_{10} to 255_{10}. Figure 5-7 illustrates this concept.

The following program demonstrates the limitations of one byte The maximum value of a **char** is 255; if it is increased by 1 from that value it resets to zero.

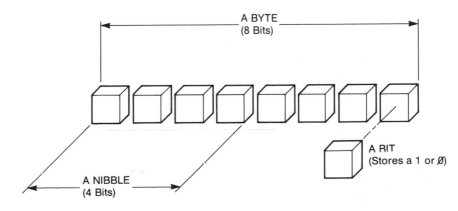

Figure 5-5   Storage Classification

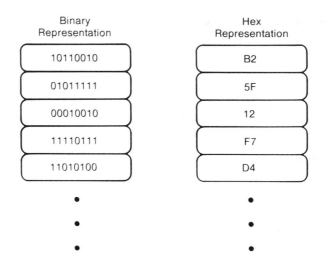

Figure 5-6   Memory Organization with 8-Bit Words

Program 5-1

```
#include <stdio.h>

main( )
{
 unsigned char  counter;

   counter = 0;

   while(counter <= 256)
   {
   counter++;
   printf(" %d - %X ",counter,counter);
   }
}
```

When the above program is executed, the output will be

```
2 - 2 3 - 3 4 - 4 5 - 5 6 - 6 7 - 7 8 - 8 9 - 9 10 - A
11 - B 12 - C 13 - D 14 - E 15 - F 16 - 10
```

This series will continue up to

```
253 - FD 254 - FE 255 - FF 0 - 0
```

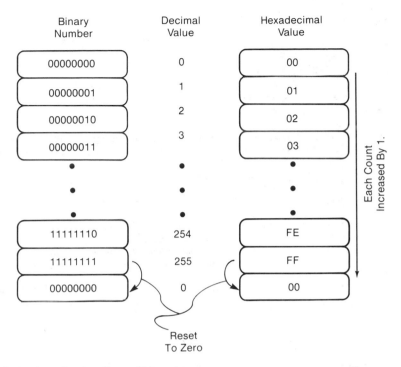

Figure 5-7   Counting in a Byte of Memory

What has happened here is that the full count for a byte has taken place (from 0 to 255). Since there are only eight bits, the maximum value that can be stored in binary is $1111\ 1111_2$. If one more is added to this value then it becomes $00\ 00\ 0000_2$ (just like the odometer on your car rolling over when it reaches its maximum count). Note that in the program a type **unsigned char** is used. You will see the reason for this shortly.

sizeof

The **sizeof** function in C will determine the size of a variable's storage in bytes. This is illustrated by the following program.

Program 5-2

```
#include <stdio.h>

main( )
{
 char character;
 int  character_size;
```

(continued)

Program 5–2 *(continued)*

```
int    integer;
int    integer_size;

   character_size = sizeof(character);
   integer_size = sizeof(integer);

printf("Size of a character on your system is %d byte.\n",character_size);
printf("Size of an integer on your system is %d bytes.",integer_size);

}
```

When executed, the above program will return

```
Size of a character on your system is 1 byte.
Size of an integer on your system is 2 bytes.
```

What this means, is that through the use of the **sizeof** function, C has determined the size, in bytes, of memory allocated to the storage of your particular variable.

Representing Signed Numbers

As you know by now, your computer can work with positive or negative numbers. However, all numbers stored in it must be represented by electrical ON's and OFF's (1s and 0s). This means that the sign of the number must somehow also be represented by a 1 or a 0. There is no way of representing a third symbol (such as a minus sign) in such a two-state system. Therefore negative numbers must somehow be distinguished from positive numbers through the use of the same 1s and 0s that are used to represent the values of the numbers themselves.

There are several ways of doing this. The most common method requires an understanding of binary addition.

Binary Addition

There are five rules you need to know for the addition of binary numbers. These are

$$
\begin{array}{ccccc}
 & & & & 1 \\
0 & 1 & 0 & 1 & 1 \\
\underline{+0} & \underline{+0} & \underline{+1} & \underline{+1} & \underline{+1} \\
0 & 1_2 & 1_2 & 10_2 & 11_2
\end{array}
$$

The first three rules are self-explanatory. The fourth rule simply states that $1 + 1 = 2_{10}$ which is 10_2 in binary. The last rule is $1 + 1 + 1 = 3_{10}$ which is 11_2 in binary.

As an example, to add $2_{10} + 3_{10} = 5_{10}$ in binary

$$1 \qquad \mathrm{<=carry}$$
$$10_2$$
$$+11_2$$
$$\overline{101_2} = 5_{10}$$

As another example, the sum of $7_{10} + 3_{10} = 10_{10}$ in binary would be

$$11 \qquad \mathrm{<= carry}$$
$$111_2$$
$$+011$$
$$\overline{1010_2} = 10_{10}$$

Notice that a 1 is carried from the addition of the first column. Thus, the addition of the second column also produces another carry.

EXAMPLE 5-2

Using binary addition, determine the sum of the following binary numbers:

A. $10_2 + 10_2$
B. $1010_2 + 1101_2$
C. $1111_2 + 1_2$

Solution

A. 10_2
$+10_2$
$\overline{100_2} = 4_{10}$
B. 1010_2
$+1101_2$
$\overline{10111_2} = 23_{10}$
C. $1111 \qquad \mathrm{<= carry}$
1111_2
$+0001_2$
$\overline{10000_2} = 16_{10}$

It is possible to perform binary subtraction in a manner similar to binary addition. However, because this would require complex logic circuits inside the computer, binary subtraction is, as you will see, actually performed by binary addition. In order to see how this is done, you first need to know how your computer represents negative binary numbers by using just 1s and 0s.

Complementing Numbers

In computer number notation, the complement of a binary number is obtained by converting all of its 1s to 0s and all of its 0s to 1s. As an example, the complement of 1011_2 is 0100_2. This seemingly useless process is the basis for obtaining the correct result of a binary subtraction. This process may also be referred to as the **ones complement** of a binary number. This is illustrated in the following example.

EXAMPLE 5-3

Determine the ones complement of the following binary numbers:

A. 101_2 B. 1111_2 C. 1000_2

Solution

A. $101_2 => 010_2$
B. $1111_2 => 0000_2$
C. $1000_2 => 0111_2$

Negative binary numbers are actually represented inside your computer by what is called the **twos complement** notation. To find the twos complement of any binary number, simply first take the ones complement and add 1 to the result. As an example, to find the twos complement of 1011_2.

1. Find the ones complement: 0100
2. Add one to the result: 0101
3. Thus the twos complement of 1011_2 is 0101.

EXAMPLE 5-4

Determine the twos complement of the following binary numbers:

A. 0110_2 B. 1001_2 C. 1111_2

Solution

A. Twos complement of $0110_2 = 1001 + 1 = 1010_2$
B. Twos complement of $1001_2 = 0110 + 1 = 0111_2$
C. Twos complement of $1111_2 = 0000 + 1 = 0001_2$

Binary Subtraction

To subtract using the twos complement notation, do the following:

1. Convert the subtrahend to twos complement notation. Add the results and ignore the final carry. As an example $(8 - 5 = 3)$ in binary is

1000_2
-0101_2

2. Convert the subtrahend to twos complement:

1000_2
1011_2

3. Add the results, and ignore the final carry:

$$
\begin{array}{r}
1 \\
1000_2 \\
+1011_2 \\
\hline
0011_2 = 3_{10}
\end{array}
$$

The resulting answer is 3_{10} which is what you would expect when subtracting 5 from 8.

Effectively what was done above was to change the value of -5 ($8 - 5$ was the problem) into twos complement notation. In this case it causes the MSB (most significant bit) of the binary number to be a 1 (0101_2 in twos complement notation is 1011_2). Thus, for signed binary numbers, anytime the MSB is a 1, the number is negative and represents the twos complement of the actual number. For example, in twos complement notation 1011_2 represents the twos complement of a negative number. To find what the value is, simply take the twos complement and prefix a $-$ sign:

$$1011_2 => 0101_2 = -5_{10}$$

This notation of representing a signed binary nibble in twos complement notation is shown in Table 5–1.

As you can see from the above table, a **nibble** in unsigned notation may represent a range of values from 0 to 15 (sixteen different values). However, in signed notation, it represents a range of values from -8 to $+7$ (still sixteen different values, including

Table 5–1 Signed and Unsigned Nibbles

Unsigned Value	Binary	Signed Value
0	0000	0
1	0001	1
2	0010	2
3	0011	3
4	0100	4
5	0101	5
6	0110	6
7	0111	7
8	1000	-8
9	1001	-7
10	1010	-6
11	1011	-5
12	1100	-4
13	1101	-3
14	1110	-2
15	1111	-1

0, but now negative numbers are represented). Thus, to find the range of values used in signed binary numbers, you can use the relationship

$$\text{Range} => -(2^{N-1}) \text{ to } +(2^{N-1}-1)$$

Where

$$N = \text{Bit size of the binary number.}$$

As an example, for the 4-bit nibble, its signed range is

$$-(2^{4-1}) \text{ to } (2^{4-1}-1) = -(2^3) \text{ to } +(2^3-1)$$
$$-8 \text{ to } (8-1) = -8 \text{ to } 7$$

EXAMPLE 5-5

Determine the range of values for a signed byte.

Solution

A byte is 8 bits. Thus, using the relationship

$$\text{Range} => -(2^{N-1}) \text{ to } +(2^{N-1}-1)$$

yields

$$-(2^{8-1}) \text{ to } +(2^{8-1}-1)$$
$$-2^7 \text{ to } +(2^7-1) = -128 \text{ to } +127$$

As illustrated in the above example, in signed notation a byte represents a range of -128 to $+127$ (256 different values, including 0, or 2^8 different values). While in unsigned notation, the same byte represents a value range of 0 to 255 (still 256 different values). This is illustrated by the following program.

Program 5-3

```
#include <stdio.h>

main( )
{
char value;

    printf("Give me a two place hex number => ");
    scanf("%X",&value);

    printf("The signed decimal value is %d",value);

}
```

The above program illustrates how negative values are stored in a single byte. For example, when this program is executed, any hex number smaller than 80_{16}

that is put in by the program user (that is a value of $7F_{16}$ or less, meaning that the MSB is not set) will produce a positive number. However, as soon as any hex value 80_{16} or larger (1000 0000 or larger, here is where the MSB is set to 1 signifying a negative number in twos complement form), then a negative decimal value is displayed. Here's an example:

```
Give me a two-place hex number => 7F
The signed decimal value is 127
```

In the above case, since the MSB was not set ($0111\ 1111_2$), this was interpreted as a positive value and equal to 127. However,

```
Give me a two-place hex number => 80
The signed decimal value is -128
```

In the above example, the MSB has been set ($80_{16} = 1000\ 0000_2$) and is interpreted as the twos complement of a negative number. Thus, taking the twos complement of 1000 0000 yields 1000 0000 which represents a -128. By the same token,

```
Give me a two-place hex number => FF
The signed decimal value is -1
```

The above example again illustrates how a negative value is stored. The hex number FF_{16} equals $1111\ 1111_2$, which represents the twos complement of a negative number. The twos complement of $1111\ 1111_2$ equals $0000\ 0001_2$ which equals -1_{10}.

This takes us back to the **unsigned char** that was introduced. When this C type is used, the setting of the MSB to 1 is no longer taken to be a negative value, and the full range of numbers from 0 to 255 can now be realized for 8 bits.

Program 5–4

```
#include <stdio.h>

main( )
{
unsigned char value;

  printf("Give me a two-place hex number => ");
  scanf("%X",&value);

  printf("The unsigned decimal value is %d",value);

}
```

When the above program is executed, the MSB will no longer be treated as a sign bit because of the **unsigned char** type:

```
Give me a two-place hex number = 7F
The unsigned decimal value is 127
```

This is no different from before, but now look what happens when the MSB is set to 1:

```
Give me a two-place hex number => 80
The unsigned decimal value is 128
```

And

```
Give me a two-place hex number => FF
The unsigned decimal value is 255
```

This illustrates the difference between signed and unsigned types in C. It relates directly to two important facts: how may bytes are used to store the data type and if the MSB is treated as a sign bit. The range of values for Learn C is given in Table 5-2.

As shown in Table 5-2, the unsigned **type** modifier gives you larger positive values for the same amount of memory space because the MSB of its binary representation is no longer used to represent a negative value.

Recall that the **type char** occupies one byte (8 bits) of memory, thus making it capable of storing a range of signed values from −128 to +127 and unsigned values of 0 to 255. The other data types have a greater range of values because they use more than one byte of memory. This is illustrated by the following program.

Program 5-5

```
#include <stdio.h>

main( )
{
int integer;
char character;
short shortnumber;
long longnumber;
unsigned char unsigned_character;
unsigned int unsigned_integer;
unsigned long unsigned_long;
float float_number;
double double_number;
long double long_double;

    printf("The size in bytes of a/an:\n");
    printf("integer => %d\n",sizeof(integer));
    printf("character => %d\n",sizeof(character));
    printf("short => %d\n",sizeof(shortnumber));
    printf("long => %d\n",sizeof(longnumber));
    printf("unsigned char => %d\n",sizeof(unsigned_character));
```

(continued)

Table 5-2 Range of Values for C Types

Identified by	Sometimes Called	Value Range
char	signed char	+128 to 127
unsigned char		0 to 255
int	signed int	+32768 to 32,767
unsigned int	unsigned	0 to 65,535
short	signed short	+32,768 to −32,767
unsigned short	unsigned short int	0 to 65,535
long	long int, signed long	+2,147,483,648 to −2,147,483,647
unsigned long	unsigned long int	0 to 4,294,967,295
float		+/− 3.4E +/− 38 (7 digits)
double		+/− 1.7E +/− 308 (15 digits)
long double		+/− 1.7E +/− 308 (15 digits)

Program 5-5 *(continued)*

```
    printf("unsigned int => %d\n",sizeof(unsigned_integer));
    printf("unsigned long => %d\n",sizeof(unsigned_long));
    printf("float => %d\n",sizeof(float_number));
    printf("double => %d\n",sizeof(double_number));
    printf("long double => %d",sizeof(long_double));

}
```

When executed, the above program will give the size in bytes used by your system to store the various C data types (such as **int, char** and so on.). As an example, Learn C on an IBM PC uses the following memory:

```
The size in bytes of a/an:
integer => 2
character => 1
short => 2
long => 4
unsigned char => 1
unsigned int => 2
unsigned long => 4
float => 4
double => 8
long double => 8
```

Conclusion

This section presented the ways in which C uses computer memory. Here you saw how values were stored in memory. Most importantly, the concept of how negative numbers are stored was presented along with programming methods that can be used to actually see how this was done. Check your understanding of this section by trying the following section review.

5-2 Section Review

1 State what is meant by a computer word.
2 What is the word size of a C **char** type?
3 How does the computer represent signed numbers?
4 Explain the difference between a signed and an unsigned data type in C.
5 What C data type uses the least amount of memory? The largest?

5-3   Pointers

Discussion

In this section you will discover the secrets of a very important programming concept commonly referred to as **pointers**. This is a powerful tool in the design of C programs for any area of technology. Take your time working with this section. You will find that working the short programs presented here will be helpful in establishing the concepts of pointers.

Basic Idea

Think of the memory in your computer as being constructed as shown in Figure 5-8.

As shown in the figure, each memory location can hold one byte of information. Since each memory location is the same size as the next, they are distinguished from each other by a number called an address. These addresses are sequentially numbered from 0 to the maximum size of the memory.

The following program uses a type **char** to simply place a value into a single byte of your computer's memory.

Program 5-6

```
#include <stdio.h>

main( )
{
char value;      /*  A memory location to hold a value.  */

  value = 97;
  printf("Value is = %d",value);

  }
```

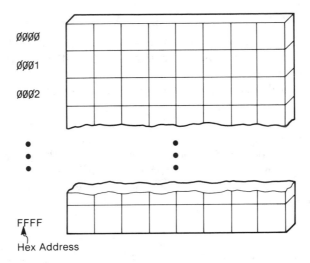

ØØØØ

ØØØ1

ØØØ2

FFFF
Hex Address

Figure 5–8 Construction of Byte-Size Memory

When the above program is executed, its output will be

```
Value is = 97
```

What the program has done is to store a number in one of the memory locations inside your computer. You know the value of the number stored is 97. However, you do not yet know at what address (exactly where in your compter's memory) this value is stored. However, using C you can find this out by using the **address operator &**:

Program 5-7

```
#include <stdio.h>

main( )
{
char value;      /*  A memory location to hold a value.  */

    value = 97;
    printf("%u => |  %d  | <=  address and data of value.",&value,value);

}
```

In the above program, there are two variables that will be displayed to your monitor screen. One is the value of 97 that you put into a particular memory location. The other will be the address (the exact location inside *your* computer) at which the value of 97 is stored. For the system I used when writing this text, the output of the above program was

```
1204 => | 97 | <= address and data of value.
```

As you can see, the address of this memory location was 1204 (this may be different on your system—it may also be different every time you run the same program—this address is assigned by the computer). The way to find out the value of the address is by using the **&** just before the variable: **&value**. This means, "Display the address of where the variable is stored in memory."

Now, look at Program 5–8. It uses a pointer. The program has a pointer declaration. A **pointer declaration** names a pointer variable and also states the type of the object to which the variable points. What happens is that a variable declared as a pointer holds a memory address. The type specifier gives the type of the object. In this program, the pointer type specifier is **char** because the object it is pointing to is of type **char**.

Program 5–8

```
#include <stdio.h>

main( )
{
char value;      /*  A memory location to hold a value.  */
char *pointer;   /*  A pointer.  */

  value = 97;
  printf("%u =>|  %d  | <=  address and data of value.\n",&value,value);

pointer = &value;
printf("%u =>|  %d  | <= address and data of pointer.",&pointer,pointer);
}
```

Execution of the above program yields

```
1204 => | 97 |  <= address and data of value.
4562 => |1204| <= address and data of pointer.
```

The above program now has two variables. One, called **value**, contains the number 97. The other, called **pointer**, contains the address where the number 97 is stored. This concept is illustrated in Figure 5–9.

Program Analysis

Program 5–8 is the first to use a pointer. The pointer is declared by using a space and the * symbol following the C **type**:

```
char *pointer;  /*  A pointer. */
```

This means that the variable **pointer** will now contain the address of a given memory location. In the case of this program, the address of the memory location is placed into **pointer** by

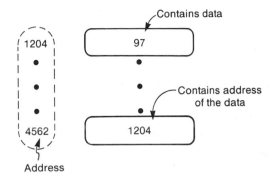

Figure 5-9 Idea of Pointer and Value

```
pointer = &value;
```

Now the `pointer` contains the address of `value`. (*Note:* A pointer may have a type specifier of **void**. Doing this essentially delays the specification of the type to which the pointer refers. When it comes time to use the pointer, you must specify the type each time you use it. This is most easily done with a C cast.)

Using the Pointer

The above program simply stores the address of a given variable in another variable. You don't need pointers to do that. The next program shows the significance of the * (called the **indirection operator**) and & (address operator).

Program 5-9

```c
#include <stdio.h>

main( )
{
char value;      /*  A memory location to hold a character.  */
char *pointer;   /*  A pointer.  */

   value = 97;
   printf("%u =>|  %d  |  <=  address and data of value.\n",&value,value);

   pointer = &value;
printf("%u =>|  %u  |<= address and data of pointer.\n",&pointer,pointer);

   printf("\n Value stored in pointer = %d\n",pointer);
   printf(" Address of pointer: &pointer = %u\n",&pointer);
   printf(" Value pointed to: *pointer = %d\n",*pointer);

}
```

Execution of the above program yields

```
1204 => | 97 | <= address and data of value.
4562 => |1204| <= address and data of pointer.
Value stored in pointer = 1204
Address of pointer: &pointer = 4562
Value pointed to: *pointer = 97
```

What is happening is illustrated in Figure 5-10.

As you can see from the figure, the variable *pointer will have exactly the same value as the address of the variable to which it is pointing. What is important is to understand the meaning of the following when it comes to using pointers:

> pointer → Contains the value stored in the variable pointer.
> &pointer → Will give the address of the variable pointer.
> *pointer → Will give the value stored at the memory location whose address is stored in the variable pointer.

Passing Variables with Pointers

You can change the value of your program variables by using pointers. This may not seem too important now, but it will become very important because, as you will see, it is a way of passing more than one variable back from a called function to the calling function. Program 5-10 illustrates how this is done.

Program 5-10

```c
#include <stdio.h>

main( )
{
char *pointer;     /* A pointer.  */
char variable;     /* A memory location to hold a value. */

    variable = 1;
    pointer = &variable;

    printf("Value stored in variable = %d\n",variable);
    printf("Value stored in pointer = %d\n",pointer);

    *pointer = 2;

    printf("New value stored in variable = %d\n",variable);
    printf("Value stored in pointer = %d\n",pointer);

}
```

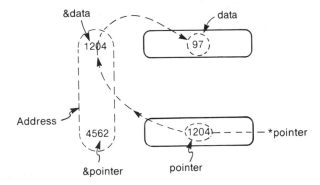

Figure 5-10 Concept of the Pointer

Execution of the above program yields

```
Value stored in variable = 1
Value stored in pointer = 7730
New value stored in variable = 2
Value stored in pointer = 7730
```

Program Analysis

The above program assigns two variables:

```
char *pointer;    /*  A pointer.  */
char variable;    /* A memory location to hold a value. */
```

The first variable is a pointer because it is preceded by the * sign. The second variable is a memory location that will hold an assigned value. The pointer is of type **char** and the variable to store a given value is of type **char** meaning that it will hold one byte of data.

The variable is assigned a value of 1, and the pointer is assigned the address of where the value of 1 is stored.

```
variable = 1;
pointer = &variable;
```

variable now contains 1, and pointer now contains the address of variable.

To demonstrate what is contained in variable and pointer, print out the values stored in variable and pointer.

```
printf("Value stored in variable = %d\n",variable);
printf("Value stored in pointer = %d\n",pointer);
```

This produces

```
Value stored in variable = 1
Value stored in pointer = 7732
```

The value of 7732 is the address assigned by the computer when the program was executed on my computer when I was writing. This means that at address 7732, the value of 1 is stored.

A value of 2 is now assigned to `*pointer`:

```
*pointer = 2;
```

Figure 5-11 shows exactly what this instruction causes to happen inside the computer.

As you can see from Figure 5-11, the value of 2 is passed to the memory location whose address is contained in `pointer`. Thus, the value of this memory location is changed from 1 to 2. This does not affect the value stored in the variable `pointer`. This is demonstrated by the following program lines:

```
printf("New value stored in variable = %d\n",variable);
printf("Value stored in pointer = %d\n",pointer);
```

The output produced by the above lines is

```
New value stored in variable = 2
Value stored in pointer = 7732
```

Note that the value in `pointer` has not changed.

Some Examples

Example 5-6 offers some examples of work with pointers.

EXAMPLE 5-6

From the following program excerpt, determine the indicated values. In each case, the declared variables are

```
char number;
int *p;
```

(Assume that the **address** of number = 7735, and the address of p = 8364.) For each case below determine the value of

> a. `number`   b. `&number`   c. `p`   d. `&p`   e. `*p`

All of the results are cumulative.

A. p = 12; number = 5
B. number = p
C. number = &p
D. p = &p
E. p = &number
F. *p = 10

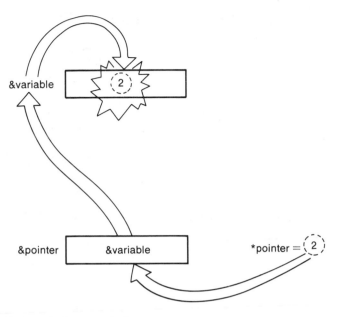

Figure 5-11 Pointer Assignment Action

Solution

A. number = 5, p = 12, &number = 7735, &p = 8364,
   *p = The data stored at memory location 12.
B. number = 12, p = 12, &number = 7735, &p = 8364,
   *p = The data stored at memory location 12.
C. number = 8364, p = 12, &number 7735, &p = 8364
   *p = The data stored at memory location 12.
D. number = 8364, p = 8364, &number 7735, &p = 8364,
   *p = The data stored at memory location 8364 which is 8364.
E. number = 8364, p = 7735, &number = 7735, &p = 8345,
   *p = The data stored at memory location 7735 which is 8364.
F. number = 10, p = 7735, &number = 7735, &p = 8345,
   *p = The data stored at memory location 7735 which is now 10.

All of the above action is illustrated in Figure 5-12.

Conclusion

This section presented an important introduction to the concept of pointers. Effectively a pointer points to a memory location. Here you saw the difference between the value of a variable, its address, and a pointer that contains the address of the variable. Here you also saw how to change the value of a memory location by using a pointer. The concepts presented here will be used in the next section. For now, test your understanding of this section by trying the following section review.

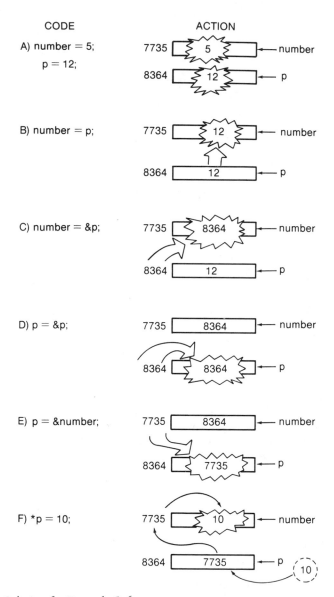

Figure 5–12   Solution for Example 5–6

5-3 Section Review

1   What is a pointer?
2   Why is a pointer called a pointer?
3   How does a pointer get the address of another memory location? Give an example.
4   How do you pass a value to a variable by using a pointer? What must you make sure of before doing this? Give an example.
5   How is a pointer declared?

5-4   Passing Variables

Discussion

This is another important section. Here you will see the application of concepts that were presented in the last section. This section demonstrates another method of passing a value from a called function to the calling function—using pointers.

This is a very powerful tool in the design and structure of C programs. Again, devote adequate time to the study of this section. As with the last section, you will find entering and trying the short programs here will help reinforce these important concepts.

Basic Idea

Program 5–11 illustrates one way of passing a value from a called function back to the calling function. Note that if the function does not have any parameters, this is indicated by type **void**.

Program 5-11

```
#include <stdio.h>

int callme(void)   /* The function to be called by main. */

main( )
{

int x;       /* Variable to receive a value from the called function.*/

    x = callme( );

    printf("Value of x is: %d",x);

    exit(0);
{

/*------------------------------------------------------------*/

int callme(void)
{
    return(5);
}
```

Execution of the above program produces

```
Value of x is: 5
```

Essentially all the above program does is to get a value from a called function. This is illustrated in Figure 5–13.

As shown in Figure 5–13, the value assigned to the variable **x** in function **main( )** is really received from the function **callme( )**. This is done by the C **return(5)** contained in **callme( )**. You have seen this method of returning a value to the calling function done before.

This procedure works fine for returning one value from a called function. It does not work for returning two or more values from a called function. As an example, suppose you were using a separate function to get the values of two or more variables from the program user and you wanted these values returned to the calling function. To return two or more values from a called function to the calling function, you need to use pointers in C.

Using Pointers

A pointer will now be used to return a single value back to the calling function. What will happen here is that a value of 5 will be assigned to a variable of the calling function by the called function as shown in Figure 5–14. The final result is no different from what was done in the last program. What happens now is that the same thing will be done using a pointer instead of the C **return( )**.

To do this the function **callme( )** will use a pointer in its formal argument:

```
void callme(int *p);
```

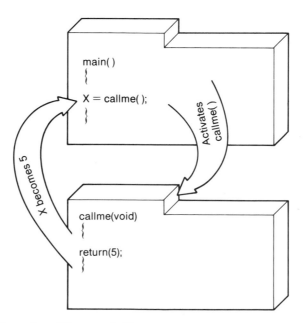

Figure 5–13   Operation of Program 5–11

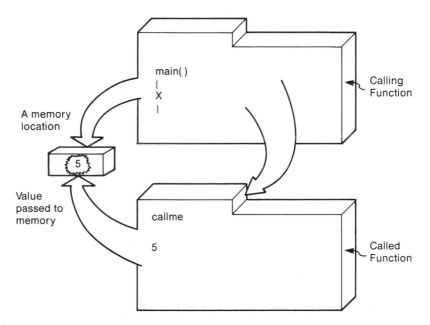

Figure 5-14   Concept of Passing a Value from a Called Function

It is helpful to think of this as if it sets up a memory location to act as a pointer—meaning that what will be stored there will be the address (memory location) of another variable. This concept is shown in Figure 5–15.

When the function **callme** is called, the address of the variable to be changed will be passed to it:

```
main( )
{
int x;
callme(&x);
```

What has happened is shown in Figure 5–16.

When the function is called, it places the value 5 at the address pointed to by the pointer:

```
void callme(int *p)
{
 *p = 5;
}
```

This is shown in Figure 5–17.

The program that makes all of this happen is Program 5–12. Since the function use_me does not return a value, it is declared as a type **void**.

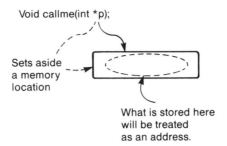

Figure 5-15   Setting a Pointer Argument

Program 5-12

```
#include <stdio.h>

void use_me(int *p);

main( )
{
int x;

    x = 0;
    printf("The value of x is %d\n",x);

    use_me(&x);
   printf("The new value of x is %d",x);

 }

void use_me(int *p)
{
    *p = 5;
}
```

Execution of the above program yields

```
The value of x is 0
The new value of x is 5
```

This seems like a lot of effort just to assign a value of 5 to the variable **x**. However, this example demonstrates the passing of a value from one function to another using a pointer. What follows is an application of this principle.

Pointer Application

Consider Program 5-13. It calculates the current in a resistor given the value of the resistor and the voltage across it. The program uses **main( )** to simply call functions.

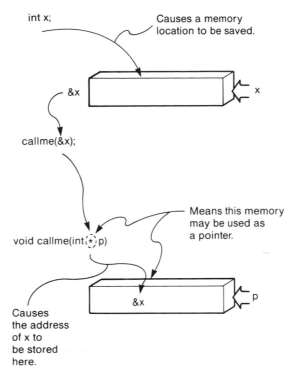

Figure 5-16   Action of Calling the Function

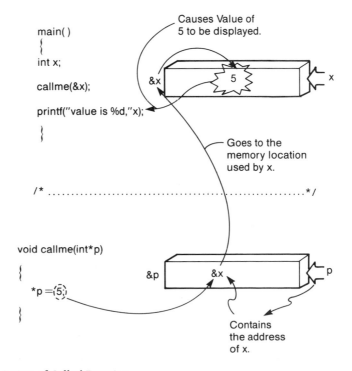

Figure 5-17   Action of Called Function

It is these called functions that do the work in the program. The function **main( )** simply shows the structure of the program.

Program 5-13

```
#include <stdio.h>

void explain_program(void);
void get_values(float *r, float *v);
float do_calculations(float resistance, float voltage);
void display_answer(float current);

main( )
{

float resistor;
float volts;
float current;

 explain_program( );
 get_values(&resistor, &volts);
 current = do_calculations(resistor, volts);
 display_answer(current);

 }

/*------------------------------------------------------------*/

 void explain_program( )
 {
 printf("This program calculates the value of the current\n");
 printf("in amps.  You need to enter the value of the resistor\n");
 printf("and the voltage in volts.\n");

 }
  /* end of explain_program( ) */

/*------------------------------------------------------------*/

void get_values(float *r, float *v)
{
 float resistance;
 float voltage;

   printf("\n\nInput the resistance in ohms = ");
   scanf("%f",&resistance);
```

(continued)

Program 5-13 *(continued)*

```
    printf("Input the voltage in volts = ");
    scanf("%f",&voltage);

    *r = resistance;
    *v = voltage;

 } /*  end of get_values( ) */

/*------------------------------------------------------------*/
_
float do_calculations(float resistance, float voltage)
{
 float current;

    current = voltage/resistance;

    return(current);

 }  /* end of do_calculations( ),  */

/*------------------------------------------------------------*/

void display_answer(float current)
{

   printf("The value of the current is %f amps.",current);

 }
```

Assuming that the program user enters a value of 10 for the voltage and 5 for the resistor, execution of the above program yields

```
This program calculates the value of the current
in amps. You need to enter the value of the resistor
and the voltage in volts.

Input the resistance in ohms = 5
Input the voltage in volts = 10
The value of the current is 2 amps.
```

Program Analysis

The first thing to note about this program is how the functions are prototyped. The design of the program follows the typical sequence of an interactive technical program:

1. Explain program to user.
2. Get values from user.
3. Do the calculations.
4. Display the answer(s).

This is reflected in the names of each of the function prototypes:

```
void explain_program(void);
void get_values(float *r, float *v);
float do_calculations(float resistance, float voltage);
void display_answer(float current);
```

Observe that the function **get_values** uses two pointers in its formal argument. It does this because these values will have to be passed back to the calling function. As demonstrated in the beginning of this section, this can be done with pointers. When two or more variables are to be passed back, it *must* be done with pointers.* Note that only one of the function prototypes is not **void**: this is **float do_calculations**. The reason for this is the function itself will return a single value—the value of the current. Since only one value needs to be returned, the C **return( )** function can be used, and no pointers are necessary (though one could have been used instead in the formal argument).

The function **main( )** declares three variables. These contain the values passed between the called functions:

```
main( )
{
 float resistor;
 float volts;
 float current;
```

The body of **main( )** now contains calls to the various functions.

```
explain_program( );
get_values(&resistor, &volts);
current = do_calculations(resistor, volts);
display_answer(current);
```

Note in the above calls that **get_values** contains the address operators in its actual arguments. This is necessary because the formal argument of this function identifies each of these to be pointers. The addresses of the two variables must be passed to this function so that the values entered by the program user can be returned back to **main( )**.

The first function called simply explains the purpose of the program to the program user:

```
void explain_program( )
{
 printf("This program calculates the value of the current\n");
 printf("in amps. You need to enter the value of the resistor\n");
```

---

*This refers to information presented up to this point in the text.

```
  printf("and the voltage in volts.\n");
}
 /* end of explain_program( ) */
```

The next function called, `get_values`, is the one that has the pointers in its formal argument. It also defines two variables that will be used to store the values entered by the program user:

```
void get_values(float *r, float *v)
{
 float resistance;
 float voltage;
```

The body of this function gets the values from the program user using the C `scanf`:

```
printf("\n\nInput the resistance in ohms = ");
scanf("%f",&resistance);
printf("Input the voltage in volts = ");
scanf("%f",&voltage);
```

Now these values are assigned to the pointers. Since these pointers contain the addresses of actual arguments used in **main( )**, the values entered by the program user are now actually being passed back to the two variables in the calling function **main( )**.

```
*r = resistance;
*v = voltage;
```

Now that **main( )** has the values entered by the program user, these are passed to the next function where the actual calculation is performed:

```
float do_calculations(float resistance, float voltage)
{
 float current;

   current = voltage/resistance;

   return(current);
}
```

Since there is only one value to return back to the calling function, this is done with the C **return( )**.

The last called function simply has the actual value of the current passed to it so it may be displayed:

```
void display_answer(float current)
{
   printf("The value of the current is %f amps.",current);
}
```

Key Points

Even though the previous program does a simple division calculation that could have easily been done in your head, it illustrates an ideal structure of a C program. First, the function **main( )** does nothing but call other functions. This is important because for more complex programs, this allows you to see exactly what the structure of the program looks like.

The next point is that each called function does one specific task. If it needed to do more than one task, then it would in turn call another function. Doing this makes a program easy to understand, debug, modify, and develop. You only need to develop one function at a time, test it, and then continue with the development of another function.

Conclusion

This section illustrated the passing of values through the use of pointers. You will use the concept of pointers throughout the remaining chapters of this text. Test your understanding of this section by trying the following section review.

5–4 Section Review

1 Name two ways of passing a value from a called function to the calling function.
2 What are the limitations of using the C **return** to pass a value back to the calling function?
3 State how more than one value may be returned to the calling function from the called function.
4 Describe the mechanism for passing values to the called function that uses pointers in its formal argument.
5 State why it is necessary to use separate functions.

5–5   Scope of Variables

Discussion

This section presents information about your C variables. Here you will be introduced to the concept of the relationships between your C functions and C variables. This information will help you understand what you should not do as well as what you should do when developing technical programs in C.

Local Variables

All of the variables declared in the programs up to this point have been local. The concept of a **local variable** is illustrated in the following program. The program is attempting to use a variable in a second function that was declared only in the calling function.

Program 5-14

```
#include <stdio.h>

void other_function(void);

main( )
{
 int a_variable;

  a_variable = 5;

  printf("The value of a_variable is %d",a_variable);
  other_function( );

}

void other_function(void)
{

   printf("The value of a_variable in this function is %d",a_variable);

}
```

The above program will not compile because the second function does not know the meaning of the variable a_variable declared in the calling function **main( )**. The reason for this is when a variable is declared by you, within a function, it is known only to that function and none of the others. This means that the variable is local to the function in which it is declared. Another way of saying this is that the **scope** of a local variable is only within the function in which it is declared and the **life** of that variable is only while the function in which it is declared is active.

Global Variables

In order to make a variable known to all the functions in your C program, it must be declared ahead of **main( )**. This is illustrated in the following program. This program will compile because the variable is now known to both functions.

Program 5-15

```c
#include <stdio.h>

void other_function(void);
int a_variable;

main( )
{

  a_variable = 5;

  printf("The value of a_variable is %d\n",a_variable);
  other_function( );

}

void other_function( )
{

    printf("The value of a_variable in this function is %d",a_variable);

}
```

For the above program, since the variable a_variable is declared outside the first function, it is now known to all the functions within the program. Execution of the above program will yield

```
The value of a_variable is 5
The value of a_variable in this function is 5
```

When a variable is declared in this fashion it is called a **global variable**. The scope of a global variable is every function within the program. The life of a global variable is as long as any part of the program is active. Figure 5–18 illustrates.

Caution with Global Variables

Using global variables is usually considered poor programming practice because any function can change the value of a global variable. Thus, when another function uses a global variable it may now have a value different from what you might expect.

It is considered good programming practice to keep your variables as local as possible. In this manner you are protecting these variables from being changed by other functions. This is why variables are passed between functions as arguments. Now the values of each of these variables are protected.

Again, it's a good rule to avoid global variables in your programs. They were introduced here so you would know what they look like. The following program illustrates one reason why you should be very cautious about using them.

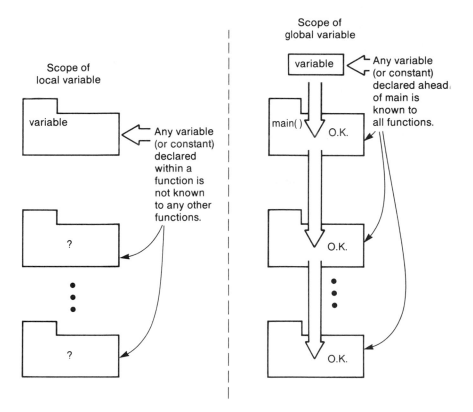

Figure 5-18 Concept of Scope

Program 5-16

```
#include <stdio.h>

void other_function(void);
int counter;

main( )
{

    counter = 0;

    while (counter < 5)
    {
    printf("The count is %d\n",counter);
    counter++;
```

(continued)

Program 5-16 *(continued)*

```
      }
   other_function( );

}

void other_function(void)
{

   while (counter < 6)
   {
   printf("The count in this function is %d\n",counter);
   counter++;
   }

}
```

When the above program is executed, the results are

```
The count is 0
The count is 1
The count is 2
The count is 3
The count is 4
The count in this function is 5
```

Observe that the output of the second function could be a surprise to the programmer (or someone else trying to read and understand the program). Since a global variable is used, it retains its counting value from the function **main( )**. (Recall that the life of a global variable is as long as any part of the program is active.) Thus the count for the second function starts off with the value of the variable `counter` equal to 5. This could have a disastrous effect in larger programs where different functions are written by different programmers in the design of a large program.

Conclusion

This section introduced the concept of the scope and life of program variables. What applies to variables also applies to program constants. You also saw the difference between local and global variables. Here you learned that it is considered poor programming practice to use global variables. Doing so could produce undesirable results during program run time.

Check your understanding of this section by trying the following section review.

5-5 Section Review

1   What is meant by a local variable?
2   State the meaning of the term scope as applied to variables in C.
3   How is a global variable declared? A local variable?
4   Is it considered good programming practice to use global variables? Explain.
5   If global variables are not used, state how values may be passed between functions.

5-6   Variable Class

Discussion

This section presents various ways of classifying your data in C. Being able to do this gives you great flexibility on how you treat your variables. The material presented in this section makes use of the concepts introduced in the last section.

Constants

You can think of one class of data as a constant. In C a constant is a value that never changes anywhere during program execution. It has a particular value assigned to it, and from then on, its value is never allowed to change.

C allows you to declare this type of data by using the keyword

```
const
```

This is illustrated in the following program.

Program 5-17

```
#include <stdio.h>

const float TWO_PI = 6.28;

main( )
{

printf("The value of 2 times pi is %f",TWO_PI);

}
```

In the above program, if you attempt to change the value of the global constant (TWO_PI) anywhere in the program, you will get a warning message during compile time. It is OK to use global constants because their values can never be changed. As a matter of fact, the use of global constants is considered good programming practice.

Automatic Variables

The local variables that you have been using, in C are called **automatic variables**. By default, any variable you have been specifying inside a function is treated as an automatic variable. If you wish to emphasize this (to show, for example, that you are overriding an external function definition) you may use the keyword

auto

This is demonstrated in the following program.

Program 5-18

```
#include <stdio.h>

main( )
{
  auto int value;

  value = 5;

  printf("The value is %d",value);

}
```

In the above program you could have left off the keyword **auto**, and nothing in the above program would have changed as far as its operation is concerned.

External Variables

Another class of variables allowed by C are external. An **external variable** is similar to a global variable except that it may also have a life between different C files. (C files are presented in Chapter 7 of this text.) For now, you may consider an external variable as one that has the entire program as its scope. The keyword of this class of variables is:

extern

Static Variables

In C a **static variable** is a variable that is local. But, unlike a local, its life is as long as the program is active—even though its scope is only within the function that declared it. What this means is a "normal" (automatic) local variable is completely forgotten by the computer once the function in which it was declared is no longer active. When the function is called again, the automatic variable no longer has the value it was left with when the function terminated.

This is not the case with a static variable. Its value is retained by the computer so that when the function that declares it is called again, its previous value is used. This concept is illustrated by the following program.

Program 5–19

```
#include <stdio.h>

void second_function(void);

main( )
{
    second_function( );
    second_function( );
    second_function( );

}

void second_function(void)
{
  static int number;

  number++;
  printf("This function has been called %d times.\n",number);

}
```

Execution of the above program yields

```
This function has been called 1 times.
This function has been called 2 times.
This function has been called 3 times.
```

The important point about the above program is that the value of the **static** variable is remembered between calls to the function. It makes no difference from how many different functions this is done within the same program, its value will still be retained as long as the program is active. However, it is still a local variable known only to the function in which it was defined.

Register Variable

Another class of variable used in C is the **register**. This type of variable tells C that you want the quickest possible speed in the use of this variable. This can be accomplished by having the variable stored inside one of the registers in your computer's microprocessor rather than inside one of your computer's memory locations. Doing this saves time because the value of the variable does not have to be exchanged between your computer's microprocessor and its memory. You can only do this with a few variables during any one program. The reason for this is your computer's microprocessor has a limited number of internal registers, and some of these will be saved for other needed types of processing. However, when you request a register variable, your computer will try to honor the request if possible; otherwise the variable will be stored in a memory location.

The keyword for this is

register

The following program illustrates the use of a register variable.

Program 5-20

```
#include <stdio.h>

main( )
{
  register counter;

    for(counter = 0; counter > 5, counter++);

}
```

The above program will execute faster with a **register** variable because there is no need to access memory each time to increment the value of the counter.

Overview

Table 5-3 presents the various classes of data available in C.

Table 5-3   Classes of Data

Type	Keyword	Where Declared	Scope	Life
Global	None	Ahead of all functions	Entire program.	Until program ends.
Global	**extern**	Ahead of all functions within a file	All files including other files where declared extern.	While any of these files are active.
Local	None or **auto**	Within the function	Only within the function where it is declared.	Until function is no longer active.
Local	**Register**	Within the function	Only within the function where it is declared.	Until function is no longer active.
Local	**Static**	Within the function	Only within the function where it is declared.	Until program ends.

Conclusion

This section presented the various classes of data available to you in C. Here you also saw the effects of different C format types. The section presented various programs to illustrate the effects of the different classes as well as the format types. Check your understanding of this section by trying the following section review.

5-6 Section Review

1   What effect does the C keyword **const** have when used to identify a data type?
2   State what is meant by an automatic variable.
3   State what is meant by a static variable.
4   What is a register type variable?

5-7   Program Debugging and Implementation

Problems in C Programming

This section presents some of the most common errors encountered by C programmers. Bringing these to your attention now may save you hours of frustration later. As you will discover, the unique characteristics of C open the door for some specific programming problems that may, at first, be difficult to spot.

Pointers

Understandably, many of those who are new to C have some difficulty with the concept and application of pointers. As a first step in understanding pointers you should make sure you understand the difference between the various forms of the pointer variable. Table 5-4 summarizes these important differences.

Table 5-4   Pointer Nomenclature

Given: `int *ptr;` `int data;` [It then follows that:]		
Operation	**Operation**	**Meaning**
Address of Operator	`&data`	Address of the variable `data`
Assignment	`ptr = &data;`	Places the address of `data` into the variable `ptr`
Variable	`ptr`	Value of the variable `ptr` which now contains the address of `data`
Indirection Operator	`*ptr`	Value in the variable `data`

The indirection operator (*) accesses a value indirectly, through a pointer. The operand must be a pointer value.

It is often helpful to try short programs that work with the concepts presented in the above table. The interactive exercise section for this chapter contains several of these.

Not Initializing Pointers

A common error made by those new to C is the failure to initialize pointers. **Initialization** means that you must know what value is in the pointer before you use it because this value will be the actual memory location that the pointer will be pointing to. As an example, the following program excerpt is not correct:

```
main( )
{
int *ptr;

    *ptr = 5;
}
```

Here the programmer is attempting to assign a value of 5 to the memory location pointed to by the pointer. However, no one (including the programmer) knows where in the computer memory the value of 5 will be stored. Doing this in small programs may have no noticeable effect. However, in larger programs, this could have a disastrous effect because the number 5 could be placed in a memory location that is being used by your program for other necessary items. The point to remember here is always know where the pointer is pointing to before using it as a pointer.

Assignment and Equality

Another area that may cause problems to new C programmers is forgetting that the = (assignment operator) does not mean equal to. The = in C means to assign the value on the right to the variable on the left. If you wish to test for equality, you must use the == (equality) symbol. For example, the following programming segment will compile and execute. However, the programmer will not get what was expected:

```
if(this = that)
    printf("This equals that!");
else
    printf("This does not equal that!");
```

All that is happening in the above program segment is the value of **that** is assigned to the variable **this**. There is no comparison being made. What should have been done is as follows:

```
if(this == that)
    printf("This equals that!");
else
    printf("This does not equal that!");
```

Forgetting to Use the Address Operator

Another common error is forgetting to pass values by address. This is illustrated in the following program excerpt:

```
main( )
{
int input;

printf("Give me a whole number => ");
scanf("%d",input);

printf("The value you entered was %d",input);

}
```

In the above excerpt, the value entered by the program user will not do what is expected. The **scanf** function requires that you use the address operator **&** in order to get the entered value returned to the calling function. For the above program excerpt, the **scanf** function must be changed to

```
scanf("%d",&input);
```

Now you will get the expected results.

The same problem occurs when you define your own functions that must return values back to the calling function. As with the **scanf** function, you must use the address operator **&** as the actual parameter.

Conclusion

This section presented some of the most common programming pitfalls encountered by beginning C programmers. A table was presented to emphasize the meaning of the different aspects of using pointers. In this section you also saw some of the more common problems encountered if a distinction is not made between the assignment and equal operators. A reminder of how to pass by address was also presented. Check your understanding of this section by trying the following section review.

5-7 Section Review

1 What is the address of operator in C? What does it do?
2 What is the indirection operator in C? What does it do?
3 State the difference between the C equality and assignment symbols.

5-8   Case Study

Discussion

The case study for this section presents program excerpts demonstrating the effect of bit manipulation in C. As you will see, **bit manipulation** is a method of working at the bit level with your machine. There are many applications for this low-level work such as in the area of hardware interfacing and using the computer as a controlling device for other systems.

Bit Manipulation

When doing bit manipulation, you are working with the individual bits of data within the computer. With bit manipulation you should think of all stored data in its binary form of 1s and 0s. Doing this is a great aid to understanding what is to follow. Think of the binary 1 as a Boolean TRUE and the binary 0 as a Boolean FALSE. Table 5-5 illustrates the meaning of various **Boolean operators**. Some of these will already be familiar to you.

The following examples give you some practice with each of the C bitwise operators presented in Table 5-5.

Bitwise Complementing

To get the **bitwise complement** of a number, convert the number to its binary equivalent, then change each 1 to a 0 and each 0 to a 1 (the same as taking the ones complement). Convert the resulting binary number back to the base of the original number. The resulting value is the bitwise complement of the number.

Table 5-5   Boolean Operations

Operation	Bitwise Operator	Meaning	With Bits
Bitwise COMPLEMENT	~	Change bit to its opposite	~1 = 0 ~0 = 1
Bitwise AND	&	The result is 1 if both bits are 1	0 & 0 = 0 0 & 1 = 0 1 & 0 = 0 1 & 1 = 1
Bitwise OR	|	The result is 0 if both bits are 0	0 | 0 = 0 0 | 1 = 1 1 | 0 = 1 1 | 1 = 1
Bitwise exclusive OR	^	The result is 1 if both bits are different	0 ^ 0 = 0 0 ^ 1 = 1 1 ^ 0 = 1 1 ^ 1 = 0

EXAMPLE 5-7

Determine the bitwise complement of the following hex numbers:

$$\text{A. } 00_{16} \quad \text{B. } FF_{16} \quad \text{C. } A3_{16}$$

Solution

A. Convert to binary: $00_{16} = 0000\ 0000_2$
     Take ones complement: $1111\ 1111_2$
     Convert back: F   F   = FF_{16}
     Thus, the bitwise complement of 00_{16} is FF_{16}
B. Convert to binary: $FF_{16} = 1111\ 1111_2$
     Take ones complement: $0000\ 0000_2$
     Convert back: 0   0 = 00_{16}
     Thus, the bitwise complement of $FF_{16} = 00_{16}$
C. Convert to binary: $A3_{16} = 1010\ 0011_2$
     Take ones complement: $0101\ 1100_2$
     Convert back: 5   C = $5C_{16}$
     Thus, the bitwise complement of $A3_{16} = 5C_{16}$

Program 5-21 is a program excerpt that illustrates the action of the bitwise complement.

Program 5-21

```
#include <stdio.h>

main( )
{
int value;

 printf("Input a hex number from 0 to FF => ");
 scanf("%X",&value);
 printf("The bitwise complement is => %X",~value);

}
```

Assuming the program user enters the value A3, execution of the above program yields

```
Input a hex number from 0 to FF => A3
The bitwise complement is => 5C
```

Note that the bitwise complement operator ~ is used on the variable `value`. This is done in the second **printf** function:

```
printf("The bitwise complement is => %X",~value);
```

Bitwise ANDing

To get the **bitwise AND** of a number, convert the number to its binary equivalent, then AND each corresponding bit of the two resulting binary numbers. Convert the resulting binary number back to the base of the original number. The resulting value is the bitwise ANDing of the two numbers.

EXAMPLE 5-8

Determine the hex results of bitwise ANDing the following hex numbers:

A. 00_{16}&FF_{16} B. FF_{16}&$A5_{16}$ C. $D3_{16}$&$8E_{16}$

Solution

A. Convert to binary: $00_{16} = 0000\ 0000_2$
$$FF_{16} = \underline{1111\ 1111_2}$$
AND each bit pair: $0000\ 0000_2$
Convert back: 0 0 $= 00_{16}$
Thus, the bitwise ANDing of 00_{16} and FF_{16} is 00_{16}

B. Convert to binary: $FF_{16} = 1111\ 1111_2$
$$A5_{16} = \underline{1010\ 0101_2}$$
AND each bit pair: $1010\ 0101_2$
Convert back: A   5 $= A5_{16}$
Thus, the bitwise ANDing of FF_{16} and $A5_{16} = A5_{16}$

C. Convert to binary: $D3_{16} = 1100\ 0011_2$
$$8E_{16} = \underline{1000\ 1110_2}$$
AND each bit pair: $1000\ 0010_2$
Convert back: 8   2 $= 82_{16}$
Thus, the bitwise ANDing of $D3_{16}$ and $8E_{16} = 82_{16}$

Program 5–22 illustrates the action of bitwise ANDing.

Program 5-22

```
#include <stdio.h>

main( )
{
int value1;
int value2;

 printf("Input a hex number from 0 to FF => ");
 scanf("%X",&value1);
 printf("Input a hex number to be bitwise ANDed => ");
 scanf("%X",&value2);
 printf("Bitwise ANDing of %X and %X produces => ",value1, value2);
 printf("%X",value1&value2);

}
```

Assuming that the program user enters the values of D3 and 8E, execution of the above program produces

```
Input a hex number from 0 to FF => D3
Input a hex number to be bitwise ANDed => 8E
Bitwise ANDing of D3 and 8E produces => 82
```

Bitwise ORing

To get the **bitwise OR** of a number, convert the number to its binary equivalent, then OR each corresponding bit of the two resulting binary numbers. Convert the resulting binary number back to the base of the original number. The resulting value is the bitwise ORing of the two numbers.

EXAMPLE 5-9

Determine the hex results of bitwise ORing the following hex numbers:

$$\text{A. } 00_{16} \mid FF_{16} \qquad \text{B. } FF_{16} \mid A5_{16} \qquad \text{C. } D3_{16} \mid 8E_{16}$$

Solution

A. Convert to binary: $00_{16} = 0000\ 0000_2$
$$FF_{16} = \underline{1111\ 1111_2}$$
   OR each bit pair: $1111\ 1111_2$
   Convert back: F   F $= FF_{16}$
   Thus, the bitwise ORing of 00_{16} and FF_{16} is FF_{16}
B. Convert to binary: $FF_{16} = 1111\ 1111_2$
$$A5_{16} = \underline{1010\ 0101_2}$$
   OR each bit pair: $1111\ 1111_2$
   Convert back: F   F $= FF_{16}$
   Thus, the bitwise ORing of FF_{16} and $A5_{16}$ is FF_{16}
C. Convert to binary: $D3_{16} = 1100\ 0011_2$
$$8E_{16} = \underline{1000\ 1110_2}$$
   OR each bit pair: $1100\ 1111_2$
   Convert back: C   F $= CF_{16}$
   Thus, the bitwise ORing of $D3_{16}$ and $8E_{16}$ is CF16

Program 5–23 illustrates the action of bitwise ORing.

Program 5–23

```
#include <stdio.h>

main( )
{
int value1;
int value2;
```

(continued)

Program 5-23 *(continued)*

```
printf("Input a hex number from 0 to FF => ");
scanf("%X",&value1);
printf("Input a hex number to be bitwise ORed => ");
scanf("%X",&value2);
printf("Bitwise ORing of %X and %X procuces => ",value1, value2);
printf("%X",value1|value2);

}
```

Assuming that the program user enters the values of D3 and 8E, execution of the above program produces

```
Input a hex number from 0 to FF => D3
Input a hex number to be bitwise ORed => 8E
Bitwise ORing of D3 and 8E produces => CF
```

Bitwise XORing

To get the **bitwise XOR** (exclusive OR) of a number, convert the number to its binary equivalent, then XOR each corresponding bit of the two resulting binary numbers. Convert the resulting binary number back to the base of the original number. The resulting value is the bitwise XORing of the two numbers.

EXAMPLE 5-10

Determine the hex results of bitwise XORing the following hex numbers:

$$\text{A. } 00_{16} \hat{\ } FF_{16} \quad \text{B. } FF_{16} \hat{\ } A5_{16} \quad \text{C. } D3_{16} \hat{\ } 8E_{16}$$

Solution

A. Convert to binary: $00_{16} = 0000\ 0000_2$
   $\qquad\qquad\qquad FF_{16} = \underline{1111\ 1111_2}$
   XOR each bit pair: $0000\ 0000$
   Convert back: $0 \quad 0 \ = 00_{16}$
   Thus, the bitwise XORing of 00_{16} and FF_{16} is 00_{16}

B. Convert to binary: $FF_{16} = 1111\ 1111_2$
   $\qquad\qquad\qquad A5_{16} = \underline{1010\ 0101_2}$
   XOR each bit pair: $0101\ 1010_2$
   Convert back: $5 \quad A \ = 5A_{16}$
   Thus, the bitwise XORing of FF_{16} and $5A_{16} = 5A_{16}$

C. Convert to binary: $D3_{16} = 1100\ 0011_2$
   $\qquad\qquad\qquad 8E_{16} = \underline{1000\ 1110_2}$
   XOR each bit pair: $0100\ 1101_2$
   Convert back: $4 \qquad D \ = 4D_{16}$
   Thus, the bitwise XORing of $D3_{16}$ and $8E_{16} = 4D_{16}$

Program 5–24 illustrates the action of bitwise XORing.

Program 5–24

```
#include <stdio.h>

main( )
{
int value1;
int value2;

 printf("Input a hex number from 0 to FF => ");
 scanf("%X",&value1);
 printf("Input a hex number to be bitwise XORed => ");
 scanf("%X",&value2);
 printf("Bitwise XORing of %X and %X produces => ",value1, value2);
 printf("%X",value1^value2);

}
```

Assuming that the program user enters the values of D3 and 8E, execution of the above program produces

```
Input a hex number from 0 to FF => D3
Input a hex number to be bitwise XORed => 8E
Bitwise XORing of D3 and 8E produces => 4D
```

Shifting Bits

C also allows for shifting bits left or right. This is accomplished with the shift left operator $<<$ or the shift right operator $>>$. To determine the result of a bitwise shift, convert the value to binary, then shift the binary bits the required number of bits in the indicated direction. Convert the resulting binary number back to the base of the original value. The following examples illustrate the result of **bit shifting**.

EXAMPLE 5-11

Determine the hex value of the following hex numbers when shifted to the right or left the number of bits indicated:

$$\text{A. } FF_{16} << 2 \quad \text{B. } 5C_{16} >> 3 \quad \text{C. } 05_{16} << 4$$

Solution

A. Convert to binary $=> FF_{16} = 1111\ 1111_2$
  Shift left 2 bits $=>$ $1111\ 1100_2$
  Convert back to hex $=>$ F   C $= FC_{16}$
  Thus $FF_{16} << 2$ is FC_{16}

B. Convert to binary $\Rightarrow 5C_{16} = 0101\ 1100_2$
   Shift right 3 bits $\Rightarrow$ $\quad\quad\quad \underline{0001\ 0111_2}$
   Convert back to hex $\Rightarrow$ $\quad\quad 1 \quad\quad 7 = 17_{16}$
   Thus $5C_{16} \gg 3$ is 17_{16}

C. Convert to binary $\Rightarrow 05_{16} = 0000\ 0101_2$
   Shift left 4 bits $\Rightarrow$ $\quad\quad\quad \underline{0101\ 0000_2}$
   Convert back to hex $\Rightarrow$ $\quad\quad 5 \quad\quad 0 = 50_{16}$
   Thus $05_{16} \ll 4$ is 50_{16}

The following program illustrates a shift left operation.

Program 5-25

```
#include <stdio.h>

main( )
{
int value1;
int shift_left;

 printf("Input a hex number from 0 to FF => ");
 scanf("%X",&value1);
 printf("Input a number of bits to be left shifted => ");
 scanf("%X",&shift_left);
 printf("Shifting %X %d places to the left produces => ",value1,shift_left);
 printf("%X",value1<<shift_left);

}
```

Assuming that the program user enters the value 5C to be shifted to the left 3 bits, execution of the above program yields

```
Input a hex number from 0 to FF => 5C
Input a number of bits to be left shifted => 3
Shifting 5C 3 places to the left produces =>2E0
```

Conclusion

This section presented the concepts of bitwise operations available with C. In this section you saw how to perform bitwise Boolean operations as well as bitwise shifting. The information presented here was in the context of a case study that presented small programs which were developed to demonstrate these concepts. Test your understanding of this section by trying the following section review.

5-8 Section Review

1 Define what is meant by a bitwise complement in C.
2 What is the meaning of the bitwise AND operation?
3 Explain what is meant by the bitwise OR operation.
4 Define what is meant by a bitwise XOR operation.
5 What is meant by a bitwise shift in C?

Interactive Exercises

DIRECTIONS

These exercises require that you have access to a computer and software that supports C. They are provided here to give you valuable experience and immediate feedback on what the concepts and commands introduced in this chapter will do. They are also fun.

Exercises

1 In the following program, which of the displayed values are predictable? Try the program to check your answer.

Program 5-26

```
#include <stdio.h>

main( )
{
 int data = 15;
 int *point_to;

    point_to = &data;

    printf("%d\n",point_to);
    printf("%d\n",&point_to);
    printf("%d\n",*point_to);

}
```

2 The next program has a global constant. Predict what the output will be, then try the program to test your prediction.

Program 5-27

```
#include <stdio.h>

const int a_value = 15;
int function(void);

main( )
{
 int a_value = 12;

   printf("%d\n",a_value);
   printf("%d\n",function( ));


}

  int function(void)
  {

    return(a_value);

  }
```

3   The next program uses bitwise operators. Note that the input is to the base 10 while the output is in hex. As before, predict what the output will be, then test it.

Program 5-28

```
#include <stdio.h>

main( )
{
 char bit1;
 char bit2;

   bit1 = 5;
   bit2 = 10;

   printf("%X\n",bit1&bit2);
   printf("%X\n",bit1|bit2);
   printf("%X\n",bit1^bit2);
   printf("%X\n",~bit2);

}
```

Self-Test

DIRECTIONS

Program 5–29 was developed to illustrate many of the key concepts presented in this chapter. The questions for this Self-Test pertain to this program.

Program 5–29

```c
#include <stdio.h>

double function_1(void);
void function_2(char *letter1, char *letter2);
int AND_bits(int this_byte, int that_byte);

const unsigned int number_1 = 57532;
int *look_at;

main( )
{
  int memory_location_1;
  char this_value;
  char that_value;
  int result;

    look_at = &memory_location_1;
    function_1( );

    this_value = 'a';
    that_value = 'b';
    function_2(&this_value, &that_value);

    result = AND_bits(15,15);

    printf("The constant is %u.\n",number_1);
    printf("The value of memory_location_1 is %d\n",memory_location_1);
    printf("The contents of this_value are = %c\n",this_value);
    printf("The result is %X.\n",result);

}
extern char new_value;

double function_1(void)
{
    *look_at = 375;
}
```

(continued)

Program 5-29 *(continued)*

```
void function_2(char *letter1, char *letter2)
{

    *letter1 = 'b';
    *letter2 = 'a';

}

int AND_bits(int this_byte, int that_byte)
{

    return(this_byte&that_byte);

}
```

Questions

1 State what the output of each of the **printf** functions in **main( )** will be when the program is executed.
2 Which data value(s) is/are global for the entire program?
3 Which data value(s) is/are global for a part of the program? What is its scope?
4 Explain how the variable **memory_location1** receives the value of 375.
5 What statement in **main( )** causes the value of **memory_location1** to become 375?
6 Explain why the output of the last **printf** function in **main( )** is F.
7 State why **function_1** does not require pointers in its argument and **function_2** does require them.
8 Does **function_1** need to be of type **double**? Explain.
9 State how **function_1** "knows" the pointer ***look_at** since it is not declared within the function.
10 Why does **function_1** cause the value of the variable **memory_location_1** to change?

End-of-Chapter Problems

General Concepts

Section 5-1

1 State a way of visualizing a computer's memory as suggested in this chapter.
2 How is one memory location distinguished from another in computer memory? What is this called?
3 What is the process of the CPU getting an instruction from memory and then executing the instruction called?
4 State the difference between immediate addressing and direct addressing.

Section 5-2

5   State some of the most common word sizes used by computers.

6   What is the purpose of the **sizeof** function in C?

7   Determine the binary sum of the following binary numers:
    A. $0101_2 + 0001_2$ B. $0111_2 + 0001_2$

8   Find the ones complement of the following binary numbers:
    A. 0001_2 B. 1111_2 C. 1010_2

9   Find the twos complement of the following binary numbers:
    A. 0001_2 B. 1111_2 C. 1010_2

10 Using the MSB as the sign bit in twos complement notation, determine the value (including the sign) of the decimal values of the following binary numbers:
    A. 0111_2 B. 1000_2 C. 1101_2

11 Using signed twos complement notation, convert the following hex values to their signed decimal equivalent.
    A. A_{16} B. 8_{16} C. $9C_{16}$

Section 5-3

12 What is the address operator in C? How is it used?

13 What is indirection operator in C? How is it used?

14 What is a pointer? How is it used?

15 Assuming the following declarations, answer the questions that follow:

```
int  data;
int *pointer;

    pointer = &data;
    *pointer = 5;
```

    A. What is the value of **pointer**?
    B. What is the value of **data**?
    C. What is the value of *****pointer**?

16 State how more than one variable is passed from a called function to the calling function.

Section 5-4

17 State why separate functions are used in a C program.

18 What are the limitations of the C **return( )** used in a called function?

19 Explain the use of the **&** operator in returning values in the arguments of a called function.

20 What must be in the formal argument of a function definition that is to return values to the calling function through its argument.

Section 5-5

21 What is the name of a variable declared within a function whose life is only when the function is active?

22 When is the life of an automatic variable the duration of the program?

23 What is the name given to a variable that is given at the beginning of the program before **main( )**? What is the scope of such a variable?

24 State what is considered good programming practice concerning the use of variables.

Section 5-6

25 What is the C keyword used to ensure that an assigned value cannot be changed during program execution?
26 Give the name of a variable that is local to the function but whose value persists during the life of the program.
27 State the name of the variable class that requests the compiler to keep the variable in one of the internal registers of the microprocessor.
28 What is the name for a variable class that has a scope of more than one function?

Section 5-7

29 State how you would obtain the address of the variable **data** in a C program.
30 Show how you would pass the address of **data** to a pointer called **ptr** in C.
31 What is the C symbol used for equality? For assignment?
32 What terminology is used when an address operator is used to assign values to function arguments?

Section 5-8

33 Determine the bitwise complement of the following hexadecimal values:
    A. C_{16} B. FE_{16} C. 50_{16}
34 Find the result of bitwise ANDing the following hexadecimal values:
    A. $3_{16} \& 5_{16}$ B. $E_{16} \& E_{16}$ C. $F_{16} \& 7_{16}$
35 Solve for the bitwise ORing of the following hexadecimal values:
    A. $3_{16} \char`^ 5_{16}$    B. $A_{16} \char`^ 5_{16}$ C. $0_{16} \char`^ F_{16}$
36 Determine the bitwise XORing of the following hexadecimal values:
    A. $3_{16} \char`^ 5_{16}$    B. $A_{16} \char`^ 5_{16}$ C. $F_{16} \char`^ F_{16}$

Program Design

In developing the following C programs, use the method developed in Program 5-13 of this chapter. This means that you will use top-down design and block structure with no global variables. The function **main( )** is to do little more than call other functions. When more than one variable is to be passed back to the calling function, then pointers are to be used. As before, be sure to include all of the documentation in your final program. This should consist of, but not be limited to, the programmer's block, function prototypes, and a description of each function as well as any formal arguments you may use.

Electronics Technology

37 Create a C program that will compute the voltage across each component of a series resonant circuit consisting of a resistor, capacitor, inductor, and AC voltage source. The program user is to input the value of each component along with the applied frequency and source voltage. The relationships are

$$X_L = 2\pi fL$$

Where
X_L = Inductive reactance in ohms.
f   = Frequency in hertz.
L   = Inductance of inductor in henrys.

$$X_C = 1/(2\pi fC)$$

Where

X_C = Capacitive reactance in ohms.
f = Frequency in hertz.
C = Capacitance of capacitor in farads.

$$Z = \sqrt{((X_L - X_C))^2 + R^2}$$

Where

Z = Circuit impedance in ohms.
X_L = Inductive reactance in ohms.
X_C = Capacitive reactance in farads.
R = Resistance in ohms.

$$I_T = V_T/Z$$

Where

I_T = Total circuit current in amps.
V_T = Applied source voltage in volts.
Z = Circuit impedance in ohms.

$$V_L = I_T X_L$$
$$V_C = I_T X_C$$
$$V_R = I_T R$$

Where

V_L = Inductor voltage in volts.
V_C = Capacitor voltage in volts.
V_R = Resistor voltage in volts.

Business Applications

38 Create a C program that represents the input of a cash register. The program user will see a display of ten items as follows:

A. Hamburger—Regular
B. Hamburger—Double
C. Hamburger—Super
D. Fries—Small
E. Fries—Medium
F. Fries—Large
G. Shake—Small
H. Shake—Medium
I. Shake—Large
T Total:

The program user selects as many items as desired by letter along with the quantity of each. When done, the letter T is entered for a total. The program then totals the order and includes a 6% sales tax. As the programmer, you enter the price of each item in your source code.

Computer Science

39 Create a C program that allows the program user to select any of the following six bitwise operations:

A. COMPLEMENT
B. AND
C. OR
D. XOR
E. Shift Left
F. Shift Right

Once the user selects the operation, then the values for the bitwise operation may be entered.

Drafting Technology

40 Create a C program so that the program user may observe the change in volume of a sphere as the value of the radius is changed. This is similar to the drafting technology problem of the last chapter except now the program must use pointers for the passing of variables between functions.

Agriculture Technology

41 Create a C program that will calculate the total expenses for planting a given crop. The program user may select among the following:

A. Corn seed
B. Wheat seed
C. Fertilizer
D. Fuel
E. Person hours
T Total

The program user selects the input by letter and is then prompted to enter the amount (with the proper units—pounds, gallons, hours, etc.). The program will then present the total when T is pressed. As the programmer, you must code in the unit cost of each item.

Health Technology

42 Develop a C program that creates the total of a patient bill. The items selected by the program user are

A. Room—Intensive Care
B. Room—Single
C. Room—Double
E. Room—4-bed
F. Room—Ward
G. Medication—cost
H. Doctor—hours
I. Days in Hospital
T Total

Manufacturing Technology

43 Create a C program that keeps information on up to ten different production schedules. The information entered by the program user is the number of items produced per working day, number of workers needed to produce each item, hourly wages of each worker, and material cost of each item. The program then assumes a 33% overhead charge on the total cost. The program user will receive the total cost of one day's production including the overhead.

Electronics Technology

44   Expand the program written for problem 37 so the program user may have any of the intermediate calculations displayed (such as the total circuit current) as selected by the program user.

Business Applications

45   Modify the program in problem 38 so that the program user may enter a secret number code (made up by you) and then change any of the prices for any item (including the amount of sales tax).

Computer Science

46   Modify the program in problem 39 so that the program user may also select the type of output to be displayed as a result of the bitwise operation (decimal, hex, octal or all three).

Drafting Technology

47   Expand problem 40 so that the program user may select a cone as well as a sphere. In both cases, the program user may observe the change in volume with a change in radius of the sphere or the base of the cone.

Agriculture Technology

48   Modify the program in Problem 41 so the program user may change the price of any item.

Health Technology

49   Expand problem 42 so that the program user may modify any of the prices provided they enter a secret code (developed by you the programmer).

6 Strings and Arrays

Objectives

This chapter provides you the opportunity to learn:

1 The relationship between characters and pointers.
2 Methods of putting characters in memory to form strings.
3 The use of strings in C.
4 What an array is.
5 Applications of arrays.
6 Passing arrays between C functions.
7 The basic concepts of sorting with arrays.
8 Methods of sorting numerical data.
9 Methods of sorting string data.
10 Other sorting methods.

Key Terms

Array
Elements
Dimension
Array Initialization
Array Passing
Multidimensional Arrays
Matrix

Row
Column
Field Width
Array Index
Bubble Sort
Rectangular Array
Ragged Array

Outline

6-1 Characters and Strings
6-2 More about Arrays
6-3 Multidimensional Arrays
6-4 Introduction to Array
       Applications

6-5 Sorting
6-6 More About Strings
6-7 Program Debugging and
       Implementation
6-8 Case Study

Introduction

This chapter will give you an opportunity to apply what you have learned about C pointers. Here you will see the relationship between pointers, strings, and arrays. Understanding how these are related will help you in developing C programs that are capable of many powerful applications including sorting.

The chapter concludes with the development of a sorting program. This is developed in the Case Study section. With the information in this chapter, you will have a solid foundation for the development of future C programs that will solve a wide variety of complex technology programs.

6-1   Characters and Strings

Discussion

This section presents the relationship between characters, pointers, and strings. Here you will see how pointers are used to store characters in memory. You will receive your first introduction to arrays.

The use of strings is an important part of any technology program. Understanding how to use strings in your programs opens another dimension of technical programming. The ability to use the names of objects, people, and data is a powerful addition to your programming skills. This section presents that important introduction.

Storing Strings

Recall from Chapter 1 that a character can be thought of as a single memory location that contains an ASCII code. A string is nothing more than an arrangement of characters. The word **array** means arrangement; therefore, a string can be thought of as an array of characters. How a string is stored in memory is illustrated in Figure 6-1.

Such a sequential arrangement of data within memory is referred to as an array. Note that the last character of the string array in Figure 6-1 consists of the C null character (represented by a /0). All of C's character strings will have the null to let C know where the string ends.

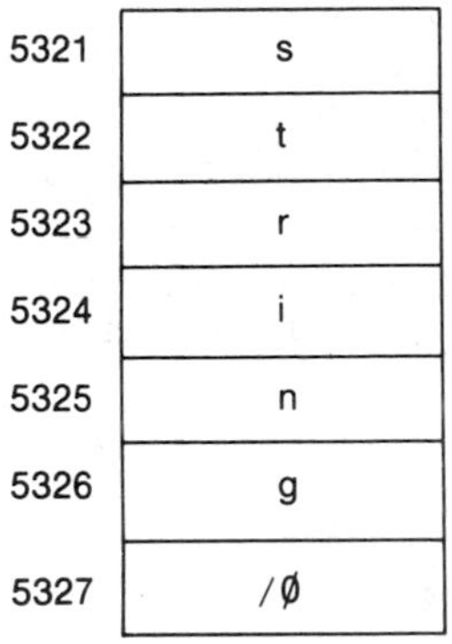

5321	s
5322	t
5323	r
5324	i
5325	n
5326	g
5327	/∅

Figure 6–1   Storing a String in Memory

To tell C you want an array of characters (a string), you must use the brackets [ ] immediately following your string identifier. This is demonstrated in the following program.

Program 6–1

```
#include <stdio.h>

 char string[] = "Tech";

main( )
{

   printf("The string is %s.",string);

}
```

When the above program is executed the output will be

`The string is Tech.`

Note the format specifier is the **%s** for string.

Program Analysis

The global constant **char string[]** = "Tech"; tells C to reserve enough consecutive memory spaces to hold the string of characters: T e c h. The variable **string** is now an array variable because it represents an arrangement of information in memory, not just one memory location.

The **%s** as the format specifier lets C know to display the array variable as a string of characters.

An Inside Look

An array is said to be made up of **elements**. As an example, in the string array of Tech, each character is an element of the array. Arrays in C start with zero, so the zero element of this array is the letter T. To represent any single element of an array, simply place the element number inside the square brackets of the array variable; string[0] in this case is the letter T. This is illustrated by the following program.

Program 6–2

```
#include <stdio.h>

 char string[] = "Tech";

main( )
{

   printf("The string is %s.\n",string);
   printf("The characters are:\n");

   printf("%c\n",string[0]);
   printf("%c\n",string[1]);
   printf("%c\n",string[2]);
   printf("%c\n",string[3]);
   printf("%c\n",string[4]);

}
```

Execution of the above program yields

```
The string is Tech.
The characters are:
T
e
c
h
```

This shows some interesting facts about string arrays in C. First, they all start with the 0 element of the array. Second, all string arrays in C require a null terminator as the last element of the array. This terminator is necessary so that the program knows when the string array ends. Going beyond the null terminator will display anything that just happens to be in that particular memory location. This concept is illustrated in Figure 6–2.

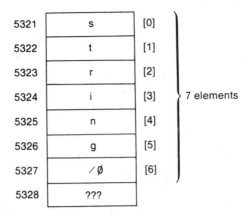

Figure 6-2 Starting and Ending Memory for String Arrays

Program Analysis

Program 6-2 again uses the global constant

char string[] = "Tech";

to define the string array **string**. The individual characters are then displayed by the **printf** functions using the character specifier **%c** along with a single array element:

printf("%c\n",string[0]);

The above statement, for example, causes the first element (the 0 element) of the string array to display its contents as a character, and hence the capital letter T is displayed. In a like manner, the other **printf** functions cause each of the other individual array elements to be displayed.

Where Are the Elements?

In a string array, each element is a character and represents a single memory location of one byte. As with other data, a pointer can also be used to access any element of the array. Consider the following program. It defines the same string constant and also a pointer (of type **char**). It then uses the *ptr to access each individual element of the array (each memory location).

Program 6–3

```
#include <stdio.h>

char string[] = "Tech";

main( )
{
  char *ptr;

    ptr = string;

  printf("The string is %s.\n",string);
  printf("The characters are:\n");
  printf("%c\n",*ptr);
  printf("%c\n",*(ptr + 1));
  printf("%c\n",*(ptr + 2));
  printf("%c\n",*(ptr + 3));
  printf("%c\n",*(ptr + 4));

}
```

Execution of the above program yields

```
The string is Tech.
 The characters are:
T
e
c
h
```

As you can see, the output of Program 6-3 is identical to that of Program 6-2. This would mean that the following are exactly equal:

```
*ptr = string[0]
*(ptr+1) = string[1]
*(ptr+2) = string[2]
*(ptr+3) = string[3]
*(ptr+4) = string[4]
```

The reason for this equality is because the variable **string** is declared as an arrayed variable (it was followed by the square brackets []). When the value of ptr is assigned to **string**

```
ptr = string;
```

this automatically places the address of the first string element (**string[0]**) in **ptr**. Thus, if you add one to the value contained in **ptr** (as with *(ptr+1)), you are adding one to the address contained there, which is the memory location of the next character. This concept is illustrated in Figure 6-3.

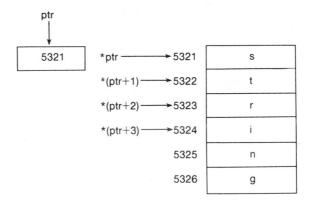

Figure 6-3   Getting Characters Using Pointers

It's important to observe the difference between

`*(ptr + 1)`

and

`ptr + 1`

In the first case, the program will be directed to the address that is one data space larger than that stored in `ptr`. In the second case, the value stored in `ptr` will have the number one added to it. This is illustrated by the following program.

Program 6-4

```
#include <stdio.h>

char string[] = "Tech";

main( )
{
 char *ptr;

    ptr = string;

 printf("The string is %s.\n",string);
 printf("The characters are:\n");
 printf("Address => %d | %c |\n",ptr, *ptr);
 printf("Address => %d | %c |\n",ptr+1, *(ptr + 1));
 printf("Address => %d | %c |\n",ptr+2, *(ptr + 2));
 printf("Address => %d | %c |\n",ptr+3, *(ptr + 3));
 printf("Address => %d | %c |\n",ptr+4, *(ptr + 4));

}
```

When the above program is executed, the output is

```
Address => 5696 | T |
Address => 5697 | e |
Address => 5698 | c |
Address => 5699 | h |
Address => 5700 |   |
```

This demonstrates three things. First, the addresses of the character array are contiguous. Second, `ptr + 1` is a value that is one larger than the value contained in `ptr`. Third, `*(ptr + 1)` represents the character stored in the memory location whose address is one larger than the address stored in `ptr`.

Conclusion

This section introduced you to using strings in C. You were also introduced to the concept of an array as a contiguous set of data elements in memory. Here you saw that the first element (number) of an array in C is 0. You also saw that a string array must be terminated with the C null.

The relationship between arrays and pointers was also demonstrated. Check your understanding of this section by trying the following section review.

6–1 Section Review

1 State what is meant by a string.
2 Explain how you indicate a **char** string in C.
3 For a string consisting of 5 characters, how many array elements are required? Explain.
4 What is the element number of the first character in a C string array?
5 Explain the relationship between pointers and string array elements.

6-2   More about Arrays

Discussion

In this section you will learn more about arrays. You will discover how to dimension an array and learn more about the relationship between arrays and pointers.

Basic Idea

To **dimension** an array, you simply place a number inside its brackets. As an example, `int array[3];` means that an array of three elements has been declared. These elements will consist of

array[0], array[1], and array[2]

The following program shows the relationship between

array and &array[0]

Program 6-5

```
#include <stdio.h>

 main( )
 {
  int array[3];

    printf("array = %d\n",array);
    printf("&array[0] = %d\n",&array[0]);

 }
```

Execution of the above program produces the same value for each of the three variables:

```
array = 7325
&array[0] = 7325
```

What this means is that the name of the array variable and the address of the first element of the array have the same value. This is the memory location of the first element of the array. This is illustrated in Figure 6–4.

Type int Array

Note that the above program uses an **int** as an array. Recall that an **int** uses *two* 8-bit memory locations (a **char** uses one). To illustrate this, the same array will be looked at again in the following program. The difference this time is the address of each array element is displayed.

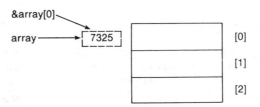

Figure 6–4   First Address of the Array

Program 6-6

```c
#include <stdio.h>

main( )
{
  int array[3];

   printf("address array[0] => %d\n",&array[0]);
   printf("address array[1] => %d\n",&array[1]);
   printf("address array[2] => %d\n",&array[2]);

}
```

Execution of the above program yields

```
address array[0] = 7325
address array[1] = 7327
address array[2] = 7329
```

Note from the output that each address is two larger than the previous one. This is because a type **int** is the arrayed variable which allocates two bytes per location. What is going on here is illustrated in Figure 6-5.

What this means is that C will automatically compensate for the required memory space depending upon the data type you are using in the array. However, the fact remains that the elements of the array are contiguous in memory.

Inside Arrays

In the above programs, you have seen the addresses of each array element. What is inside each of these elements (what is stored at these addresses)? Program 6-7 gives you an idea of what is there:

Program 6-7

```c
#include <stdio.h>

main( )
{
  int array[3];

   printf("contents of array[0] => %d\n",array[0]);
   printf("contents of array[1] => %d\n",array[1]);
   printf("contents of array[2] => %d\n",array[2]);

}
```

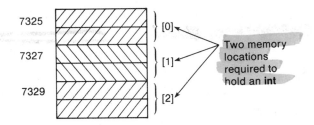

Figure 6-5   Address with Type **int** Arrays

Execution of the above program produces

```
contents of array[0] = 0
contents of array[1] = -5672
contents of array[2] = 58
```

Where did these values come from? They are just what happened to be in those memory locations at the time the program was executed. What has happened is that the program has set aside enough memory space for your three element array, but it hasn't put anything in these memory locations. Now consider a slight variation on the same program. This time, however, the array is being declared as a global variable (Program 6-8).

Program 6-8

```
#include <stdio.h>

int array[3];

main( )
{

    printf("contents of array[0] => %d\n",array[0]);
    printf("contents of array[1] => %d\n",array[1]);
    printf("contents of array[2] => %d\n",array[2]);

}
```

Look at what happens when the above program is executed:

```
contents of array[0] = 0
contents of array[1] = 0
contents of array[2] = 0
```

Now all of the elements of the array have been set to zero. Setting the elements of an array to a known quantity is called **array initialization**.

The only difference between the last two programs is that in Program 6-7, the array was declared automatic (local) and in Program 6-8 the array was declared global (external). This illustrates an important point in C. Arrays declared global are initialized to zero by default, while local variables are not.

Putting in Your Own Values

You can place your own values in an array. One method of doing this is shown in Program 6–9.

Program 6–9

```
#include <stdio.h>

main( )
{
  int array[3];

     array[0] = 10;
     array[1] = 20;
     array[2] = 30;

    printf("contents of array[0] => %d\n",array[0]);
    printf("contents of array[1] => %d\n",array[1]);
    printf("contents of array[2] => %d\n",array[2]);

}
```

Execution of the above program produces

```
contents of array[0] = 10
contents of array[1] = 20
contents of array[2] = 30
```

As you can see from the above program, each element of the array now contains a value entered by you. You could also have used pointers to get the contents of each of your array elements in the above program. Program 6–10 shows how to do this.

Program 6–10

```
#include <stdio.h>

main( )
{
  int array[3];
  int *ptr;
```

(continued)

Program 6-10 *(continued)*

```
        array[0] = 10;
        array[1] = 20;
        array[2] = 30;

    ptr = array;

   printf("contents of array[0] => %d\n",*ptr);
   printf("contents of array[1] => %d\n",*(ptr+1));
   printf("contents of array[2] => %d\n",*(ptr+2));

}
```

Execution of this program yields exactly the same results:

```
contents of array[0] = 10
contents of array[1] = 20
contents of array[2] = 30
```

This demonstrates the following equalities:

```
array[0] = *ptr;
array[1] = *(ptr + 1);
array[2] = *(ptr + 2);
```

Note that *ptr is declared as type **int**. This lets C know that every increment of *ptr is to be two bytes and not one (because type **int** uses two memory locations).

Passing Arrays

You can pass arrays from one function to the other. This is shown in Program 6-11.

Program 6-11

```
#include <stdio.h>

void function1(int this[]);

main( )
{
  int array[3];
```

(continued)

Program 6-11 *(continued)*

```
        array[0] = 10;
        array[1] = 20;
        array[2] = 30;

    function1(array);

    }

    void function1(int this[])
    {
        printf("contents of this [0] => %d\n",this[0]);
        printf("contents of this [1] => %d\n",this[1]);
        printf("contents of this [2] => %d\n",this[2]);

    }
```

Execution of the above program produces

```
contents of this[0] = 10
contents of this[1] = 20
contents of this[2] = 30
```

Program Analysis

The above program declares a function prototype of type **void** that contains an array argument:

void function1(**int** this[]);

Notice that the array is called **this[]**. The indirection operator * is not used (it could have been) because **this[]** is a pointer that will point to the first element of the array.

The function **main( )** also defines an array type consisting of three elements; then it initializes each element:

int array[3];

```
array[0] = 10;
array[1] = 20;
array[2] = 30;
```

After this initialization, the function **function1** is then called:

function1(array);

Note that the actual argument did not use the **&** operator. (It could have used &array[0].) However, **array** is the beginning address of the array. This is what is passed to **function1**.

Next, `function1` receives the value of the address of the first array element and then proceeds to display the values of the first three elements.

```
printf("contents of this[0] => %d\n",this[0]);
printf("contents of this[1] => %d\n",this[1]);
printf("contents of this[2] => %d\n",this[2]);
```

What is happening is shown in Figure 6–6.

The next program illustrates the passing of an array from a called function.

Passing Arrays Back

Program 6–12 illustrates **array passing**. Data is passed from a called function back to the calling function. This is just the opposite of the previous program. In Program 6–12, the called function initializes the array elements, and the calling function displays the contents. This is possible because the called function again has the starting address of the array passed to it.

Program 6–12

```
#include <stdio.h>

void function1(int this[]);

main( )
{
  int array[3];

    function1(array);
    printf("contents of array[0] = %d\n",array[0]);
    printf("contents of array[1] = %d\n",array[1]);
    printf("contents of array[2] = %d\n",array[2]);
}

  void function1(int this[])
  {
    this[0] = 20;
    this[1] = 40;
    this[2] = 60;
  }
```

The output of the above program is

```
contents of array[0] = 20
contents of array[1] = 40
contents of array[2] = 60
```

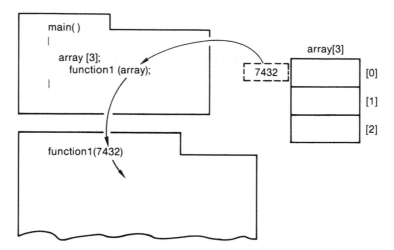

Figure 6–6   Passing an Array to a Called Function

Again, the starting address of the array is passed to the called function by `function1(array);` in **main( )**. Since the starting address is known by `function1`, each element of the array can be initialized. Figure 6–7 illustrates the process.

As you can see from the above programs, arrayed variables can be easily passed between functions in a C program.

What you can do with arrays, you can just as easily do with strings. This is illustrated in the remainder of this section.

Revisiting Strings

Program 6–13 asks program users to input their first name. The program will then display the name in a sentence.

Program 6–13

```
#include <stdio.h>

main( )
{
  int stringarray[10];

      printf("Enter your first name = ");
      scanf("%s",stringarray);

      printf("Hello there %s!",stringarray);

}
```

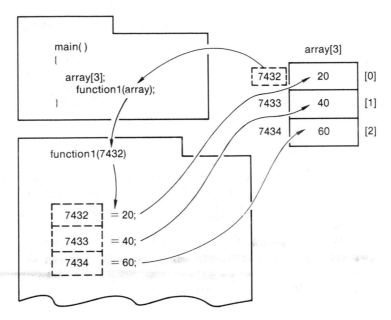

Figure 6-7   Process of Initialization in Called Function

Assume that the program user enters Tom as the first name. Then execution of the above program will yield

```
Enter your first name = Tom
Hello there Tom!
```

Note that in this case the size of **stringarray** is 10. Since the name is only three characters, the remaining elements are never initialized. To see what was in each element of the entered string, look at the next program.

Program 6-14

```
#include <stdio.h>

main( )
{
 char stringarray[10];
     printf("Enter your first name = ");
     scanf("%s",stringarray);

     printf("Hello there %s!\n\n",stringarray);
     printf("stringarray[0] => %c\n",stringarray[0]);
     printf("stringarray[1] => %c\n",stringarray[1]);
     printf("stringarray[2] => %c\n",stringarray[2]);
     printf("stringarray[3] => %c\n",stringarray[3]);

}
```

Again assume that the program user enters the name Tom. Execution of the above program will produce

```
Enter your first name = Tom
Hello there Tom

stringarray[0] = T
stringarray[1] = o
stringarray[2] = m
stringarray[3] =
```

Note that C stores the entered string starting at element 0 and then uses an extra element to store the null (to indicate where the string ends). This brings up a very important point. When reserving elements in a string array, make sure that you save enough room for the null character!

When the **scanf** function is used to input the name of the program user, the **&** (address operator) is not used with the array variable. Because **stringarray** is defined as an array, **stringarray** contains the address of the first element of the array. Thus, from the above program

```
scanf("%s",stringarray);
```

is all that is necessary to get the user's name into the array.

Passing Strings

Since strings are nothing more than character arrays, you can pass strings between functions in the same manner that you can pass other arrays. This is demonstrated by Program 6–15 and Program 6–16.

Program 6–15

```c
#include <stdio.h>

void function1(char name[]);

main( )
{
  char string[20];

   printf("What is your name = ");
   scanf("%s",string);

   function1(string);

}
```

(continued)

Program 6–15 *(continued)*

```
void functionl(char name[])
{

 printf("Hello there %s!",name);
}
```

The above program illustrates the passing of a string to a called function. Execution of the program produces

```
What is your name = Tom
Hello there Tom!
```

Program 6–16 shows the string being passed back to the calling function.

Program 6–16

```
#include <stdio.h>

void functionl(char name[]);

main( )
{
  char string[20];

  functionl(string);
  printf("Hello there %s",string);

 }

 void functionl(char name[])
 {

  printf("What is your name = ");
  scanf("%s",name);
  }
```

The output of Program 6–16 is identical to that of Program 6–15.

Conclusion

This section presented a closer look at arrays and strings. You saw the relationship between pointers and arrays as well as how to pass arrays between functions. Passing strings between functions was also demonstrated. Check your understanding of this section by trying the following section review.

6-2 Section Review

1   How can you let C know how many elements an array will have?
2   What is always the index of the first element of a C array?
3   If the variable `int value[5];` is declared in a function, then what is the relationship between `value`, and `&value[0]`?
4   What is the difference between declaring a local array or a global array?
5   How many elements are needed in a string array? Why?

6-3   Multidimensional Arrays

Discussion

In the last section you worked with arrays that had a single dimension. This means that there was only one index. In technology applications, it is common to find arrays with more than one dimension. These are called **multidimensional arrays**. This section introduces two-dimensional arrays and shows an application with them. The material presented in this section is applied too in the case study section of this chapter.

Basic Idea

The only kinds of arrays you can initialize are static and external. In working with strings, you could initialize a string array as shown in the following program.

Program 6-17

```
#include <stdio.h>

main( )
{

  static char array[7] = {'H','e','l','l','o'};

    printf("%s",array);

}
```

Execution of the above program will produce

`Hello`

The array initialization is performed using the { }. Inside these, separated by commas, are the individual characters to be used in the array. All arrays may be initialized this way. This is not the most efficient method for strings, but it is quite easy for numbers. Consider the one-dimensional array of the next program.

Program 6–18

```
#include <stdio.h>

main( )
{

  static int array[3] = {10,20,30};
  int index;

        for(index = 0; index < 3; index++)
        printf("array[%d] = %d\n",index,array[index]);

}
```

Execution of the above program produces

```
array[0] = 10
array[1] = 20
array[2] = 30
```

To initialize this array, the { } are again used. Values assigned to each element of the array are again separated by commas: {10,20,30}.

Arrays of more than one dimension may be initialized as shown in the next program. This program has a two-dimensional array. You can visualize a two-dimensional array as a checkerboard pattern where each square is identified by a unique row and column number. Each square contains data. The concept is shown in Figure 6–8. Such an arrangement is usually referred to as a **matrix**.

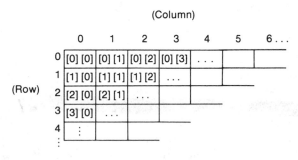

Figure 6–8 Concept of a Matrix

Program 6-19

```c
#include <stdio.h>

main( )
{

  static int array[2][3] = {
                            {10,20,30},
                            {11,21,31}
                           };
  int row;
  int column;

        for(row = 0; row < 2; row++)
        {
         for(column = 0; column < 3; column++)
           printf("%5d",array[row][column]);
         printf("\n\n");
        }

}
```

Execution of the above program produces

10   20   30

11   21   31

Program Analysis

First a two-dimensional array is declared and initialized in

```c
main( ):
 static int array[2][3] = {
                           {10,20,30},
                           {11,21,31}
                          };
```

The indexes [2][3] indicate that the array is a two-dimensional array containing two **rows** (numbered 0 and 1) and 3 **columns** (numbered 0 through 2). Note that each dimension of the array uses a new set of [ ]. The way in which C lays out memory is to lay out one row at a time with a given number of columns. This arrangement in memory is illustrated in Figure 6-9.

The method of initializing each element is done by specifying each element in the first row, then each element in the second row. This could have been done in the above declaration by

{{10,20,30},{11,21,31}}

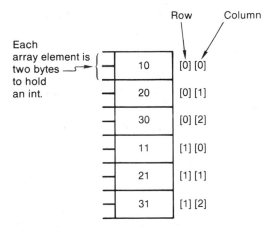

Figure 6–9 Memory Allocation with Rows and Columns for a Two-Dimensional Array

However, it is more descriptive of an actual two-dimensional matrix to lay it out as shown.

Next two variables are declared:

```
int row;
int column;
```

These two variables will be used to step through each element of the array so its value can be displayed. This is accomplished by the nested C **for** loops:

```
for(row = 0; row < 2; row++)
{
 for(column = 0; column < 3; column++)
  printf("%5d",array[row][column]);
printf("\n\n");
}
```

Remembering that the index of `row` starts with 0 and ends with 1, while the index for `column` starts with 0 and ends with 2, the two loops cause each element value in the array to be displayed through the variable

```
array[row][column]
```

To see how this is done, the first time through the loop, `row` = 0 and `column` = 0; thus, the first element to be displayed by the **printf** function is `array[0][0]` which has the value of 10. The next element to be displayed is `array[0][1]` which has a value of 20. This process continues until the nested **for** loop reaches the count of 3 and then the **printf("\n\n");** function causes two spaces, and the outer loop increments and the process begins again, this time displaying the values of each column in the second row.

There is a **field width** specifier used with the format specifier in the **printf** function of this program:

```
printf("%5d",array[row][column]);
```

Note the **%5d**. The 5 means to use 5 characters when printing the output. Thus, when the first column in displayed, it produces:

```
10    20    30
  12345          <= indicates number of spaces.
```

Thus, when the output is displayed there are five spaces reserved for each output number (if the number to be displayed happened to be larger and require more spaces, C would automatically adjust for that).

Array Applications

Arrays have many applications when solving technical problems. Many times the use of arrays requires manipulation of the array elements. Hence it's important to develop programming skills that allow calculations with arrays.

One of the simplest processes to perform with array arithmetic is to determine the sum of any column or any row. Program 6–20 determines the sum of the first column of the two-dimensional array of the last program.

Program 6–20

```c
#include <stdio.h>

int add_column(int arrayin[][3]);

main( )
{
 static int array[2][3] = {
                             {10,20,30},
                             {11,21,31}
                           };
 int row;
 int column;
 int first_column_sum;

 for(row = 0; row < 2; row++)
 {
 for(column = 0; column < 3; column++)
  printf("%5d",array[row][column]);
  printf("\n\n");
 }

     first_column_sum = add_column(array);
     printf("The sum of the first column is %d",first_column_sum);

}
```

(continued)

Program 6–20 *(continued)*

```c
int add_column(int arrayin[][3])
{
 int row;
 int column_sum;
      column_sum = 0;

 for(row = 0; row < 2; row++)
    column_sum += arrayin[row][0];

 return(column_sum);

}
```

Why – dimension
or function prototype

Execution of the above program produces

```
10    20    30

11    21    31
The sum of the first column is 21
```

This seems like a lot of work just to add two numbers. However, this is just a simple example to demonstrate a powerful concept—arithmetic operations using arrays.

Program Analysis

The program starts by first declaring a function prototype:

```c
int add_column(int arrayin[][3]);
```

Since an array of more than one dimension will be used, C requires that the limits of other dimensions be specified within the formal parameter list. The reason for this is the way arrays are stored in C. In this case, for the two-dimensional array, C needs to know how many columns will be required for each of the rows. In this manner it can properly set its array pointer.

The array elements are declared as in the previous program:

```c
static int array[2][3] = {
                          {10,20,30},
                          {11,21,31}
                         };
```

And the variables for **main( )** are declared:

```c
int row;
int column;
int first_column_sum;
```

Note the addition of the new variable `first_column_sum` that will be used to store the value of the sum for the first column.

Next, the array is displayed as in the previous program:

```
for(row = 0; row < 2; row++)
    {
    for(column = 0; column < 3; column++)
    printf("%5d",array[row][column]);
    printf("\n\n");
    }
```

Next, the function that will take the sum of the first column and return the value is called:

```
first_column_sum = add_column(array);
```

Note as before, when passing the address of the array to the called function it is only necessary to pass the identifier for the array variable. This, of course, is the starting address of the array.

This is a good time to present the details of the called function.

The function is of type **int** and defines two variables:

```
int add_column(int arrayin[][3])
    {
    int row;
    long int column_sum;
```

The variable `column_sum` will be used to store the value of the required sum. It is first initialized to `0`.

```
column_sum = 0;
```

Then it is used in a **for** loop that uses the += to generate the sum of `arryin[0][0]` + `arryin[1][0]`. This is accomplished by the following loop:

```
for(row = 0; row < 2; row++)
    column_sum += arrayin[row][0];
```

The final value is returned to the calling function:

```
return(column_sum);
```

After this, the final answer is displayed:

```
printf("The sum of the first column is %d",first_column_sum);
```

The next program illustrates a method of adding both columns of the matrix.

Adding More Columns

Consider Program 6–21. It expands on the previous program and adds all columns of the given matrix.

Program 6-21

```c
#include <stdio.h>

void add_columns(int arrayin[][3], int column_value[]);
main( )
{

static int array[2][3] = {
                             {10,20,30},
                             {11,21,31}
                          };
  int row;
  int column;
  int column_value[3];

        for(row = 0; row < 2; row++)
{
for(column = 0; column < 3; column++)
 printf("%5d",array[row][column]);
 printf("\n\n");
}

        add_columns(array, column_value);

     for(column = 0; column < 3; column++)
      printf("The sum of column %d is %d\n",column,column_value[column]);

}

void add_columns(int arrayin[][3], int column_value[])
{
 int row;
 int column;

  for(column = 0; column < 3; column++)
  {
   column_value[column] = 0;
    for(row = 0; row < 2; row++)
      column_value[column] += arrayin[row][column];
  }
}
```

When the above program is executed the output will be

```
   10    20    30

   11    21    31
The sum of column 0 is 21
The sum of column 1 is 41
The sum of column 2 is 61
```

Note now that each column in the matrix is totaled. Being able to do this has applications in the technology of business spread sheets.

Program Analysis

Program 6–21 incorporates a few additions to Program 6–20. First, the function prototype is changed to a type **void**, and its argument is expanded to include an array that will contain the sum of each column:

```
void add_columns(int arrayin[][3], int column_value[]);
```

The function definition is also expanded:

```
void add_columns(int arrayin[][3], int column_value[])
{
 int row;
 int column;

  for(column = 0; column < 3; column++)
  {
   column_value[column] = 0;
    for(row = 0; row < 2; row++)
      column_value[column] += arrayin[row][column];
  }
}
```

Note now that there is a C **for** loop that sums each column of the array using the += operator. Also note that within the loop, the array variable column_value[column] is initialized to zero each time. This is to ensure that the += operation starts with a zero value in this variable. The answers for the sum of each column are stored in the arrayed variable column_value[column]. It is this array that is returned to the calling function (**main( )**):

```
add_columns(array, column_value);
```

Observe that **main( )** has defined a new array variable:

```
int column_value[3];
```

This is used as the actual parameter of the called function.

What is important to realize is that two things are happening when this function is being called. First, the starting address of the arrayed variable **array** that contains the values of the array is being passed to the called function. Secondly, the arrayed variable column_value is being used to return the starting address of a second array that contains the values of the required column sums. Lastly, the output values are displayed:

```
for(column = 0; column < 3; column++)
printf("The sum of column %d is %d\n",column,column_value[column]);
```

The last program for this section computes the sum of each column and each row of the matrix. It uses two separate functions to do this. One function calculates the column sum; the other calculates the row sum.

Program 6–22

```
#include <stdio.h>

 void add_columns(int arrayin[][3], int column_value[]);
 void add_rows(int arrayin[][3], int row_value[]);

main( )
{

 static int array[2][3] = {
                            {10,20,30},
                            {11,21,31}
                           };
 int row;
 int column;
 int column_value[3];
 int row_value[2];

      for(row = 0; row < 2; row++)
 {
 for(column = 0; column < 3; column++)
  printf("%5d",array[row][column]);
  printf("\n\n");
 }

      add_columns(array, column_value);

   for(column = 0; column < 3; column++)
    printf("The sum of column %d is %d\n",column,column_value[column]);

      add_rows(array, row_value);

   for(row = 0; row < 2; row++)
    printf("The sum of row %d is %d\n",row,row_value[row]);

}

 void add_columns(int arrayin[][3], int column_value[])
 {
  int row;
  int column;

   for(column = 0; column < 3; column++)
   {
    column_value[column] = 0;
     for(row = 0; row < 2; row++)
       column_value[column] += arrayin[row][column];
   }
 }
```

(continued)

Program 6-22 *(continued)*

```
void add_rows(int arrayin[][3], int row_value[])
{
 int row;
 int column;

  for(row = 0; row < 2; row++)
  {
    row_value[row] = 0;
     for(column = 0; column < 3; column++)
     row_value[row] += arrayin[row][column];
  }
}
```

Execution of the above program produces

```
10    20    30

11    21    31
The sum of column 0 is 21
The sum of column 1 is 41
The sum of column 2 is 61
The sum of row 0 is 60
The sum of row 1 is 63
```

Conclusion

This section presented the concept of arrays with more than one dimension. Specifically a two-dimensional array was presented. You learned how to initialize the array and to pass values between functions. You saw an application that did arithmetic operations on the array. Check your understanding of this section by trying the following section review.

6-3 Section Review

1   State a way of initializing an array.
2   How do you declare an array with more than one dimension in C?
3   Explain how to declare a formal parameter in C that uses more than one dimension.
4   State the method used in the programs of this section that allows arrayed values to be added.
5   What must you do before using the += operator with an uninitialized array?

6-4   Introduction to Array Applications

Discussion

This section introduces you to the fundamental concepts of applying arrays in technical programs. Here you will see how to use arrays to rearrange a list of numbers. This will be developed into a program that will then be able to find the smallest value from a list of user input values. The material presented here lays the foundation for the information in the next section of this chapter.

Working with the Array Index

The one idea behind array applications is the ability to work with the index of the array. It is important not to think of the value of the **array index** as a fixed number. These values can be manipulated in any manner you choose. Understanding this concept can produce many powerful technical programs. Program 6-23 illustrates this concept.

Program 6-23

```c
#include <stdio.h>

main( )
{
 int number_array[7];
 int index;

   /* Place values in the array.  */

        number_array[0] = 0;
        number_array[1] = 2;
        number_array[2] = 4;
        number_array[3] = 8;
        number_array[4] = 16;
        number_array[5] = 32;
        number_array[6] = 64;

  for(index = 1; index <= 5; index++)
    {
    printf("number_array[%d] = %d",index, number_array[index]);
    printf(" number_array[%d + 1] = %d",index, number_array[index+1]);
    printf(" number_array[%d - 1] = %d\n",index, number_array[index-1]);
    }

 }
```

Execution of the above program produces

```
number_array[1] = 2 number_array[1 + 1] = 4 number_array[1 - 1] = 0
number_array[2] = 4 number_array[2 + 1] = 8 number_array[2 - 1] = 2
number_array[3] = 8 number_array[3 + 1] = 4 number_array[3 - 1] = 4
number_array[4] = 16 number_array[4 + 1] = 32 number_array[4 - 1] = 8
number_array[5] = 32 number_array[5 + 1] = 64 number_array[5 - 1] = 16
```

As you can see from the above output, changing the array index value causes a corresponding change in the program output value. In the above program, values are assigned to each element of the array. Then, the output of the array elements is displayed. However, to demonstrate the effect on the output of changing the index values, each index value is increased by one and then decreased by one. This is done in order to demonstrate the resulting output. Thus, number_array[3] contains the same value as number_array[2+1] and number_array[4-1].

Changing the Sequence

Program 6–24 further illustrates the manipulation of the array index. This program allows the program user to enter nine numbers. The program will then display the numbers in the opposite order in which they were entered.

Program 6-24

```
#include <stdio.h>
#define maxnumber 9   /* Maximum number of array elements.   */

void run_backwards(int user_array[]);

main( )
{
 int number[maxnumber];
 int index;

    printf("Give me nine numbers and I'll print them backwards.\n");
     for(index = 0; index < maxnumber; index++)
     {
      printf("Number[%d] = ",index);
      scanf("%d",&number[index]);
     }
    printf("Thank you...\n");

    run_backwards(number);

}
```

(continued)

Program 6-24 *(continued)*

```
void run_backwards(int user_array[])
{
 int index;

   printf("\n\nHere are the numbers you entered displayed\n");
   printf("in the reverse order of entry:\n");
    for(index = maxnumber-1; index >= 0; index--)
     printf("number[%d] = %d\n",index, user_array[index]);

}
```

Assume that the program user enters the numbers shown below. Then execution of the above program will produce

```
Give me nine numbers and I'll print them backwards.
Number[0] = 1
Number[1] = 2
Number[2] = 3
Number[3] = 4
Number[4] = 5
Number[5] = 6
Number[6] = 7
Number[7] = 8
Number[8] = 9
Thank you...

Here are the numbers you entered displayed
in the reverse order of entry:
Number[8] = 9
Number[7] = 8
Number[6] = 7
Number[5] = 6
Number[4] = 5
Number[3] = 4
Number[2] = 3
Number[1] = 2
Number[0] = 1
```

Program Analysis

The program first defines the maximum size of the array with a **#define**:

```
#define maxnumber 9   /* Maximum number of array elements.   */
```

A function prototype that uses an array in its formal argument is then declared:

```
void run_backwards(int user_array[]);
```

Next a C **for** loop gets input from the program user. Note that in this loop, the array index is being incremented, and it goes from 0 (first array element) to one less than maxnumber (last array element). This is an efficient method of getting array element values from the program user

```
printf("Give me nine numbers and I'll print them backwards.\n");
  for(index = 0; index < maxnumber; index++)
  {
   printf("Number[%d] = ",index);
   scanf("%d",&number[index]);
  }
```

It then calls on the function that will now display the array in reverse order. Note that the pointer of the starting address of the array is used in the actual parameter.

```
run_backwards(number);
```

The called function uses another C **for** loop, but this time the loop starts the array index at maxnumber − 1 (the maximum index value for the array) and ends at 0, doing a C −− on the variable index.

```
for(index = maxnumber−1; index >= 0; index−−)
 printf("number[%d] = %d\n",index, user_array[index]);
```

In each case, the value of the variable is printed out.

Finding a Minimum Value

The previous program illustrates changing the order of how an array is displayed by reording the array index. This concept will be developed further in Program 6–25 where the smallest value from a list of numbers is extracted. The program user enters the numbers, and the program searches through the list and extracts the smallest value and displays it on the screen.

Program 6–25

```
#include <stdio.h>
#define maxnumber 9   /* Maximum number of array elements.   */

int minimum_value(int user_array[]);

main( )
{
 int number[maxnumber];
 int index;

    printf("Give me nine numbers and I'll find the minimum value:\n");
     for(index = 0; index < maxnumber; index++)
       {
```

(continued)

Program 6–25 *(continued)*

```
            printf("Number[%d] = ",index);
            scanf("%d",&number[index]);
          }
      printf("Thank you...\n");

      printf("The minimum value is %d\n",minimum_value(number));

   }

int minimum_value(int user_array[])
{
 register int index;
 int minimum;

    minimum = user_array[0];

      for(index = 0; index < maxnumber; index++)
       if(user_array[index] < minimum)
          minimum = user_array[index];

             return(minimum);

   }
```

Assuming the program user enters the series of numbers shown below, execution of the above program yields

```
Give me nine numbers and I'll find the minimum value:
Number[0] = 12
Number[1] = 21
Number[2] = 58
Number[3] = 3
Number[4] = 5
Number[5] = 8
Number[6] = 19
Number[7] = 91
Number[8] = 105
Thank you...
The minimum value is 3
```

Program Analysis

Program 6–25 is different from 6–24 in the called function. First Program 6–25 does a function prototype for the new function:

```
int minimum_value(int user_array[]);
```

The function **main( )** still gets nine values from the program user in the same manner as before. The difference is function `minimum_value`. This function defines a variable called `minimum`. This will hold the minimum value of the array. First `minimum` is initialized to the value of the first element of the array:

```
minimum = user_array[0];
```

This is done so that the variable `minimum` has a value from the array to compare the other values to. This comparison is done in a C **for** loop.

```
for(index = 0; index < maxnumber; index++)
  if(user_array[index] < minimum)
    minimum = user_array[index];
```

What happens is the next element of the array is compared to `minimum`. If this element is less than minimum, then `minimum` is given this value. If this is not the case, then minimum retains its previous value. It is in this manner that the smallest value of the array is selected. This value is then returned to the calling function:

```
return(minimum);
```

Conclusion

In this section you were introduced to the concept of accessing arrayed data. Here you saw how to modify the index of an array in order to accomplish this. This section presented a method of rearranging how an array is displayed to show a minimum value from a list of values entered by the program user. Test your understanding of this section by trying the following section review.

6–4 Section Review

1 State the basic idea behind array applications.
2 Explain what programming method was used in order to get arrayed values from the program user.
3 State the method used to cause a series of entered values to be displayed in the opposite order from which they were entered.
4 What method was used in order to extract a minimum value from a list of entered values?

6-5   Sorting

Basic Idea

There are many different methods of sorting a list of numbers. The method presented in this section is called a **bubble sort**. It is called this because numbers are bubbled to the top of the list as the sorting is processed.

As an example, consider the following list of numbers.

6

3

6

8

Suppose it is your job to arrange the above list in ascending order (smallest number first, largest last). Assume that you are to use the bubble sorting technique to accomplish this. Here are the rules you would use:

Bubble Sorting Rules (to sort in ascending order):

1. Test only two numbers at a time, starting with the first two numbers.
2. If the top number is smaller, leave as is. If the top number is larger, switch the two numbers.
3. Go down one number and compare that number to the number that follows it. These two will be a new pair.
4. Continue this process until no switch has been made in an entire pass through the list.

To sort in descending order, simply change rule 2 as follows:

2. If the top number is larger, leave as is. If the top number is smaller, switch the two numbers.

To sort the above list of data, start with the first rule. Test only two numbers at a time, starting with the first two numbers. So, using the above list, start with the top two numbers:

6

3

The top number is larger, so using rule 2, switch the two numbers:

3

6

The list now looks like this:

3

6

6

8

Go down one number down and compare it to the number that follows (a new number pair):

6

6

These are both the same. Since they are equal, it makes no difference what you do to them!

Go to the next new number pair:

$$6$$
$$8$$

Since the smallest number is already on the top, leave them as is. You have completed the list, but a switch was made on this pass through the list; therefore, you must make another pass through the list:

Testing the first two numbers

$$3$$
$$6$$

no switch is necessary. Now the next two:

$$6$$
$$6$$

Again, no switch is necessary. The next two:

$$6$$
$$8$$

Still no required switch. Since there were no switches in this pass through the list, the sorting is completed and the resulting list is

$$3$$
$$6$$
$$6$$
$$8$$

Sample Program

A sample program that implements a bubble sort is shown in Program 6-26.

Program 6-26

```
#include <stdio.h>

#define maxnumber 9   /* Maximum number of array elements.   */

void bubble_sort(int user_array[]);
void display_array(int sorted_array[]);

main( )
{
  int number[maxnumber];
  int index;
```

(continued)

Program 6–26 *(continued)*

```
        printf("Give me nine numbers and I'll sort them:\n");
          for(index = 0; index < maxnumber; index++)
          {
           printf("Number[%d] = ",index);
           scanf("%d",&number[index]);
          }
        printf("Thank you...\n");

        bubble_sort(number);

        display_array(number);

       }

  void bubble_sort(int user_array[])
  {
   int index;
   int switch_flag;
   int temp_value;

      do
       {
        switch_flag = 0;

          for(index = 0; index < maxnumber; index++)
          {
          if((user_array[index] > user_array[index+1])&&(index!=maxnumber-1))
           {
             temp_value = user_array[index];
             user_array[index] = user_array[index+1];
             user_array[index+1] = temp_value;
             switch_flag = 1;
           }  /* end of if */
          }  /* end of for */
        }  /* end of while */
      while (switch_flag);

   }

  void display_array(int sorted_array[])
  {
   int index;

      printf("\n\nThe sorted values are: \n");

      for(index = 0; index < maxnumber; index++)
       printf("Number[%d] = %d\n",index,sorted_array[index]);

   }
```

Assuming that the program user enters the numbers as shown, execution of the above program produces

```
Give me nine numbers and I'll sort them:
Number[0] = 6
Number[1] = 7
Number[2] = 5
Number[3] = 8
Number[4] = 4
Number[5] = 9
Number[6] = 3
Number[7] = 0
Number[8] = 2
Thank you...

The sorted values are:
Number[0] = 0.00
Number[1] = 2.00
Number[2] = 3.00
Number[3] = 4.00
Number[4] = 5.00
Number[5] = 6.00
Number[6] = 7.00
Number[7] = 8.00
Number[8] = 9.00
```

Program Analysis

After the #define maxnumber 9 which is used to set the maximum size of the array, the program defines two function prototypes:

```
void bubble_sort(int user_array[]);
void display_array(int sorted_array[]);
```

Both of these functions are of type **void** since the functions themselves will not be used to return any values. However the formal parameters of both functions declare **int** arrays. The declaration user_array[] is a pointer just as *user_array is a pointer. It will be used to store the address of the first element ([0]) of the array. The first function bubble_sort will be used by **main( )** to do the actual sorting. The next function display_array is used to display the sorted array.

Function **main( )** simply gets nine numbers from the program user:

```
int number[maxnumber];
int index;

    printf("Give me nine numbers and I'll sort them:\n");
      for(index = 0; index < maxnumber; index++)
      {
       printf("Number[%d] = ",index);
       scanf("%d",&number[index]);
      }
    printf("Thank you...\n");
```

The above program excerpt defines an array `number` consisting of `maxnumber` elements. This means that the first element of the array will be `number[0]` and the last element `number[8]`. When the program user has entered all nine numbers the program displays

```
Thank you...
```

The program then calls the two functions, one to sort the given numbers, the other to display them. Note that the actual parameter passed is the address of the first element. Recall that this contains the address of the first element of the array. It is the same as using `&number[0]` as the parameter.

```
bubble_sort(number);
display_array(number);
```

The function `bubble_sort` declares three variables: `index`; `switch_flag`; and `temp_value`;. The variable `index` will be used as the index for each of the array elements. `switch_flag` will let the program know when the sorting of the array is completed, and `temp_value` will temporarily store the value of one array element while it is being switched with another array element. You can think of a flag as a variable that is either up (ON) or down (OFF). Thus, for this example, `switch_flag` has just one of two values.

```
int index;
int switch_flag;
int temp_value;
```

In the body of the `bubble_sort` function, a C **do** loop is used to repeatedly go through the array to check if a switch is needed between array elements. The loop starts by setting the `switch_flag` to 0 (thus making `switch_flag` false).

```
do
{
  switch_flag = 0;
```

Next there is a nested **for** loop that causes a scan through each element of the array.

```
for(index = 0; index < maxnumber; index++)
```

Inside the C **for** loop, a comparison of each array element is made, and a check is made to ensure that no more than the maximum number of array elements is tested (`user_array[8]` is the largest array element).

```
if((user_array[index] > user_array[index+1])&&(index!=maxnumber-1))
```

If a switch is needed, the following then takes place:

```
{
temp_value = user_array[index];
user_array[index] = user_array[index+1];
user_array[index+1] = temp_value;
switch_flag = 1;
}
```

Notice how the variable `temp_value` is used. This is similar to having a cup of milk and a glass of orange juice and your task is to put the milk in the glass and the orange juice in the cup. One method of doing this is to get a third container (call it `temp_container`). Pour the milk into the temporary container, pour the juice into the cup, then pour the milk from the temporary container into the glass. `temp_value` is used in this manner. If a switch is made, then `switch_flag` is set to 1 (making it true). This means that the outer do loop will have to be repeated

```
while (switch_flag);
```

Recall that the way of ensuring that the sorting of the loop is completed is by not having a switch on a comparison through the loop.

The display function simply uses an index counter in a C **for** loop to cause the now sorted values to be displayed on the monitor:

```
for(index = 0; index < maxnumber; index++)
printf("Number[%d] = %d\n",index,sorted_array[index]);
```

Conclusion

Bubble sorting was presented here as a simple method of sorting numbers. In the next section, bubble sorting will be used to sort string variables. Test your understanding of this section by trying the following section review.

6-5 Section Review

1   Briefly describe the process of a bubble sort.
2   For any list of data, what is the minimum number of times a sorting program must go through a list of numbers before the sort is considered completed. Under what circumstances would this happen?
3   What determines if a list of data will be sorted in descending or ascending order?
4   In order to make the following list into ascending order, how many passes through the list will a bubble sort require?

8 7 3 1

6-6   More about Strings

Discussion

This section will prepare you for sorting strings. In the last section you saw how to sort numbers. You did this by manipulating the index of the array. When you sort strings in C you will again be manipulating an index, but this time it will be the index of pointers to strings. What you will discover is a method of sorting strings by sorting the pointers to these strings—not the actual strings themselves. Doing this makes for a more efficient and faster operating sorting program. It also sets the foundation for arranging more complex data structures that will be presented in the next chapter.

Before diving into an actual string sorting program, you need some more information about strings and their pointers. That's the purpose of this section.

An Array of Characters

Recall that a string is nothing more than an array of characters. This means that if you have three different strings you can envision them as a two-dimensional array. This is shown in the following program.

Program 6-27

```
#include <stdio.h>

main( )
{
static char string[3][10] = {
                             {'T','h','r','e','e'},
                             {'l','i','t','t','l','e'},
                             {'w','o','r','d','s'}
                             };

    printf("%s %s %s.",string[0], string[1], string[2]);

}
```

Execution of the above program produces

```
Three little words.
```

Program Analysis

The main point of Program 6-27 is to demonstrate the initialization of a two-dimensional array of characters. This is done by using a C static char called string. It is given two dimensions—three rows by ten columns:

```
static char string[3][10] = {
                             {'T','h','r','e','e'},
                             {'l','i','t','t','l','e'},
                             {'w','o','r','d','s'}
                             };
```

Figure 6-10 shows how such an array is stored in memory. Such an arrangement of data is called a **rectangular array**. Note that the null terminator is automatically placed at the end of the character array by C.

The **printf** function is then used to print out the string. This is done by using the **s** format type. Recall that this format type only needs the beginning address

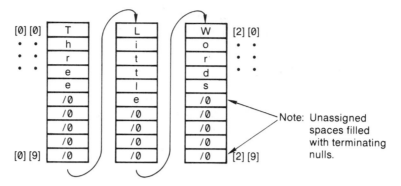

Figure 6-10   A Rectangular Array of Characters

of the string. It will then print out the series of characters starting at this address until it reaches the C null character.

```
printf("%s %s %s.",string[0], string[1], string[2]);
```

Another Way

Another way of arranging the same array of characters is to arrange them as three separate strings with three pointers pointing to the beginning address of each string. This is illustrated in Program 6-28. Here the same thing is accomplished as in Program 6-27, but in a slightly different (and more efficient) manner.

Program 6-28

```
#include <stdio.h>

main( )
{
static char *string[3] = {
                        {"Three"},
                        {"little"},
                        {"words"}
                      };

    printf("%s %s %s.",string[0], string[1], string[2]);

}
```

What the above program has done is to create another array of characters. But this time, C automatically takes care of how much space is needed for each row of strings. The resulting array is called a **ragged array** and is illustrated in Figure 6-11.

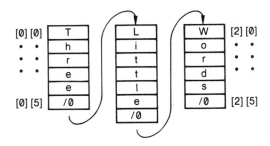

Figure 6-11   A Ragged Array of Characters

With the exception of the dimension restriction of 10 for the row size, a comparison of the two above programs shows that

`string[3][10]`

and

`*string[3]`

are similar data structures. Because of this, the **printf** function for Program 6–28 is the same as that of Program 6–27:

`printf(" %s %s %s.",string[0], string[1], string[2]);`

A Simple Way

A very simple and direct way of having a string in your program is to use the **#define** as shown in Program 6–29.

Program 6–29

```
#include <stdio.h>
#define sentence "Three little words."

main( )
{
    printf(sentence);
}
```

Execution of this program produces

`Three little words.`

However, defining strings this way is very static, especially when you want to do things like sorting them.

A Closer Look

To see how C keeps track of strings, consider the following program. It uses a pointer to point to the starting address of three strings. The values of the three pointers are displayed.

Program 6-30

```
#include <stdio.h>

main( )
{
 char *string[3];

  string[0] = "Resistor";
  string[1] = "Capacitor";
  string[2] = "Inductor";

      printf("%d %d %d",string[0],string[1],string[2]);

}
```

Using the **%s** format type in the same program produces the strings themselves:

Program 6-31

```
#include <stdio.h>
main( )
{
 char *string[3];

  string[0] = "Resistor";
  string[1] = "Capacitor";
  string[2] = "Inductor";

      printf("%s %s %s",string[0],string[1],string[2]);
}
```

It should come as no surprise to you that the output of the above program is

```
Resistor Capacitor Inductor
```

To illustrate what each of the three pointers—**string[0]**, **string[1]** and **string[2]**—are actually pointing to, look at the next program.

Program 6-32

```
#include <stdio.h>

main( )
{
  char *string[3];

    string[0] = "Resistor";
    string[1] = "Capacitor";
    string[2] = "Inductor";

          printf("%c %c %c",*string[0],*string[1],*string[2]);

  }
```

In the above program the %c format type is used to print the character that is contained in the memory location pointed to by the respective pointers. The result is,

R C I

the first letter of each string, confirming that the pointers are indeed pointing to the first character of each string. Thus what you have are three pointers, each of which contains the starting address of each of the three strings.

Pointing to Pointers

Suppose you want to create a separate function that will display the strings defined in the calling function. All you need to do is to pass the starting address of the pointers that point to each string. To get the concept of an array of pointers, consider the following program.

Program 6-33

```
#include <stdio.h>

main( )
{
  char *string[3];

    string[0] = "One";
    string[1] = "Two";
    string[2] = "Last";
```

(continued)

Program 6-33 *(continued)*

```
printf("\n          string => | %d |\n\n",string);

printf("&string[0] = %d | %d | => %o\n",&string[0], string[0], *string[0]);
printf("&string[1] = %d | %d | => %o\n",&string[1], string[1], *string[1]);
printf("&string[2] = %d | %d | => %o\n",&string[2], string[2], *string[2]);

}
```

Execution of the above program produces

```
            string => | 7842 |
&string[0] = 7842 | 5696 | => O
&string[1] = 7844 | 5700 | => T
&string[2] = 7846 | 5704 | => L
```

What is happening in the above program is illustrated in Figure 6-12.

The method of passing a string array to another function is to pass the starting address of the pointer array that contains the starting address of each string. In this manner, all of the necessary information about each of the strings may be accessed by using just one number.

The next program illustrates how to pass the addresses of a pointer array to a called function.

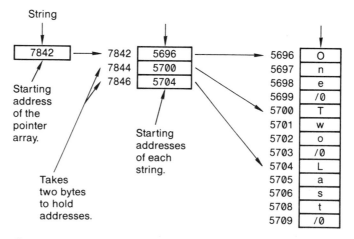

Figure 6-12   Illustration of Pointer Arrays

Program 6–34

```c
#include <stdio.h>

void display_string(char *ptr[]);

main( )
{
  char *string[3];

  string[0] = "Resistor";
  string[1] = "Capacitor";
  string[2] = "Inductor";

        display_string(string);

}

void display_string(char *ptr[])
{

        printf("%s %s %s",ptr[0],ptr[1],ptr[2]);

}
```

Figure 6–13 illustrates what is happening in Program 6–34.

Program Analysis

The program starts by defining a function prototype:

```c
void display_string(char *ptr[]);
```

This is the function that will be used to actually display the strings defined in **main( )**. Note that the pointer it uses in its formal parameter is defined to be an array of pointers. This is indeed what will happen because the value of the starting address of the three pointers will be sent to this function.

The function **main( )** simply defines the three strings as before:

```c
string[0] = "Resistor";
string[1] = "Capacitor";
string[2] = "Inductor";
```

The function to display the strings is then called. Note what is passed as the actual parameter. It is the name of the array variable **string** which is exactly the

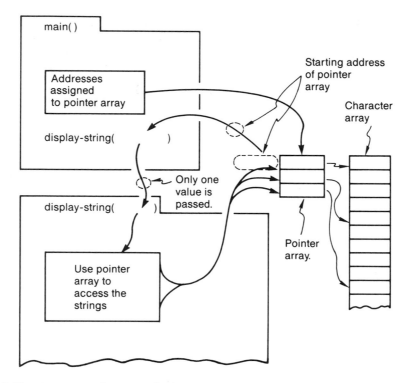

Figure 6-13   Operation of Program 6-34

same as **&string[0]** which is the starting address of where the addresses of the three strings are stored.

The called function now displays the three output strings:

void display_string(**char** *ptr[])

{

      printf("%s %s %s",ptr[0],ptr[1],ptr[2]**);**

The result is

Resistor Capacitor Inductor

Manipulating the Pointer Indexes

Since the called function in the above example is really working with an array of pointers, you can now change the index of each pointer and produce the effect of changing the ordering of the strings. In actuality, the addresses of the strings themselves will not change, just the index of the pointers. This is what Program 6-35 does. It has the user enter three strings, and then displays the entered strings in reverse order. It does this by reordering the indexes of the pointer arrays.

Program 6–35

```c
#include <stdio.h>

void display_string(char *ptr[]);

main( )
{
 char instring[3][10];
 char *string[3];
 int index;

  printf("Enter three strings and see them in reverse order: \n");
    for(index = 0; index < 3; index++)
      {
       printf("String[%d] = ",index);
       scanf("%s",instring[index]);
       string[index] = instring[index];
      }

       printf("Thank you...\n\n");

 printf("%s %s %s\n",instring[0],instring[1],instring[2]);

        display_string(string);

}

void display_string(char *ptr[])
{
int index;

       printf("Here are your strings in reverse order:\n");
       for(index = 2; index >= 0; index--)
       printf("String [%d] = %s\n",index,ptr[index]);

}
```

Assuming that the program user enters the strings as shown below, execution of the above program will yield:

```
Enter three strings and see them in reverse order:
String[0] = Jane
String[1] = was
String[2] = here
Thank you...
```

```
Jane was here.
Here are the strings in reverse order:
String[2] = here
String[1] = was
String[0] = Jane
```

As before, this seems like a lot of programming to do such an apparently simple task. But again keep in mind that the simplest possible examples are being used to illustrate new concepts. Once these concepts are understood, you can proceed to apply them in a richer and more useful way. The case study section of this chapter will demonstrate a practical application of what you are learning here.

Program Analysis

The program declares a function prototype. This function will be called by **main( )** and will reverse the order of the inputted strings and display the result. Note that the formal parameter is defined as a pointer array, meaning that the starting address of a pointer array will be passed to it from the calling function.

```
void display_string(char *ptr[]);
```

The program declares a two-dimensional array of characters and a pointer to an array of pointers as well as an index. This is done inside **main( )**:

```
main( )
{
 char instring[3][10];
 char *string[3];
 int index;
```

Next, three strings are entered by the program user. This is achieved by using the **scanf** function inside a C **for** loop. For each string, the pointer **string** is equated to the starting address of the inputted string. This is done to create the required array of three pointers to be passed to the called function.

```
printf("Enter three strings and see them in reverse order: \n");
  for(index = 0; index < 3; index++)
    {
    printf("String[%d] = ",index);
    scanf("%s",instring[index]);
    string[index] = instring[index];
    }
```

The program then displays the three entered strings using the pointers for each of the character arrays:

```
printf("%s %s %s\n",instring[0],instring[1],instring[2]);
```

Next, the function is called that displays a new order. Note that the actual parameter is the name of the array of pointers. This is the starting address of this pointer array:

```
display_string(string);
```

The called function definition declares one local variable `index` to be used for indexing the arrayed pointers

```
void display_string(char *ptr[ ])
{
int index;
```

It is important to note that the positions of the actual strings in memory are not changed. What changes is the displayed order of the arrayed pointers that point to these strings! For large programs of this type it is much faster to change simple pointers than large groups of data:

```
printf("Here are your strings in reverse order:\n");
for(index = 2; index >= 0; index--)
printf("String [%d] = %s\n",index,ptr[index]);
```

Doing It with Pointers

Here is a look at a different approach to the `Resistor Capacitor Inductor` string program.

Program 6-36

```
#include <stdio.h>

void display_string(char *ptr[]);

main( )
{
  char *string[3];

  string[0] = "Resistor";
  string[1] = "Capacitor";
  string[2] = "Inductor";

        display_string(string);

}

void display_string(char *ptr[])
{

        printf("%s %s %s",*ptr,*(ptr+1),*(ptr+2));

}
```

Look at the **printf** function in the called function `display_string`. It is now incrementing the pointer array to display each of the strings! This can be done because the starting address of this array is passed to it from **main( )**. This is the first address of a sequence, in memory, of three pointers each of which points to a different string (the first points to **Resistor**, the second to **Capacitor**, and the third to **Inductor**).

To emphasize this point even further, if you use the declared pointer as a pointer to a pointer and recode the **printf** function from the above program as follows

```
printf("%c %c %c",**ptr,**(ptr+1),**(ptr+2));
```

then the output will be

```
R C I
```

or the contents of the first three memory locations of each string.

Conclusion

This was an important section. You saw the relationship between pointers and strings in a much deeper way than before. You discovered that the beginning address of an array of pointers can be passed to a called function. Once this was done, the index of the pointer array can be manipulated to cause the original data to be displayed in different ways. The concepts you learned here will be applied to the case study section of this chapter as well as following chapters. For now, test your understanding of this section by trying the following section review.

6-6 Section Review

1 What is another name for an array of characters?
2 Explain what is meant by a rectangular array.
3 Explain what is meant by a ragged array.
4 What must be used to cause the **%s** format specifier to print out a string?
5 State what is used as the actual argument when passing information about a string array to a called function.

6-7 Program Debugging and Implementation

I/O Problems

This section presents information concerning I/O (input/output) problems that you may encounter when developing a C program that requires user input from the keyboard. This kind of problem is very common in programs that require user interaction.

Some technology programs you develop may require that the program user enter a certain kind of number (such as a whole number or a number with no commas). The same program may also require that the user enter a string that does not consist of any numbers (such as the name of a person). Problems may arise when your program expects a specific kind of data (such as a number) and the program user enters something else (such as an alphabetical character). Unless you make specific programming arrangements for this, your program could act in some very unpredictable ways that would be confusing or discouraging to the program user. This section will demonstrate some methods of making your C programs more "user friendly."

scanf( ) Again

You have used the built-in C function **scanf( )** many times in your C programs. There are some hidden problems with this function. One of these is demonstrated by the following program.

Program 6–37

```
#include <stdio.h>

main( )
{
 char string[80];

   printf("Give me an input string => ");
   scanf("%s",string);

   printf("The input string was => %s.\n",string);

}
```

The above program simply asks the program user to input a string and then displays the string. There are, however, two possible conditions for the above program. One condition is if the string entered by the user does not contain any blank spaces, the other if the string does contain spaces. As an example, consider the first case

```
Give me an input string => Hello!
The input string was => Hello!
```

Now consider what happens when the user inputs a string containing a blank space:

```
Give me an input string => Hello there!
The input string was => Hello
```

As you can see, one of the limitations of the **scanf( )** function is that it stops reading the input string after the first blank space. Thus, if you want program users to enter their full name, you could not use the **scanf( )** function.

The **gets( )**

For string input, a better choice is the built-in C function **gets( )**. Basically this function will continue to read a string of input until it encounters a newline character or an end-of-file marker. When this happens, it will replace the newline character with a null character ($\backslash$0), thus creating a string. The use of this function is illustrated in the following program. This program preforms exactly the same function as the previous program. The difference is that the **scanf( )** function has been replaced by the **gets( )** function.

Program 6–38

```
#include <stdio.h>

main( )
{
 char string[80];

  printf("Give me an input string => ");
  gets(string);

  printf("The input string was => %s.\n",string);

}
```

Now execution of the above program will produce

```
Give me an input string => Hello, my name is George!
The input string was => Hello, my name is George!
```

Note that now all of the string was accepted—including the blank spaces.

Long Strings

C provides an easy way of outputting long strings. This method is demonstrated in Program 6–39.

Program 6–39

```
#include <stdio.h>

main( )
{
 char *bigstring;
```

(continued)

Program 6-39 *(continued)*

```
    bigstring = "To connect the circuit, make sure that the power\n"
               "has not been applied. Then check your parts list\n"
               "for a complete set of materials.\n";

    printf("%s",bigstring);

}
```

Execution of the above program produces

```
To connect the circuit, make sure that the power
has not been applied.  Then check your parts list
for a complete set of materials.
```

A Character at a Time

There is another useful built-in C function that gets one character at a time from the input. This function is called **getchar( )**. As you will see, the advantage of doing this is that each individual character may be tested. Doing this will determine if a character is a letter of the alphabet, a number, or some other kind of input such as a punctuation mark. The use of this function is illustrated in Program 6-40.

Program 6-40

```
#include <stdio.h>

main( )
 {
 char ch;

     printf("Give me a single input => ");
     ch = getchar( );

 }
```

The function **getchar( )** reads a single character from input. For example, a string of characters may be entered by the program user until a newline marker is encountered (meaning the program user has pressed the carriage return). This is demonstrated in Program 6-41.

Program 6-41

```
#include <stdio.h>

main( )
{
char ch;

    printf("Give me a number => ");

    while((ch = getchar( )) != '\n');

}
```

In the above program, the program user will be able to continually input characters from the keyboard until the return key is depressed. When this happens, the C **while** will terminate.

Checking Characters

There are several different types of built in C functions that will check the type of character that is entered into a C program. Table 6-1 lists them.

Table 6-1   C Character Classifications in `<ctype.h>`

Function	Meaning (Returns a nonzero value if character meets the test, otherwise zero.)	Example (ch is character being tested)
`isalnum( )`	alphanumeric	`if(isalnum(ch) !=0)` `printf(%c is alphanumeric\n",ch);`
`isalpha( )`	alphabetic	`if(isalpha(ch) !=0)` `printf(%c is alphabetical\n",ch);`
`iscntrl( )`	control character	`if(iscntrl(ch) !=0)` `printf(%c is a control ch\n",ch);`
`isdigit( )`	digit	`if(isdigit(ch) !=0)` `printf(%c is a digit.\n",ch);`
`isgraph( )`	Checks if character is printable. Excludes the space character.	`if(isgraph(ch) !=0)` `printf(%c is printable.\n",ch);`
`islower( )`	Checks if character is lower case.	`if(islower(ch) !=0)` `printf(%c is lower case\n",ch);`
`isprint( )`	Checks if character is printable. Includes the space character.	`if(isprint(ch) !=0)` `printf(%c is printable.\n",ch);`

(continued)

Table 6–1   *(Continued)*

Function	Meaning (Returns a nonzero value if character meets the test, otherwise zero.)	Example (ch is character being tested)
ispunct( )	Checks if character is punctuation.	if(ispunct(ch) != 0) printf(%c is punctuation.\n",ch);
isspace( )	space	if(isspace(ch) != 0) printf(%c is a space.\n",ch);
isupper( )	Checks if character is upper case.	if(isupper(ch) != 0) printf(%c is upper case.\n",ch);
isxdigit( )	Hexadecimal digit.	if(isxdigit(ch) != 0) printf(%c is hex digit.\n",ch);

Program 6–42 illustrates an application of one of the built-in C character checking functions.

Program 6–42

```
#include <stdio.h>
#include <ctype.h>

main( )
 {
 char ch;

     printf("Give me a sentence => \n");

     while((ch = getchar( )) != '\r')
     printf("%d",isalpha(ch));

 }
```

Execution of the above program using Learn C on an IBM PC produces

```
Give me a sentence =>
Learn C
1222201
```

As you can see from the above output, the **isalpha( )** function returns a 0 when the character is not a letter of the alphabet (such as the blank space).

More Character Checking

Just as Program 6–42 checks for alphabetical characters, you can also check for numerical characters or any other type of character presented in Table 6–1. C contains another useful built-in function that can help you with user data input. This is the **strlen( )** function. This function returns the length of a given string excluding the terminating null character.

Program 6–43 contains a new function that will read a string of characters and ensure that each input character is a letter of the alphabet, a space or a period. The program asks users to input their name. The program will only accept input such as

```
A. Good Programmer
Mary C. Learner
Big Ed
```

It will not accept

```
4 Me
Big Ed!
Edna + George
```

Program 6–43

```c
#include <stdio.h>
#include <ctype.h>

int readstring(char *string);

main( )
{
  char name[50];

      do
       printf("Input your name => ");
      while(readstring(name));
       printf("Your name is %s",name);

}

int readstring(char *string)
{
  int index;
      gets(string);
       for(index = 0; index < strlen(string); index++)
      {
 if((isalpha(string[index])==0)&&(string[index]!=' ')&&(string[index]!='.'))
             return(1);
      }
       return(0);
}
```

Execution of the above program produces

```
Input your name => Harold C. Thompson
```

If the user inputs any other characters then the prompt will repeat:

```
Input your name => 2 Good to be True!
Input your name => Big Bad John!
Input your name => You Win
```

Program Analysis

The above program can best be analyzed by first looking at the user defined function readstring( ).

```
int readstring(char *string)
```

Note that the function is of type **int**. This means that the function itself will return a value to the calling function. Its formal argument is a character pointer. This function first uses the C **gets( )** to get the user input string.

```
gets(string);
```

Next, a C **for** loop is used to check each character of the input string. Note the use of the **strlen( )** function to test when the loop should end.

```
for(index = 0; index < strlen(string); index++)
```

This loop will continue for the length of the input string. Next, the body of the C **for** loop tests each character of the input string to see if it meets the three conditions—that it is a letter of the alphabet, a space, or a period. If it isn't one of these three conditions, then a value of 1 is returned to the calling function. If none of the characters causes the C **if** to become true, then a value of 0 is returned. Recall that a value of 0 represents a FALSE while any other value represents a TRUE.

```
   {
if((isalpha(string[index])==0)&&(string[index]!=' ')&&(string[index]!='.'));
        return(1);
   }
     return(0);
   }
```

Note the manner in which the C **if** is constructed. If a zero is returned from isalpha( ) and the character is not a space and is not a period, then and only then will the C **if** be TRUE, causing the loop to end by a **return(1)**.

The calling function **main( )** uses a C **do** loop.

```
do
 printf("Input your name => ");
while(readstring(name));
 printf("Your name is %s",name);
```

As long as the C **while** condition is TRUE (and it will be if any character of the input string is not a letter of the alphabet, a space, or a period) then the prompt Input your name => will be repeated. In this manner the program user is reminded that only certain keyboard characters will be accepted as input. Just as important, your program input is now protected, assuring that the accepted data is what your program expects.

Looking for Numbers

There is another built-in C function that will take an input string consisting of digits and convert this to a number. This is called **strtod( )** and will convert a string of digits to a double-precision value. The input string may be of the form

```
[whitespace][sign][digits.digits]
```

or in the form

```
[exponent_letter][sign][digits]
```

The term white space refers to optional blanks and tabs preceding the numbers. Digits are the decimal digits from 0 to 9, and the **exponent_letter** may be d, D, e, or E, all assumed to be to the power of ten. This function will end its conversion process when it encounters the first character in the string that does not contain any of the above. Program 6–44 illustrates its use. Note the use of the **stdlib.h** file which contains this function definition.

Program 6–44

```c
#include <stdio.h>
#include <stdlib.h>

main( )
{
 char *string;
 char **endptr;
 double value;

  printf("Give me a number => ");
  gets(string);

   value = strtod(string, &endptr);

   printf("The value is %lg.",value);

}
```

Execution of the above program could produce

```
Give me a number => 123e-3
The value is 0.123.
```

The variable **char **endptr** is a required character pointer to a pointer. It must be used in the **strtod( )** argument (where it will point to the remaining string if the function terminates because it encounters characters other than the expected numerical format).

The advantage of this approach to user input, where numerical data is first accepted as a string and then converted to a number, is that it provides you with a method of ensuring that any keyboard input from the user will be accepted by the program. Furthermore, you can then test each character of the input string to ensure that what was entered is limited to the type of data expected by your program.

Conclusion

This section presented various ways of protecting user input to your programs. Here you saw the disadvantage of the **scanf( )** function and the advantage of the **gets( )** function. You were also introduced to methods of testing each individual input character and prompting the program user for input until satisfactory data is entered. Use the information presented here in your programs. Doing so will result in a more user-friendly program which will encourage people to use it. Check your understanding of this section by trying the following section review.

6–7 Section Review

1 State one problem of using the **scanf( )** function.
2 What is a better function to use in place of the **scanf( )** for string input? Explain.
3 Explain the function of **getchar( )**.
4 What is meant by the C character classifications? Give an example.
5 State how you can allow a program user to enter numerical data as a string thus assuring that any keyboard input will not disrupt the operation of your program. State what you must test for.

6-8   Case Study

Discussion

The case study for this section traces the logic and the steps involved in the development of a string sorting program. In the previous sections of this chapter, any sorting was done with numerical values. The sorting techniques presented there are applied in this program. However, special consideration must be made because string sorting involves sorting pointers that point to other pointers. Remember that a string in

C is really treated as an array of characters and that the string variable is a pointer that points to the starting address of this array. Also recall that all strings must terminate with the null so that C string functions will know where the string ends.

The Problem

Create a C program that will sort user entered strings up to a maximum of ten strings. The sorting is to be done according to the ASCII code sequence of each character in the string. This means that the program does not require any special instructions concerning upper and lowercase letters. (Uppercase letters in the ASCII code have a lower numerical value than do the lowercase letters. This means that the uppercase "Z" will be placed ahead of the lowercase "a".)

First Step—Stating the Problem

Stating the case study in writing yields:

- Purpose of the Program: Get up to ten strings from the program user. Sort the strings alphabetically (using the string ASCII code). Display the sorted string to the user. Make provisions for the user to input fewer than 10 strings.
- Required Input: From 0 to 10 strings.
- Process on Input: Alphabetically sort the strings (using their ASCII values).
- Required Output: The sorted list of entered strings.

Developing the Algorithm

The steps in this program are the same as for most technology problems:

1. Explain the program to the user.
2. Get strings from the user.
3. Sort the strings.
4. Display the sorted output.

As with any program, there are many different approaches that could have been taken in the development of this program. The approach for this program (as with others in this text) is to develop the program code so that it is easily understood by anyone reading it. As long as program speed or memory use is not important in the development of the program, then the program should be developed in such a way as to make it easy to understand and modify. As with other case studies in this text, the program was developed using a block structure with top-down design.

Before starting the design, it is first necessary to be introduced to some of the built-in string functions used by C.

Built-in C String Functions

Table 6–2 lists the built-in C functions.

Table 6–2   Built-in C Functions `<string.h>`

Function	Meaning	Example
		`char *string1;`
		`char *string2;`

`strcat( )`	string concatenate (join strings into one) First character of `string2` overwrites null character of `string1`.	`strcat(string1, string2);`

`strchr( )`	string character Searches for the first occurrence of a character in a string and if found returns pointer to that character.	`new_string=strchr("Total:$2", $);`
		`printf("%s",new_string);`
		
		This will output: $2

`strcmp( )`	string compares Compares two strings according to the ASCII values of their characters.	`value = strcmp(string1, string2);` value = 0: strings equal. value = 1: string1 > string2. value = −1: string 1 < string 2.

`strlen( )`	string length Returns the length of the string not counting the null character.	`length = strlen(string1)`
		
		`length = Number of string characters.`

`strncmp( )`	string compare Returns an integer whose value depends upon the relationship of the first n characters of two strings.	`value = strncmp(string1,"stop",4);` value = 0; stop in first 4 characters of string 1. value = 1; "stop" > string 1. value = −1; "stop" < string1.
`strpbrk( )`	string pointer break Returns a pointer to the first occurrence of any character from `string2` in `string1`.	`bracket = strpbrk(string1, "[]");` bracket => Pointer value of first occurrence of "[" or "]" in string 1. bracket => NULL if not found.

`strrchr( )`	string return character Looks for the last occurrence of a character in a given string. This may include the null character.	`string1[] = "Total $ = $3.56";`
		`total = strrchr(string1, '$');`
		
		total => "$3.58"

`strstr( )`	string to string Locates the first occurrence of `string2` in `string1`.	`string1[]="Student number: 45-C10";`
		`number= strstr(string1, "45-C");`
		
		number => "45-C10"

Program Development

As in the development of other programs, the required programmer's block was first written. This is shown in Program 6–45.

Program 6–45

```
#include <stdio.h>

#define MAXSTRINGS 10     /* Maximum number of input strings. */
#define STRINGSIZE 81     /* Maximum string size.             */
/*****************************************************************/
/*                     String Sorter                            */
/*****************************************************************/
/*                 Developed by: This C. Student                */
/*****************************************************************/
/*                  Date: October 10, 1992                      */
/*****************************************************************/
/*       This program will take ten different input strings     */
/*   of up to 80 characters long and sort them alphabetically   */
/*       This sorting is accomplished by using the ASCII        */
/*   code for each character of the input string.  Hence this   */
/*   program does not perform a correct numerical sort for      */
/*   numerical characters. This program will accept any         */
/*   printable keyboard character.                              */
/*****************************************************************/
/*                 Function Prototypes                          */
/*--------------------------------------------------------------*/
void screen_scroll(int lines);
/*     lines = The number of lines to scroll up.                */
/*         This function scrolls the screen up 25 lines.        */
/*--------------------------------------------------------------*/
void explain_program(void);
/*         This function explains the operation of the program  */
/*   to the program user.                                       */
/*--------------------------------------------------------------*/
void sort_string(char *ptr[], int maxnumber);
/*     *ptr[] = Pointer to point to the starting address of     */
/*              the sorted pointers which in turn point to      */
/*              the starting address of each input string.      */
/*     maxnumber = The maximum number of strings inputted by    */
/*                 the program user.                            */
/*       This function sorts the input strings according to     */
/*   their ASCII values.                                        */
/*--------------------------------------------------------------*/
void display_string(char *ptr[], int maxnumber);
/*     *ptr[] = Same purpose as in previous function.           */
/*     maxnumber = Same purpose as in previous function.        */
/*        This function displays the sorted output string.      */
/*****************************************************************/

main( )
{

}    /* End of main. */

/*----------------------------------------------------------------*/
```

(continued)

Program 6-45 *(continued)*

```
void explain_program(void)              /* Explain program to user.  */
{

}    /*  End of explain_program.  */
/*-------------------------------------------------------------------*/
void sort_string(char *ptr[], int maxnumber)    /* Sort the strings.  */
{

}    /* End of sort_string.  */
/*-------------------------------------------------------------------*/
void display_string(char *ptr[], int maxnumber) /*Display sorted strings.*/
{

}   /* End of display_string. */
/*-------------------------------------------------------------------*/
void screen_scroll(int lines)      /* Scroll the screen up.  */
{

}    /*  End of screen_scroll.  */
/*-------------------------------------------------------------------*/
```

From the above program, you can see that the first step in its development is to complete the programmer's block. Then just enough information is given in the function definitions to allow the program to compile. This process is in keeping with the idea of top-down design where the most general concepts of the program are coded first and the details are filled in later.

Function Analysis

The first function void screen_scroll(int lines); is intended to act as a screen clearing function. What it will do is to cause the screen to scroll up the number of lines indicated by the value of its argument.

The function **void** explain_program(void); is your old friend, that necessary process of ensuring that the program user understands what the program is to do.

Function **void** sort_string(char *ptr[], int maxnumber); is the function that will actually do the sorting of the strings. This function will be called from **main( )** and must therefore be able to return the starting address of the sorted pointers back to **main( )**. This is the reason for a **char** pointer to an array. The **int** maxnumber is needed so that this sorting function knows how many strings the program user entered into the program.

The last function, **void** display_string(**char** *ptr[], **int** maxnumber); will also be called from **main( )** to display the sorted list of entered strings.

Final Program Phase

The final completed program is shown in Program 6–46. Note that each function has a function prototype with the necessary explanation for each function variable and purpose of the function.

Program 6–46

```
#include <stdio.h>
#include <string.h>
#define MAXSTRINGS 10    /* Maximum number of input strings. */
#define STRINGSIZE 81    /* Maximum string size.             */

/***************************************************************/
/*                    String Sorter                         */
/***************************************************************/
/*                Developed by: This C. Student             */
/***************************************************************/
/*                Date: October 10, 1992                    */
/***************************************************************/
/*      This program will take ten different input strings  */
/*  of up to 80 characters long and sort them alphabetically*/
/*        This sorting is accomplished by using the ASCII    */
/*  code for each character of the input string.  Hence this*/
/*  program does not perform a correct numerical sort for    */
/*  numerical characters.  This program will accept any      */
/*  printable keyboard character.                            */
/***************************************************************/
/*                 Function Prototypes                      */
/*---------------------------------------------------------*/
void screen_scroll(int lines);
/*    lines = The number of lines to scroll up.             */
/*        This function scrolls the screen up 25 lines.     */
/*---------------------------------------------------------*/
void explain_program(void);
/*        This function explains the operation of the program*/
/*  to the program user.                                    */
/*---------------------------------------------------------*/
void sort_string(char *ptr[], int maxnumber);
/*    *ptr[] = Pointer to point to the starting address of  */
/*             the sorted pointers which in turn point to    */
/*             the starting address of each input string.   */
/*    maxnumber = The maximum number of strings entered by  */
/*                the program user.                         */
/*        This function sorts the input strings according to */
/*  their ASCII values.                                     */
/*---------------------------------------------------------*/
```

(continued)

Program 6–46 *(continued)*

```c
void display_string(char *ptr[], int maxnumber);
/*      *ptr[] = Same purpose as in previous function.       */
/*      max_number = Same purpose as in previous function.   */
/*          This function displays the sorted output string. */
/************************************************************/


main( )
{
 char *stringptr[MAXSTRINGS]; /* Pointer for each input string.*/
 char instring[MAXSTRINGS][STRINGSIZE]; /* Array of 10 strings of 80 characters
                                         max. */
 int index;                 /* Array index.                    */
 int maxnumber;             /* Number of strings entered by user. */

      explain_program( );   /* Explain program to user. */

  printf("Give me up to 10 strings and I'll sort them.\n");
  printf("Just hit enter when you want to quit.\n");
  screen_scroll(10);  /* Scroll up 10 lines. */
  index = 0;

      do
      {
        printf("Word %d => ",index);
        gets(instring[index]);
        stringptr[index] = instring[index];
        index++;
      }
  while((index < MAXSTRINGS)&&(strcmp(instring[index-1],"") != 0));
  maxnumber = index-1;

  sort_string(stringptr, maxnumber);       /* Sort the strings. */
  display_string(stringptr, maxnumber);  /* Display the sorted strings.*/

}    /* End of main. */

/*----------------------------------------------------------------*/

  void explain_program(void)              /* Explain program to user. */
  {
   char *string;            /* String to hold program explanation. */

  string = "\n\n\n"
          " This program will sort strings entered by you\n"
          "for a maximum of up to 10 strings.\n"
          "\n"
          " Simply enter the desired string when prompted to\n"
          "do so.  Press the return key at any prompt to begin\n"
          "the sorting process if you input less then 10 strings.\n"
          "\n"
```

(continued)

Program 6-46 *(continued)*

```
                    "  The sorting is done according to the ASCII code and\n"
                    "will not perform a correct numerical sort.\n";

        printf("%s",string);     /* Print the above string. */

        string = "\n\n\n\n"
                "                    -Press RETURN/ENTER to continue- ";

        printf("%s",string);     /* Prompt user to continue. */

        getchar( );              /* Hold program for user. */

        screen_scroll(25);       /* Scrolls the screen up 25 lines. */

    }    /*  End of explain_program.  */
/*------------------------------------------------------------------*/
  void sort_string(char *ptr[], int maxnumber)  /* Sort the strings. */
  {
   int index;
   int flag;
   char *temp;

     do
      {
       flag = 0;
        for(index = 0; index <= maxnumber; index++)
         {
         if((strcmp(ptr[index],ptr[index+1]) > 0)&&(index<maxnumber))
          {
          temp = ptr[index];
          ptr[index] = ptr[index+1];
          ptr[index+1] = temp;
           flag = 1;
          }   /* end of if */
         }   /* end of for */
      }   /* end of do */
    while(flag);

   }    /* End of sort_string. */

/*------------------------------------------------------------------*/
  void display_string(char *ptr[], int maxnumber) /*Display sorted strings.*/
  {
   int index;

   index = 0;

   printf("\n\nThe sorted strings are:\n");
```

(continued)

Program 6–46 *(continued)*

```
        do
        {
        printf("String[%d] => %s\n",index, ptr[index]);
        index++;
        }
        while(index <= maxnumber);

   }  /* End of display_string. */
/*---------------------------------------------------------------*/

   void screen_scroll(int lines)      /* Scroll the screen up.  */
   {
   int counter;                       /* Counts the number of lines. */

      for(counter = 0; counter <= lines; counter++)
        printf("\n");

   }  /* End of screen_scroll.  */
/*---------------------------------------------------------------*/
```

The details of the above program are reserved for the self-test section of this chapter.

Conclusion

This case study presented the development of a string sorting program. Here you saw the development of this program using top-down design and resulting in a program using block structure. The program uses a programmer's block where documentation is given concerning the program. Function prototypes are used along with the necessary explanation of each function variable along with the purpose of each function. Check your understanding of this section by trying the following section review.

6–8 Section Review
1 State the purpose of the program used in this case study.
2 Explain how the strings are sorted in this program.
3 May the program user input fewer than 10 strings for this program? Explain how this is done.
4 How has top-down design been implemented in this program?

Interactive Exercises

DIRECTIONS

These exercises require that you have access to a computer and software that supports C. They are provided here to give you valuable experience and immediate feedback on what the concepts and commands introduced in this chapter will do. They are also fun.

Exercises

1   Program 6–47 will display uninitialized variables. What kind of output will you get?

Program 6–47

```
#include <stdio.h>

main( )
{
 char array[3];
 int index;

    for(index = 0; index < 3; index++)
      printf("%c\n",array[index]);

}
```

2   Look at the array indexes of Program 6–48. Do you think this program will compile and execute? What happens when you try it?

Program 6–48

```
#include <stdio.h>

#define MAXVALUE 1000

main( )
{
 char array[MAXVALUE];
 int index;

    for(index = 0; index > -3; index--)
      printf("%c\n",array[index]);

}
```

3 Does Program 6–49 execute? How many dimensions does it have? What is the maximum number of dimensions you can assign to an array for your system? How did you determine this?

Program 6–49

```
#include <stdio.h>

main( )
{
  char array[1][1][1][1][1][1][1][1][1][1][1][1][1][1];
  int index;

}
```

4 Predict what Program 6–50 will output. Was your prediction correct?

Program 6–50

```
#include <stdio.h>
#include <string.h>

  char *string1 = "Resistor";
  char *string2 = "Capacitor";

main( )
{
  char *newstring;

    newstring = strchr("Resistor", 'i');
    printf("%s",newstring);

}
```

Self-Test

DIRECTIONS

Answer the following questions by referring to Program 6–46 in the case study section of this chapter:

1 State the purpose of the **char** variable ***string** in function **explain_program( )**.
2 Explain the meaning of the statement **strcmp(instring[index-1],"")**

3 Where is the above statement used?
4 Explain the operation of the function **screen_scroll( )**.
5 State how the screen could be cleared using the **screen_scroll( )** function.
6 What is the purpose of the variable **flag** in function **sort_string( )**?
7 State the purpose of the **getchar( )** function used in **explain_program**.
8 Does the program really sort strings? Explain.
9 Explain the purpose of the statement:

    if((strcmp(ptr[index],ptr[index+1]) > 0)&&(index<maxnumber);

10. State the purpose of the variable **int temp** in the function **string_sort( )**.

End-of-Chapter Problems

General Concepts

Section 6–1

1 What is the index of the first character in a C string?
2 State how many array elements are required in a C string consisting of 8 characters. Explain.
3 Explain what constitutes a C string.
4 Give a method of declaring a string variable in C.

Section 6–2

5 What is meant by the word array?
6 State how you declare an array of 10 elements in a C program.
7 For the array in problem 6 above, what is the index of the first element? The last?
8 State the difference between a global and a local array.

Section 6–3

9   Show how you would declare the array variable **array** as a two-dimensional array.
10 Explain how an array may be initialized when it is declared as a local variable.
11 State the rule concerning the declaration of a formal array parameter containing more than one dimension.

Section 6–4

12 Explain the basic idea behind the applications of arrays in a C program.
13 State how a program could be developed that would display a series of entered values in the opposite order from which they were entered.
14 Explain how a program in C could be developed so that the minimum value of an input list could be extracted.
15 What is an easy method to use to get program user input into an array?

Section 6–5

16 Explain the processes of a bubble sort.
17 State why the term bubble sort is used.
18 How many times must a sorting program go through a list to be sorted?

Section 6–6

19 What is meant by a ragged array?

20 Explain the meaning of a rectangular array.

21 When passing arrays between functions, what is passed in the parameter?

22 State what is actually used with the **%s** operator in a **printf( )** function to print a given string.

Section 6–7

23 Explain the purpose of the **getchar( )** function.

24 State one of the problems of the **scanf( )** function.

25 What is a good function to use for entering strings? Explain.

26 State the purpose of the built-in C character classification functions.

Program Design

In developing the following C programs, use the method described in the case studies chapter sections. For each program you are assigned, document your design process and hand it in with your program. This should include the design outline, process on the input, and required output as well as the program algorithm. Be sure to include all of the documentation in your final program. This should consist of, but not be limited to, the programmer's block, function prototypes, and a description of each function as well as any formal parameters you may use.

Electronics Technology

27 The circuit in Figure 6–14 is a parallel series circuit. Develop a C program that will compute the total resistance of any branch selected by the program user. The total resistance of any one branch is:

$$R_T = R_1 + R_2 + R_3$$

Where

R_T = Total branch resistance in ohms.

R_1, R_2, R_3 = Value of each resistor in ohms.

28 Modify the above program so that the total resistance of any combination of branches may be found by the program user. The total resistance of any parallel branches is found

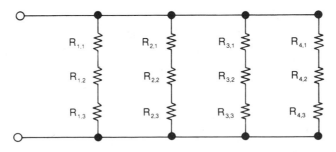

Figure 6–14   Circuit for Problems 27, 28, and 29

by first determining the total resistance of that branch, and then using the parallel resistance formula for finding the total resistance. The parallel resistance formula is

$$R_T = 1/(1/R_{T1} + 1/R_{T2} ... + 1/R_{TN})$$

Where

R_T = Total resistance in ohms.
$R_{T1}, R_{T2}, ... R_{TN}$ = Total branch resistance in ohms.

29   Expand the above program so that the program user can enter the value of each resistor and the resistors will be displayed in numerical order.

30   Develop a C program that computes the power dissipation of each resistor in a series circuit. The program user may select how many resistors there will be in the circuit. The program will sort the resistors by their power dissipation and display their value and subscript number (the subscript number represents the order they appear in the circuit). The program user enters the value of the voltage source in volts. Power dissipation in a resistor may be determined by

$$P = I^2R$$

Where

P = Power dissipation in watts.
I = Current in the resistor in amps.
R = Value of resistor in ohms.

The current in each resistor in a series circuit is the same. It may be determined from

$$I = V_S/R_T$$

Where

I = Circuit current in amps.
V_S = Source voltage in volts.
R_T = Total circuit resistance in ohms.

Business Applications

31   Develop a C program that will display the amount of money in any safety deposit box that is contained in a wall which has 10 rows by 8 columns of these boxes.

Computer Science

32   Create a C program that will take a 4 × 5 array and multiply the first column by any other column and display the sum of the products.

Drafting Technology

33   Create a C program that will give the program user the color of an area of the grid system shown in Figure 6–15. The program user must enter the row and column number.

Agriculture Technology

34   Develop a C program that will rate dairy cows according to their age, milk producing ability (quarts per day), and cost of feed per day. The program user must be able to access the name of any dairy cow by milk producing ability, age, or cost of feed per day.

Health Technology

35   A pharmacist needs a C program that will allow her to find any prescription by (1) name of medication, (2) name of patient, or (3) cost of medication. The program user

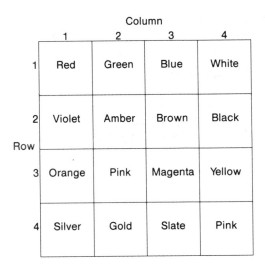

Figure 6-15   Grid System for Problem 33

must be able to locate all of this information by entering any one of the three items above.

Manufacturing Technology

36   A robotics company requires a C program that will produce the following information about factory workers on an assembly line: (1) name of worker, (2) hourly wages, (3) hours worked per week, and (4) total weekly pay. The program user must be able to locate all of the above information by entering any one of the above items.

Business Applications

37   Modify the program of problem 31 so the program user may enter the amount of money in any vault. Have the program display the amounts in each vault by displaying a matrix on the screen replicating the structure of the vault (use the gotoxy(X,Y) or similar command).

Computer Science

38   Change the program in problem 32 so the program user may select any two columns to be multiplied.

Drafting Technology

39   Modify the program in problem 33 so that the coordinates of all the areas of the same color will be displayed by the program user. Make the program so the program user may input the color of each area. The program must be able to sort the colors in order according to their frequency on the matrix.

Agriculture Technology

40   Modify the program in problem 34 so that the program will display the names of cows sorted according to any of the program parameters selected by the program user.

Health Technology

41 Change the program in problem 35 so the program user may have all the medications for any one patient sorted alphabetically and the total price displayed.

Manufacturing Technology

42 Modify the program in problem 36 so the program user may find the total wages paid for the week to any factory worker. The program must allow the program user to locate any factory worker by weekly income earnings, by those who have made less than a specified amount, or by those who have made more than a specified amount.

7   Structuring and Saving Data

Objectives

This chapter provides you the opportunity to learn:

1   The meaning of enumerated data types in C.
2   How to define your own data types.
3   The meaning of data structures.
4   Why data structures are useful.
5   Arranging data structures as arrays.
6   How to develop complex data structures.
7   How to use C unions.
8   Methods of saving user data to the disk.
9   Methods of retrieving data from the disk.
10   The use of file redirection, piping, and filtering.
11   How to implement command line arguments.

Key Terms

Enumerated
Type Definition
Structure Member
Structure
Member of Operator
Template
Structure Tag

Member Declaration List
union
Structure Arrays
DOS
File Pointer
Character Stream
Text File

Text Mode	Filtering
Binary Mode	Standard I/O
Binary File	System I/O
Buffer	Sequential Access Files
Buffered Stream	Random Access Files
Flush	File Pointer
Standard Files	**offset**
File Redirection	Command Line Arguments
Piping	

Outline

7-1　Enumerating Types

7-2　Naming Your Own Data Types

7-3　Introduction to Structures

7-4　More Structure Details

7-5　The Union and Structure Arrays

7-6　Ways of Representing Structures

7-7　Disk Input and Output

7-8　More Disk I/O

7-9　Program Debugging and Implementation

7-10 Case Study

Introduction

This chapter is about how to use C to handle data: how data can be structured, how to save it on your disk, and how to get it back again.

Up to this point, you have been primarily concerned with working with only one data type at a time. For example, when you worked with arrays, the array was of a single type (such as **int** or **char**). A single data type has a simple data structure. In complex applications, you may use more than one data type at a time. This chapter will show you how to do that.

In previous chapters, any data entered by the program user was lost once the program was terminated. This chapter will introduce you to ways to allow the program user to save data to the disk so that it may be used again at another time.

This is a very useful chapter. Essentially this chapter will show you how to develop C programs that will handle more complex technology problems.

7-1　Enumerating Types

Discussion

This section presents another way of expressing data in C. Up to this point you have had the **#define** available. Now you will have the opportunity to learn another method for expressing data that will give your programs even greater readability. The material you learn here will help set the stage for the remaining sections of this chapter.

Expressing Data

There is another way of expressing data in C called the **enum** (for **enumerated**). What this does is to hold one integer value from a fixed set of identified integer constants.

This has this form:

```
enum tag { enumeration-list }
```

The above enumeration declaration presents the name of an enumeration variable and then defines the names within the enumeration list. The declaration begins with the keyword **enum**. The resulting **enum** variable can then be used anywhere an **int** type is used. An example is illustrated in Program 7-1.

Program 7-1

```
#include <stdio.h>
  enum color_code {black, brown, red, orange, yellow,}

main( )
{
  enum color_code color;
  char value;

      printf("Input an integer from 0 to 4 => ");
      value = getchar( );

      switch (value)
         {
            case '0' : color = black;
                       break;
            case '1' : color = brown;
                       break;
            case '2' : color = red;
                       break;
            case '3' : color = orange;
                       break;
            case '4' : color = yellow;
                       break;

         }  /* End of switch */
```

(continued)

Program 7-1 *(continued)*

```
    switch (color)
        {
        case black : printf("Color is black.");
                     break;
        case brown : printf("Color is brown.");
                     break;
         case red : printf("Color is red.");
                     break;
        case orange : printf("Color is orange.");
                     break;
        case yellow : printf("Color is yellow.");
                     break;

        }  /* End of switch */

}
```

The above program asks the program user to input an integer from 0 to 4. The program will then give the program user the equivalent resistor color code for that number. As an example

```
Input an integer from 0 to 4 => 3
Color is orange.
```

The important point of this program is that it uses the **enum** data type to make the source code more readable.

Program Analysis

First, an enumerated data type is declared outside of **main( )**:

```
enum color_code {black, brown, red, orange, yellow,}
```

What this does is declare **color_code** to be an enumerated type. The list within the braces shows the constant names that are valid values of the **enum color_code** variable.

Next, inside **main( )** is the declaration of an enumerated variable of type **color_code**:

```
main( )
{
 enum color_code color;
```

This means that the variable **color** may have any of these values: black, brown, red, orange, or yellow!

Look at what the first C **switch** does. From the user input

```
switch (value)
    {
      case '0' : color = black;
                 break;
      case '1' : color = brown;
                 break;
      case '2' : color = red;
                 break;
      case '3' : color = orange;
                 break;
      case '4' : color = yellow;
                 break;
    }   /* End of switch */
```

it assigns color value to the variable color. Hence, if the user selects 3 for an input, the variable color will be assigned the value of orange. This data is then used in the next C **switch**:

```
switch (color)
    {
      case black : printf("Color is black.");
                   break;
      case brown : printf("Color is brown.");
                   break;
        case red : printf("Color is red.");
                   break;
      case orange : printf("Color is orange.");
                   break;
      case yellow : printf("Color is yellow.");
                   break;
    }   /* End of switch */
```

This results in a very descriptive C **case**. You should note that **enum** does not introduce a new basic data type. Variables of the **enum** type are treated as if they are of type **int**. What it does is improve the readability of your programs.

Another Form

Another method of displaying an **enum** declaration is in the vertical form as follows:

```
enum color_code
    {
     black,
     brown,
     red,
     orange,
     yellow
    }
```

Enumeration Example

To demonstrate exactly what is going on in an **enum** data type, look at the following program:

Program 7–2

```
#include <stdio.h>

enum color_code {black, brown, red, orange, yellow}

main( )
{
 enum color_code color;

     for(color = black; color <= yellow; color++)

     printf("Digital value of enum type color => %d\n",color);

}
```

Execution of the above program yields

```
Digital value of enum type color = 0
Digital value of enum type color = 1
Digital value of enum type color = 2
Digital value of enum type color = 3
Digital value of enum type color = 4
```

Note that the **enum** type is used in the C **for** loop just as if it were of type **int**.

Assigning **enum** Values

The C **enum** type may also be assigned values—as long as they are integer values. This is demonstrated in the following program. The program solves for the total current for a given voltage source when the Thevenin* equivalent voltage and resistance are known. In this program, the Thevenin equivalent voltage is 12 volts and the Thevenin equivalent resistance is 150 ohms. Note how these values are set by the C **enum** type.

---

*The Thevenin equivalent form of any resistive circuit consists of an equivalent voltage source and an equivalent resistance. These values depend upon the values of the original circuit. This method is used to simplify resistive circuits to a single voltage source and single resistor.

Program 7-3

```
#include <stdio.h>

enum thevenin {source = 12, resistance = 150}

main( )
{
 float load;
 float current;

    printf("Input value of load resistor => ");
    scanf("%f",&load);

    current = source/(resistance + load);

    printf("The circuit current is %f amps.",current);

}
```

In the above program, the **enum** data type thevenin is assigned value of source (which is set equal to 12) and resistance (set equal to 150). These constant names are then used inside the program to do the necessary calculations. Again, the purpose here is to make the program more readable and allow easy changes to the constant assignments used in the program.

Conclusion

This section introduced the C **enum**. Here you discovered another method of using data in your C program to improve its readability. You also saw that the use of the C **enum** is compatible with the C **int**. Check your understanding of this section by trying the following section review.

7-1 Section Review

1   In C, what does the keyword **enum** mean?
2   State what an **enum** data type does in C.
3   For the following code, what is the integer value of **first**?

   enum numbers {first, second, third}

4   What is the main purpose of using a C **enum**?
5   Can a C **enum** constant be set to a given integer value? Give an example.

7-2   Naming Your Own Data Types

Discussion

In the last section you were introduced to a method of making your program easier to read. This section elaborates on that theme. However, unlike the previous section, this section will show you how to name your own data types.

Inside typedef

The use of the **typedef** (type definition) in C allows you to define names for a C data type. A **typedef** declaration is similar to a variable declaration except that the keyword **typedef** is used. The form is

```
typedef type-specifier declarator
```

For example, you could have:

```
typedef char LETTER;
```

This now declares LETTER as a synonym for **char**. This means that LETTER can now be used as a variable declaration such as

```
LETTER character;
```

instead of

```
char character;
```

As an example, suppose you have a program that is part of an automated testing system. In this program the user inputs the status of an indicator light and the program gives a response depending on the given status. The program could be written as shown in Program 7–4.

Program 7–4

```
#include <stdio.h>

void test_it(void);
int check_lights(char condition);

main( )
{

    test_it( );

}
```

(continued)

Program 7-4 *(continued)*

```c
void test_it(void)
{
 char input;

    printf("\n\n1] Red 2] Green 3] Off \n\n");
    printf("Select light conditions by number => ");
    input = getchar( );

      check_lights(input);

 }   /* End of test_it( )  */

int check_lights(char condition)
{

    switch (condition)
         {
             case '1' : printf("Check system pressure.");
                     break;
             case '2' : printf("System OK.");
                     break;
             case '3' : printf("Check system fuse.");
                     break;
         }     /* End of switch  */

 }  /* End of check_lights( ) */
```

Execution of the above program produces

```
1] Red  2] Green  3] Off
Select light conditions by number => 2
System OK.
```

However, the program could be made more descriptive by defining a data type called light_status. The same program using this new type definition is illustrated by Program 7-5.

Program 7-5

```c
#include <stdio.h>

typedef enum light_status {Red, Green, Off};
```

(continued)

Program 7–5 *(continued)*

```c
void test_it(void);
int check_lights(light_status condition);

 main( )
 {

     test_it( );

 }

 void test_it(void)
 {
  char input;
  light_status reading;

     printf("\n\n1] Red 2] Green 3] Off \n\n");
     printf("Select light conditions by number => ");
     input = getchar( );

       switch (input)
           {
             case '1' : reading = Red;
                        break;
             case '2' : reading = Green;
                        break;
             case '3' : reading = Off;
                        break;
           }   /* End of switch */

       check_lights(reading);

 }   /* End of test_it( )  */

 int check_lights(light_status condition)
 {
       switch (condition)
           {
             case Red : printf("Check system pressure.");
                        break;
           case Green : printf("System OK.");
                        break;
             case Off : printf("Check system fuse.");
                        break;
           }   /* End of switch */

 }  /* End of check_lights( ) */
```

The output of Program 7-5 is the same as that of Program 7-4. The difference is that a new C type name is created for an existing data type. This allows new C identifiers to be defined as this data type. This can be done in the body of functions as well as in the function parameters. It's important to note that the **typedef** declaration does not create types. It simply creates synonyms for existing types.

Program Analysis

The C keyword **typedef** is used to indicate that a new definition is to be used for a data type.

```
typedef enum light_status {Red, Green, Off};
```

In this case, the data type is an **enum** type called `light_status`. It consists of the enumerated constants: `Red`, `Green` and `Off`. Now other variables may be defined in terms of this new data type. Look at the formal parameter of the function prototype:

```
int check_lights(light_status condition);
```

It declares a new variable `condition` as a type `light_status`. It is used again in the definition of function `test_it`.

```
void test_it(void)
{
 char input;
 light_status reading;
```

This means that the variable `reading` is the data type `light_status` and can assume the values `Red`, `Green`, and `Off`.

After the variable `reading` is assigned one of these three values by the C **switch**, it is then passed to the called function `check_lights( )`.

```
switch (input)
    {
    case '1' : reading = Red;
              break;
    case '2' : reading = Green;
              break;
    case '3' : reading = Off;
              break;
    }  /* End of switch  */

check_lights(reading);
```

Next it is used in the called function as part of the C **switch** to make the code more readable:

```
int check_lights(light_status condition)
{
```

```
switch (condition)
    {
      case Red : printf("Check system pressure.");
                 break;
    case Green : printf("System OK.");
                 break;
     case Off : printf("Check system fuse.");
                 break;
    }    /* End of switch */

}  /* End of check_lights( ) */
```

Other Applications

Some other applications of the C **typedef** are:

typedef char *STRING;

This allows you to now use the new word STRING as a pointer to **char**

STRING inputname;

Here, inputname is defined as a type STRING which is really a pointer to **char** (**char** *).

You will see even more powerful applications of the C **typedef** in the next sections of this chapter.

Conclusion

This section introduced you to a method naming data types using C. Here you saw how to implement this in a program where you could create new identifiers in terms of this new data type. Check your understanding of this section by trying the following section review.

7-2 Section Review

1　State what the C **typedef** does.
2　For the following code, what is the resulting new name of the data type?

   typedef char name[20];

3　What is the main purpose of using the C **typedef**?

7-3　Introduction to Structures

Discussion

Up to now, you have been using arrays and pointers to create items of the same data type. You have never mixed a type **char** and a type **int** within a single data type. Understandably, you may not have even considered the possibility. However, most of the everyday "keeping track of things" requires the use of more than one data type. Consider your checking account. Each check you write must have your name on it (a type **char**), the amount of the check (a type **float**) and the check number (a type **int**). All of this information (plus a lot more) is contained on this one item called a check. Since this is a very natural way of arranging information, C has provided a method for you to accomplish the same thing. Doing this makes your program more readable and makes it easier to handle complex data consisting of many different types that are logically related.

The C Structure

A structure declaration names a structure variable and then states a sequence of variable names—called **structure members**—which may have different types. The basic form is

```
struct
  {
    type    member_identifierl;
    type    member_identifier2;
      .
      .
      .
    type    member_identifierN;
  } structure_identifier;
```

Note that structure declarations begin with the keyword **struct**.

As an example of the use of a C structure, consider a box of parts as shown in Figure 7–1.

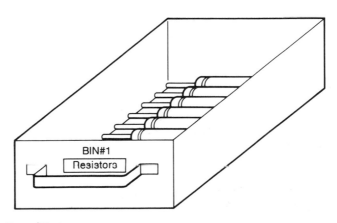

Figure 7–1 Box of Resistors

Assume that this box of parts consists of one type of resistor. Assume for the moment that there are three things you wish to keep track of concerning these resistors: the name of the manufacturer, the quantity of resistors in the box, and the price of each resistor. You could develop a C program that would easily do this without using the concept of a structure, but to keep things simple for now, the C **structure** will be used to keep this inventory. Program 7–6 illustrates the construction of a C structure.

Program 7–6

```
#include <stdio.h>

main( )
{

 struct
  {
   char    manufacturer[20];    /* Resistor manufacturer.  */
   int     quantity;            /* Number of resistors left. */
   float   price_each;          /* Cost of each resistor.  */

  } resistors; /* Structure variable.  */
}
```

Figure 7–2 illustrates the key points for the structuring of a C structure.

Note that the structure in Program 7–6 consists of a collection of data elements, called members, which may consist of the same type or different types that are logically related. The data types are **char** for the variable **manufacturer**, **int** for the variable **quantity**, and **float** for the variable **price_each**. The general form is

```
struct
 {
    type      member_identifier1;
    type      member_identifier2;
       .
       .
       .
    type      member_identifierN;
 }  structure_identifier;
```

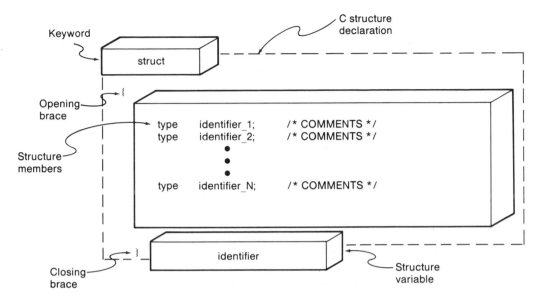

Figure 7-2 Key Points of a C Structure

Where
> **struct** = Keyword indicating that a structure follows.
> { = Necessary opening brace to indicate that a list
> structure elements is to follow.
> **type** = The C type for each element.
> member_identifier = The variable identifier for the structure member.
> { = Necessary closing brace.
> structure_identifier = Defines the structure variable.

Putting Data into a Structure

Now that you've seen what a C structure looks like, you need to know how to get values into the structure members. Program 7-7 shows you how this is done. The program gets information from the program user concerning the manufacturer of the resistors, the quantity of resistors in the bin, and the unit price of the resistors. It then calculates the total value of all the resistors, displaying this along with the information put in by the program user.

The method of getting data into and out of a structure member is by using the **member of operator**. The member of operator specifies the name of the structure member and the structure of which it is a member. For example, in Program 7-7

```
resistors.manufacture
```

represents the variable **manufacture** that is a member of the structure **resistors**. Note that the member of operator is represented by the period (.).

Program 7-7

```c
#include <stdio.h>

main( )
 {

 struct
  {
   char    manufacturer[20];    /* Resistor manufacturer.  */
   int     quantity;            /* Number of resistors left. */
   float   price_each;          /* Cost of each resistor.  */

  } resistors; /* Structure variable. */

 float total_value;     /* Total value of parts. */

  /* Get name of manufacturer: */

     printf("Name of manufacturer => ");
     gets(resistors.manufacture);

  /* Get number of parts left: */

     printf("Number of parts left => ");
     scanf("%d",&resistors.quantity);

  /* Get cost of each part: */

     printf("Cost of each part => ");
     scanf("%f",&resistors.price_each);

  /* Calculate total value: */

     total_value = resistors.quantity * resistors.price_each;

   /* Display results: */

     printf("\n\n");
     printf("Item:         Resistors\n\n");
     printf("Manufacturer: %s\n",resistors.manufacture);
     printf("Cost each:    $%f\n",resistors.price_each);
     printf("Quantity:     %d\n",resistors.quantity);
     printf("Total value:  $%f\n",total_value);

 }
```

Assuming the user inputs the following, execution of the above program results in

```
Name of manufacturer => Ohmite
Number of parts left => 10
Cost of each part => 0.05

Item:          Resistors

Manufacturer:  Ohmite
Cost each:     $0.050000
Quantity:      10.000000
Total value:   $0.500000
```

Program Analysis

The program starts with the same C structure as before and also declares another variable, total_value.

```
main( )
 {

 struct
  {
    char   manufacturer[20];    /* Resistor manufacturer.   */
    int    quantity;            /* Number of resistors left. */
    float  price_each;          /* Cost of each resistor.    */

  } resistors; /* Structure variable. */

   float total_value;    /* Total value of parts. */
```

The variable total_value will be used to hold the value that represents the total value of all the resistors in the parts bin.

Next the program gets the name of the resistor manufacturer from the program user.

```
/* Get name of manufacturer: */

   printf("Name of manufacturer => ");
   gets(resistors.manufacture);
```

Note how this is done. The name of the resistor manufacturer must be placed in the member char manufacturer[20];. Since the variable manufacturer[20] is contained in the structure referred to by the variable resistors, this must somehow be shown in the program. This is done by the member of operator ".".

```
resistors.manufacture
```

The member of operator specifies the name of the structure member and the structure of which it is a member. This C member of operator is used again to get the data for the remaining two structure member variables:

```
/* Get number of parts left: */

    printf("Number of parts left => ");
    scanf("%d",&resistors.quantity);

 /* Get cost of each part: */

    printf("Cost of each part => ");
    scanf("%f",&resistors.price_each);
```

Note that this time the address of **&** operator is needed because of the use of the **scanf( )** function. But the member of operator is still the same:

```
structure_name.member_name
```

Next, a calculation is performed using the member of operator:

```
/* Calculate total value: */

    total_value = resistors.quantity * resistors.price_each;
```

Again, the name of the structure and the name of the member are used to identify the structure variables. The results are then displayed. Note that the variable `total_value` does not require a member of operator because it is not a member of any structure.

```
/* Display results: */
  printf("\n\n");
  printf("Item:          Resistors\n\n");
  printf("Manufacturer:  %s\n",resistors.manufacture);
  printf("Cost each:     $%f\n",resistors.price_each);
  printf("Quantity:      %d\n",resistors.quantity);
  printf("Total value:   $%f\n",total_value);
```

Conclusion

This section introduced you to the concept of a C structure. In the next section you will see other more powerful ways of applying a C structure. For now, test your understanding of this section by trying the following section review.

7–3 Section Review

1   State how the information on a check from a checking account could be treated as a structure.
2   What is a C structure?
3   Describe the structure presented in this section.

4   State how a structure member variable is programmed for getting data and displaying data.
5   Give an example of problem 4 above.

7-4   More Structure Details

Discussion

In the last section you were introduced to the concept and form of a C structure. In this section, you will see different ways of letting your C program know that you are constructing a structure. You will also discover how to name function types, declare functions as structures, and pass structures between functions.

The Structure Tag

In the last section, the structure for an inventory of resistors of the same type contained in a storage box was demonstrated. Take a closer look at where the structure was defined:

```
main( )
{

struct
  {
    char    manufacturer[20];    /* Resistor manufacturer.  */
    int     quantity;            /* Number of resistors left. */
    float   price_each;          /* Cost of each resistor.  */

  } resistors; /* Structure variable. */
```

It was defined inside **main( )** and as a consequence will only be known to **main( )**. Another method of developing a structure is to define the structure before **main( )**. This method creates a global variable.

C has a method of announcing the construction of a structure that will serve as a **template** that can then be used by any function within the program. This is illustrated in Program 7–8. The output of this program is exactly the same as Program 7–7. The difference is the way the C structure is placed in the program.

Program 7-8

```
#include <stdio.h>

struct parts_record    /* Structure tag. */
  {
    char    manufacturer[20];    /* Resistor manufacturer.  */
    int     quantity;            /* Number of resistors left. */
```

(continued)

Program 7-8 *(continued)*

```
      float   price_each;          /* Cost of each resistor.  */

      };

  main( )
   {
     struct parts_record resistors; /* Structure variable. */
     float total_value;     /* Total value of parts. */

     /* Get name of manufacturer: */

     /* Get number of parts left: */

      /* Get cost of each part: */

      /* Calculate total value: */

      /* Display results: */

      }
```

Program Analysis

The program declares a structure type called `parts_record`. There is no variable identifier; `parts_record` is called a **structure tag**. The structure tag is an identifier that names the structure type defined in the **member declaration list**. Note that this was done after the keyword **struct**. It is called `parts_record`.

```
struct parts_record     /* Structure tag. */
 {
   char    manufacturer[20];   /* Resistor manufacturer.  */
   int     quantity;           /* Number of resistors left. */
   float   price_each;         /* Cost of each resistor.  */
   };
```

Now any function within the program can use this structure by making reference to the structure tag. This was done in **main( )**.

```
main( )
 {
   struct parts_record resistors; /* Structure variable. */
```

Note the format of the above declaration. It uses the keyword **struct** and the tag identifier `parts_record` which defines the variable `resistors` as the structure tagged `parts_record`. The general form of the declaration when a structure tag is used is

```
struct tag variable_identifier;
```

The same method is used to access the individual members of the structure as before, using the member of operator ".". For example, `resistors.quantity` identifies the structure member `quantity`. Using the tag method and declaring the structure outside of **main** sets a template that may now be used by any function within the program.

Naming a Structure

Another method of identifying a structure is to use the C **typedef**. This can be accomplished as shown in Program 7-9.

Program 7-9

```
#include <stdio.h>

typedef struct
  {
  char    manufacturer[20];    /* Resistor manufacturer.   */
  int     quantity;            /* Number of resistors left. */
  float   price_each;          /* Cost of each resistor.   */

  } parts_record;    /* Name for this structure. */

main( )
  {
  parts_record resistors; /* Structure variable. */
  float total_value;      /* Total value of parts. */

  /* Get name of manufacturer: */

  /* Get number of parts left: */

   /* Get cost of each part: */

   /* Calculate total value: */

   /* Display results: */

  }
```

The above program does exactly the same as the previous one. The difference is in how the C structure is declared. This time, the C **typedef** is used to name the structure type `parts_record`.

```
typedef struct
  {
    char    manufacturer[20];   /* Resistor manufacturer.  */
    int     quantity;           /* Number of resistors left. */
    float   price_each;         /* Cost of each resistor.   */

  } parts_record;     /* Name for this structure. */
```

Note that a structure tag is not used. Instead the structure variable `parts_record` is now being named as a data type by the keyword **typedef**. The advantage of doing this is that (as before) you can use the defined structure in any function. However, you only need to declare local structure variables in terms of this new type definition. This is done inside **main( )** to keep the record variable `resistors` local.

```
main( )
  {
    parts_record resistors; /* Structure variable. */
```

Note that the structure variable `resistors` is used as its data type `parts_record`. Again, you still access each member of the structure by the member of operator; for example, `resistors.quantity` identifies the structure member `quantity`.

The various ways of declaring structures are illustrated in Figure 7–3.

Structure Pointers

You can declare a structure variable to be of type `pointer`. This is illustrated in the following program. Notice that the C **typedef** is used to define the structure type. The program does exactly the same thing as before; however it is now doing it by using a structure pointer.

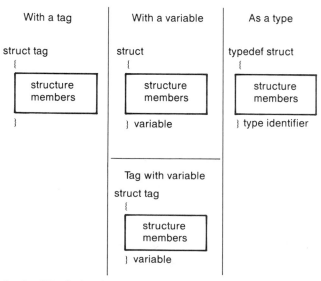

Figure 7–3   Methods of Declaring Structures

Program 7-10

```c
#include <stdio.h>

typedef  struct
  {
  char    manufacturer[20];    /* Resistor manufacturer.   */
  int     quantity;            /* Number of resistors left. */
  float   price_each;          /* Cost of each resistor.   */

  } parts_record;    /* Name for this structure. */

main( )
  {
  parts_record *rcd_ptr; /* Structure pointer. */
  float total_value;        /* Total value of parts. */

    /* Get name of manufacturer: */

      printf("Name of manufacturer => ");
      gets(rcd_ptr -> manufacturer);

    /* Get number of parts left: */

      printf("Number of parts left => ");
      scanf("%d",&rcd_ptr -> quantity);

    /* Get cost of each part: */

      printf("Cost of each part => ");
      scanf("%f",&rcd_ptr -> price_each);

    /* Calculate total value: */

      total_value = rcd_ptr -> quantity * rcd_ptr -> price_each;

     /* Display results: */

      printf("\n\n");
      printf("Item:            Resistors\n\n");
      printf("Manufacturer:    %s\n",rcd_ptr -> manufacture);
      printf("Cost each:       $%f\n",rcd_ptr -> price_each);
      printf("Quantity:        %d\n",rcd_ptr -> quantity);
      printf("Total value:     $%f\n",total_value);

  }
```

Note that the structure pointer is declared the same way a pointer is normally declared:

```
main( )
  {
    parts_record *rcd_ptr; /* Structure pointer. */
```

However, rcd_ptr is now a pointer to a structure and as such, when an individual member of the structure is to be accessed, a special C symbol —> must be used:

```
/* Get name of manufacturer: */

    printf("Name of manufacturer => ");
    gets(rcd_ptr —> manufacturer);
```

The symbol —> refers to the member of a structure pointed to by a pointer. Effectively, this replaces the member of operator "." used in previous examples.

Since a structure may be defined as a pointer, it may also use the address of & operator as with any other variable. This means that the address of a structure may be assigned to a structure pointer.

A Function as a Structure

You may declare a C function to be of a structure type. This is demonstrated in the next program. Again, the program does exactly the same thing as previous ones; the difference is that now **main( )** is serving as the function that simply calls other functions. It is the other functions that do all the work. Note in the program that the function get_input( ) is declared as a type defined by **parts_record**. Thus, this function itself now has a record structure! Note also that a structure variable is used as a function argument to pass the structure data to the function display_output( ).

Program 7-11

```
#include <stdio.h>

typedef struct
  {
    char    manufacturer[20];    /* Resistor manufacturer.  */
    int     quantity;            /* Number of resistors left. */
    float   price_each;          /* Cost of each resistor.   */

  } parts_record;      /* Name for this structure. */

parts_record get_input(void);    /* Get user input. */
  void display_output(parts_record resistor_record); /* Display output. */
```

(continued)

Program 7–11 *(continued)*

```c
main( )
{
  parts_record resistor_box;   /* Declares a structure variable. */

  resistor_box = get_input( );   /* Get user input. */
  display_output(resistor_box);  /* Display output. */

}   /* End of main( ) */

/*---------------------------------------------------------*/

parts_record get_input(void)   /* Get user input.  */
{
 parts_record resistor_information;

  /* Get name of manufacturer: */

    printf("Name of manufacturer => ");
    gets(resistor_information.manufacturer);

  /* Get number of parts left: */

    printf("Number of parts left => ");
    scanf("%d",&resistor_information.quantity);

   /* Get cost of each part: */

    printf("Cost of each part => ");
    scanf("%f",&resistor_information.price_each);

    return(resistor_information);

}   /* End of get_input( ) */

/*--------------------------------------------------------*/


  void display_output(parts_record resistor_information) /*display output.*/

  {

    /* Display results: */

      printf("\n\n");
      printf("Item:          Resistors\n\n");
      printf("Manufacturer:  %s\n",resistor_information.manufacturer);
      printf("Cost each:     $%f\n",resistor_information.price_each);
      printf("Quantity:      %d\n",resistor_information.quantity);

  }   /* End of display_output( ) */

/*---------------------------------------------------------*/
```

Program Analysis

The first new material in the above program is in the function prototypes:

```
parts_record get_input(void);    /* Get user input. */
```

The C `typedef` has been used to give the name `parts_record` to the structure. Now, the function `get_input` is being defined as a type `parts_record`.

The next item is the use of a structure type as a formal parameter:

```
void display_output(parts_record resistor_record); /* Display output. */
```

Here, the formal parameter `resistor_record` is of type `parts_record` (which is the name for the previously defined structure type). This means that the argument passed to this function must be of the same type. This is what happens in the body of **main( )**.

```
main( )
 {
  parts_record resistor_box;   /* Declares a structure variable. */

  resistor_box = get_input( );   /* Get user input. */
  display_output(resistor_box);   /* Display output. */

 }  /* End of main( ) */
```

The variable `resistor_box` is defined as the type `parts_record` (as were the other structure variables). Now, since `resistor_box` and `get_input( )` are both of the same type of structure, one may be assigned to the other.

Next, in the definition of the `get_input` function, the whole structure is returned back using the defined variable `resistor_information` which is again of type `parts_record`.

```
parts_record get_input(void)   /* Get user input.   */
 {
  parts_record resistor_information;

   /* Get name of manufacturer: */

   /* Get number of parts left: */

    /* Get cost of each part: */

      return(resistor_information);

 }  /* End of get_input( ) */
```

The last function, `display_output`, simply takes the actual parameter from **main( )** (which now contains the structure data) and uses it to display the output data).

Conclusion

This section demonstrated various forms of the C structure. Here you saw different ways of letting your C program know that you are constructing a structure. You also saw how to name function types, declare functions as structures, and pass structures between functions.

7-4 Section Review

1 Describe what is meant by a structure tag.
2 Give an example of a structure tag.
3 Can a structure be a variable type? Explain.
4 How is a structure member pointed to when a structure pointer is used?
5 What are the three operations that are allowed with structures?

7-5 The **union** and Structure Arrays

Discussion

This section presents the C **union** which you will see is similar to a C structure in many ways but with an important difference. You will also be introduced to a method of bringing together the power of a C structure with your old friend, the C array. This combination will give you tremendous programming potential for handling many different data types. This section brings together much of the material that has been presented up to this point.

The C union

The C **union** is similar to the C structure. The difference is that a C **union** is used to store different data types in the same memory location.

The C **union** is

```
union tag
    {
        type    member_identifier₁
        type    member_identifier₂
            .
            .
            .
        type    member_identifierₙ
    }
```

As you can see a union declaration has the same form as the structure declaration. The difference is the keyword **union**. A union declaration names a union variable and states the set of variable values (members) of the union. These members can have different types. What happens in a **union** is that a union type variable will store one of the values defined by that type. Program 7–12 demonstrates the action of a C union.

Program 7–12

```
#include <stdio.h>

main( )
{
  union      /*  Define union.   */
      {
        int integer_value;
        float float_value;
      } integer_or_float;

  printf("Size of the union => %d bytes.\n", sizeof(integer_or_float));

  /*  Enter an integer and display it:  */

  integer_or_float.integer_value = 123;
  printf("The integer value is %d\n",integer_or_float.integer_value);
  printf("Starting address is => %d\n",&integer_or_float.integer_value);

  /*  Enter a float and display it:  */

  integer_or_float.float_value = 123.45;
  printf("The float value is %f\n",integer or float.float value);
  printf("Starting address is => %d\n",&integer_or_float.float_value);

}
```

Execution of the above program produces

```
Size of the union => 4
The integer value is 123
Starting address is => 7042;
The float value is 123.45
Starting address is => 7042
```

What you see from the above output is that the size of the union is 4 bytes. This is because it takes 4 bytes to store a type **float**. First, an integer value is stored and then retrieved from this memory location (shown with a starting address of 7042). Then a float value is stored and retrieved from the same memory space.

As shown in Program 7–12, the union declarations have the same form as structure declarations. The difference is that the keyword **union** is used in place of the keyword **struct**. The rules covered up to this point for structures also apply to unions. The amount of storage required for a union variable is the amount of storage required for the largest member of the union. All members are stored in the same memory space (but not at the same time) with the same starting address.

Initializing the C Structure

A C structure may be initialized if it is a global or static variable. This is illustrated in Program 7-13.

Program 7-13

```
#include <stdio.h>


    typedef struct
        {
        char        part[20];        /* Type of part.  */
        int         quantity;        /* Number of parts left. */
        float       price;           /* Cost of each part. */

        } parts_record;

    main( )
    {
     static parts_record  bin_1_contents =
         {
          "Resistors",
           25,
           0.05
          };
     static parts_record  bin_2_contents =
         {
           "Capacitors",
            37,
            0.16
          };

    printf("Contents of bin #1:\n");
    printf("Item => %s\n",bin_1_contents.part);
    printf("Quantity => %d\n",bin_1_contents.quantity);
    printf("Cost each => $%f\n",bin_1_contents.price);

    printf("\nContents of bin #2:\n");
    printf("Item => %s\n",bin_2_contents.part);
    printf("Quantity => %d\n",bin_2_contents.quantity);
    printf("Cost each => $%f\n",bin_2_contents.price);

    }
```

Execution of the above program produces

```
Contents of bin #1:
Item => Resistors
Quantity => 25
Cost each => 0.05

Contents of bin #2
Item => Capacitors
Quantity => 37
Cost each => 0.16
```

Program Analysis

The program starts by naming a structure type called `parts_record`.

```
typedef struct
  {
  char       part[20];      /* Type of part.  */
  int        quantity;      /* Number of parts left. */
  float      price;         /* Cost of each part. */

  } parts_record;
```

Now `parts_record` means a structure that contains three members. Each member will contain information about parts in a storage bin. In the function **main( )**, the data for each of the three members is assigned. Note that the new variable `bin_1_contents` is defined as a **static** of type `parts_record`:

```
main( )
{
  static parts_record  bin_1_contents =
      {
        "Resistors",
        25,
        0.05
      };
```

Another variable, `bin_2_contents`, is also defined as a **static** of type `parts_record`. Again each member is assigned data. Note that each assignment corresponds to the member type.

```
static parts_record  bin_2_contents =
      {
        "Capacitors",
        37,
        0.16
      };
```

There are now two structures of the same type, `bin_1_contents` and `bin_2_contents`. Their structure members are the same, but the data assigned to each member is different.

The remainder of the program uses the member of operator to display the data contained in each structure:

```
printf("Contents of bin #1:\n");
printf("Item => %s\n",bin_1_contents.part);
printf("Quantity => %d\n",bin_1_contents.quantity);
printf("Cost each => $%f\n",bin_1_contents.price);

printf("\nContents of bin #2:\n");
printf("Item => %s\n",bin_2_contents.part);
printf("Quantity => %d\n",bin_2_contents.quantity);
printf("Cost each => $%f\n",bin_2_contents.price);
```

Structure Arrays

The real power of using C structures comes when there are **structure arrays**. Consider an inventory program where there are many different storage bins for parts. For the contents of each storage bin you may want to know the name of the part, the quantity, and the price of each part. Thus you would want to know the same structure of information about many storage bins. This concept is illustrated in Figure 7–4.

Since the structure of information for each of these is the same, you can create a C program that is an array of the same kind of structure. This concept is illustrated in Figure 7–5.

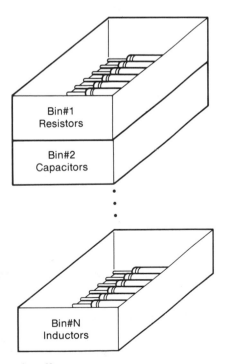

Figure 7–4   Storage Bins with Different Parts

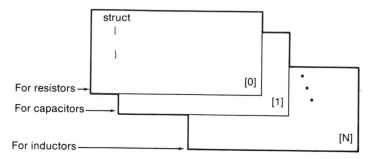

Figure 7-5   Creating an Array of Structures

Program 7-14 is a C program that contains an array of three structures. This program will allow the program user to input—for each of three part bins—the name of the part, the number of parts in the bin, and the cost of each part. The program will then display what the user has entered. This is achieved by defining one structure type and then making an array of this type.

Program 7-14

```
#include <stdio.h>
#include <string.h>

  typedef struct
  {
    char      part[20];            /* Type of part.  */
    int       quantity;           /* Number of parts left. */
    float     price;              /* Cost of each part. */
    char      initialized;        /* Test if record exists. */
  } parts_record;

main( )
{
 parts_record  bin_contents[3];
 int record;

  do
  {
  /* Get bin number: */

    printf("Enter bin number from 1 to 3 (0 to quit) => ");
    scanf("%d",&record);

    --record;   /* Start at bin 0 */
```

(continued)

Program 7-14 *(continued)*

```
     if (record < 0) continue;

   /* Get name of part:  */

      printf("Part name => ");
      scanf("%s",bin_contents[record].part);

   /* Get number of parts left: */

      printf("Number of parts left = ");
      scanf("%d",&bin_contents[record].quantity);

   /* Get cost of each part: */

      printf("Cost of each part => ");
      scanf("%f",&bin_contents[record].price);

   /*  Set record initialization TRUE.  */

bin_contents[record].initialized = 'T';  /* Show data entered. */

  } while(record >= 0);

  /*  Output information: */

    for(record = 0; record <= 2; record++)
      {
        if(bin_contents[record].initialized == 'T')
          {
          printf("Bin number %d contains:\n",record+1);
          printf("Part => %s\n",bin_contents[record].part);
          printf("Quantity => %d\n",bin_contents[record].quantity);
          printf("Cost each => $%f\n",bin_contents[record].price);
          }
      } /* End for. */

}
```

Assuming the following program user inputs, execution of the above program produces

```
Enter bin number from 1 to 3 (0 to quit) => 1
Part name => Inductor
Number of parts left = 12
Cost of each part => 0.18
```

```
Enter bin number from 1 to 3 (0 to quit) => 3
Part name => Resistor
Number of parts left = 25
Cost of each part => 0.04

Enter bin number from 1 to 3 (0 to quit) => 0

Bin number 1 contains:
Part => Inductor
Quantity => 12
Cost each => $0.18

Bin number 3 contains:
Part => Resistor
Quantity => 25
Cost each => $0.14
```

Note that no data was entered for bin #2, and no data was displayed for it.

Program Analysis

The program defines a type called `parts_record` that is a structure of four members.

```
typedef struct
    {
    char        part[20];           /* Type of part.  */
    int         quantity;           /* Number of parts left. */
    float       price;              /* Cost of each part. */
    char        initialized;        /* Test if record exists. */
    } parts_record;
```

The important point is what follows next. A new variable is defined as an array (consisting of three elements). The new variable is called `bin_contents[3]` and is defined as type `parts_record`. This means that you now have a three-element array that is a structure consisting of the four members defined in the `typedef`.

```
main( )
{
  parts_record  bin_contents[3];
```

This sets aside an array of three structures, all of which have the same number and type of members. The next part of the program asks the program user to input the bin number. This value will set the array number. However, since the C language starts its arrays at 0, the actual array number entered by the user has one subtracted from it. This is done using the C —record operation. User input is part of a C **do** loop that continues while the user input is not less then 0. (If the user inputs a 0 to stop input, one is subtracted from it making the final input less than zero.)

```
do
{
  /* Get bin number: */

    printf("Enter bin number from 1 to 3 (0 to quit) => ");
      scanf("%d",&record);

      --record;   /* Start at bin 0 */

   if (record < 0) continue;
```

When input is taken, the format for selecting a specific structure from the array and a specific member of that structure is `structure_variable[N].structure_member`, where N is the number of the array.

```
/* Get name of part:   */

    printf("Part name => ");
    scanf("%s",bin_contents[record].part);

    printf("Number of parts left = ");
    scanf("%d",&bin_contents[record].quantity);

    printf("Cost of each part => ");
    scanf("%f",&bin_contents[record].price);
```

In C, an initialized record will contain some random data. (For example, in this program, if the random contents are T for the structure member initialization, then the program will fail.) In order to avoid displaying this information when an initialized variable record is accessed, the last structure member (`char initialized`) is used to indicate if data has been placed in the array by the user. This is done by setting this member to the letter T:

```
    /*  Set record initialization TRUE.  */
  bin_contents[record].initialized = 'T';   /* Show data entered. */
```

The program uses a C **for** loop to output the data. It's important to note that no data will be displayed for a structure that has not been initialized (when `bin_contents[record].initialized` does not equal T:

```
for(record = 0 record <= 2; record++)
  {
    if(bin_contents[record].initialized == 'T')
```

The output now simply uses C **printf** functions and identifies each member of a specific structure from the structure array:

```
{
  printf("Bin number %d contains:\n",record+1);
  printf("Part => %s\n",bin_contents[record].part);
```

```
printf("Quantity => %d\n",bin_contents[record].quantity);
printf("Cost each => $%f\n",bin_contents[record].price);
}
```

Conclusion

This section presented the C **union**. You also saw the C structure and the C array brought together. This combination is a powerful new tool for handling complex data structures. In the next section, you will see how structures and arrays may be mixed to handle even more complex data structures. Check your understanding of this section by trying the following section review.

7-5 Section Review

1 Describe the purpose of a C **union**.
2 How is a C **union** declared?
3 Can a C structure be initialized by the program? Explain.
4 Explain how an array of structures may be declared.
5 Show how an individual member of an array of structures is accessed.

7-6   Ways of Representing Structures

Discussion

C structures are rich in the variety of ways they may be used. This section will demonstrate some of them. You will find many potential technical applications for this powerful feature of C.

Structures within Structures

You can have a C structure within another structure. Consider Program 7-15.

Program 7-15

```
#include <stdio.h>

typedef struct
   {
      int          member_1;
      char         member_2;
      float        member_3;
      } first_structure;
```

(continued)

Program 7–15 *(continued)*

```
struct second_structure
   {
     first_structure  second_member_1;
     int               second_member_2;
     char              second_member_3;
   }

main( )
{

}
```

As you can see from Program 7–15, the C structure `second_structure` contains the structure `first_structure` in the member called `second_member`. This is a case of a C structure containing another structure. The parts inventory program could use a structure within a structure such as this. For example, the manufacturer of the part could in itself be a structure that would contain the address and telephone number as members.

Arrays within Structures

You can also have C arrays within C structures. For example, consider Program 7–16. Here, one member of the structure is a two-dimensional array. Recall that you used character arrays to represent strings in previous C structures. This is no different.

Program 7–16

```
#include <stdio.h>

struct structure_tag
   {
     float       array_variable[5][6];
     int         second_member_2;
     char        second_member_3;
   }

main( )
{

}
```

As you can see from the above program, the member `array_variable[5][6]` is a two-dimensional array contained within the C structure.

Multidimensional Structure Arrays

You may also have multidimensional arrays that are C structures. Consider Program 7–17. Here you have a two-dimensional array that is a structure which contains three members, one of which is itself a two-dimensional array.

Program 7–17

```
#include <stdio.h>

    typedef struct
      {
      float      array_variable[5][6];
      int        second_member_2;
      char       second_member_3;
      } record_1;

    typedef record_1 complex_array[3][2];

    main( )
    {

    }
```

As shown in the above program, `complex_array[3][2]` is a two-dimensional array of type `record_1`. Thus you have a data structure where there are actually $3 \times 2 = 6$ structures, each of which contains three members, one of which is an array itself.

Conclusion

A rich variety of structures and arrays are available with C. In this section you saw a few examples of the possibilities. Check your knowledge of this section by trying the following section review.

7–6 Section Review

1 Can a structure be an array? Explain.
2 Can a structure contain another structure? Explain.
3 Is it possible for an array to be a structure type that contains an array as one of its members? Explain.

7-7   Disk Input and Output

Discussion

The C programs you have developed up to this point did not allow the program user to save any entered data. As an example, the parts inventory programs used in this chapter do not save any of the data concerning the parts once the computer is turned off. This is a severe restriction for many different kinds of technology programs. In this section, you will discover how to save data to the disk and how to get it back. Doing this will allow you to create C programs that will let the program user store data entered into the program as well as get the data back when the program is used again.

Creating a Disk File with C

In order to store user input data on the computer disk, a file must be created on the disk for storing the data. This can be done in many different ways. One way is to have the C program do this automatically for the program user. In order to do this, you must observe the rules for the Disk Operating System (**DOS**) used by your system. The DOS used by the IBM PC and compatibles contains specific rules concerning the naming of a disk file. What follows is a summary of legal DOS file names. A more in-depth coverage is presented in Appendix A.

A DOS Review

A legal DOS file name has the form

`[Drive:]FileName[.EXT]`

Where

Drive = The optional name of the drive that contains the disk which you want to access.

FileName = The name of the file (up to eight characters).

.EXT = An optional extension to the file name (up to three characters).

Both the `Drive:` and `.EXT` are optional. If `Drive:` is not specified, then the active drive will be used. The colon (:) following the drive letter is necessary. For example,

`B:MYFILE.DAT`

is the name of a disk file on drive `B:` with the file name `MYFILE` and the extension `.DAT`. It makes no difference if you enter a file name with uppercase or lowercase letters; DOS will always show all disk files in uppercase.

A First File Creation Program

Program 7–18 shows a C program that creates a disk file called MYFILE.DTA on the active drive. The program does not store any information in the file. However, when the program is executed, a new file MYFILE.DTA will now exist on the disk in the active drive.

Program 7–18

```
PROGRAM 7-18
#include <stdio.h>

main( )
{
 FILE *file_pointer;    /* This is the file pointer.  */

    /* Create a file called MYFILE.DTA and
       assign its address to the file pointer:  */

       file_pointer = fopen("MYFILE.DTA", "w");

    /* Close the created file.  */

       fclose(file_pointer);

 }
```

Program Analysis

The program starts by creating a **file pointer** that is of a data type called FILE:

```
FILE *file_pointer;    /* This is the file pointer.  */
```

This is a data type defined in both Turbo and Quick C.

Next, the file pointer is set equal to the built-in C **fopen( )** function. In order to use this function, the legal DOS name of the file and a file code string are required in the argument (**w** means the file will be opened for writing into—it is a string because of the double quotes).

```
file_pointer = fopen("MYFILE.DTA", "w");
```

The above will actually create a file named MYFILE.DTA on the active drive and open it for writing data into it.

The next statement is necessary when you are finished with the open files. If you don't use it, you may damage any data in your opened files

```
fclose(file_pointer);
```

This is the built-in C **fclose( )**. Its argument is the file pointer.

Putting Data into the Created File

Program 7–18 creates a disk file using a C program. However, nothing was put into the file. The program illustrates the minimum requirements for creating a disk file, but the file is empty. Program 7–19 illustrates the creation and opening of the same file, but this time the character C is put into the file.

Program 7–19

```
#include <stdio.h>

main( )
{
  FILE *file_pointer;      /* This is the file pointer.  */

    /* Create a file called MYFILE.DTA for writing and
       assign its address to the file pointer:   */

       file_pointer = fopen("MYFILE.DTA", "w");

    /* Put a letter into the opened file: */

       putc('C', file_pointer);

    /* Close the created file.  */

       fclose(file_pointer);
}
```

Program Analysis

Program 7–19 does the same thing as Program 7–18 with one exception; it puts the letter C into the opened file. This is done by the built in function **putc( )**. This function puts a single character into an opened file. Its argument requires the character and the name of the file pointer.

You now have a file called **MYFILE.DTA** that contains the character C.

Reading Data from a Disk File

Program 7–19 shows you how to put a single character into a file created by a C program. What you now need is a method of allowing the C program to get data from an existing disk file. Program 7–20 shows the basic requirements to do just that.

Program 7-20

```c
#include <stdio.h>

main( )
{
 FILE *file_pointer;     /* This is the file pointer.  */
 char file_character;    /* Character to be read from the file. */

   /* Open the existing file called MYFILE.DTA for reading and assign
      its address to the file pointer:  */

      file_pointer = fopen("MYFILE.DTA", "r");

   /* Get the first letter from the opened file: */

      file_character = getc(file_pointer);

   /* Display the character on the monitor. */

     printf("The character is %c\n",file_character);

   /* Close the created file.  */

       fclose(file_pointer);
}
```

For the file created in Program 7-20, when the above program is executed, it will display

```
The character is C
```

Program Analysis

This program for reading data from an existing file is very similar to the program for creating and writing data into a file. As before, a pointer of type FILE is declared:

```c
FILE *file_pointer;     /* This is the file pointer.  */
```

This is the same requirement as for creating and inputting data to a disk file. For this program, a variable to hold the character to be received from the file is declared:

```c
char file_character;    /* Character to be read from the file. */
```

The file is opened using the same built-in C function **fopen( )**. Again, its argument contains the file name and a different file operation command. This time **r** is used to mean the file is being opened for reading!

```c
file_pointer = fopen("MYFILE.DTA", "r");
```

The file pointer is evaluated from this function as before.

Next, the file character variable is evaluated to the built-in C function **getc( )** where the argument is the name of the file pointer.

```
file_character = getc(file_pointer);
```

Now the character read from the file is displayed on the monitor:

```
printf("The character is %c\n",file_character);
```

Last is the important step of closing the opened file when you are finished with it:

```
fclose(file_pointer);
```

Saving a Character String

The previous programs on disk access, save, and retrieve only a single character. However, they illustrate the fundamental requirements for a disk file with C.

The next program illustrates a method of saving a character string to a disk file. In this case, the program user may input any string of characters from the keyboard and have them automatically saved to a file called **MYFILE.DTA**. The program will continue getting characters from the program user until a carriage return is entered.

Program 7–21

```
#include <stdio.h>

main( )
{
 FILE *file_pointer;    /* This is the file pointer.  */
 char file_character;   /* Character to be read from the file. */

   /* Create a file called MYFILE.DTA for writing and assign
      its address to the file pointer:  */

      file_pointer = fopen("MYFILE.DTA", "w");

   /* Put a stream of characters into the file: */

     while((file_character = getc( )) != '\r')
       file_character = putc(file_character, file_pointer);

   /* Close the created file.  */

       fclose(file_pointer);
}
```

Note that like the other file programs, this one uses the built-in C function **fopen( )** which includes the **w** character meaning that the file is being opened for writing into. Again, when all file activity is completed, the file is closed with the built in C function **fclose( )**.

The main difference between this program and the others is the line

```
file_character = putc(file_character, file_pointer);
```

The built-in C function **putc** inputs one character at a time into the file pointed to by the file pointer:

```
putc('C', file_ptr);
```

where C is the character and **file_ptr** is the file pointer.

Note that this is included in a C **while** loop:

```
while((file_character = getc( )) != '\r')
  file_character = putc(file_character, file_pointer);
```

Here the loop will continue as long as the program user does not press the Return key (\r).

Reading a Character Stream

The concept of a **character stream** is a sequence of bytes of data being sent from one place to another (such as from computer memory to the disk).

To read a character stream from a disk file you use the same tactics. The difference is that your program will continue to read a stream of characters until an end-of-file marker (EOF). This is automatically placed at the end of a disk file when it is created. When this marker is read by the program, it indicates that the end of the file has been reached and there is nothing further to read. Program 7–22 illustrates:

Program 7–22

```
#include <stdio.h>

main( )
{
 FILE *file_pointer;    /* This is the file pointer.  */
 char file_character;   /* Character to be read from the file. */

   /* Open an existing file called MYFILE.DTA for reading and assign
      its address to the file pointer:  */

      file_pointer = fopen("MYFILE.DTA", "r");

   /* Get a stream of characters from the file
      and display them on the monitor.  */
```

(continued)

Program 7-22 *(continued)*

```
    while((file_character = getc(file_pointer)) != EOF)
    printf("%c",file_character);

/* Close the created file.  */

    fclose(file_pointer);
}
```

Program 7-22 is very similar to the other file programs. The **fopen( )** function is used with the **r** character directive, meaning that the file is being opened for reading. The main difference is in the C **while** loop:

while((file_character = getc(file_pointer)) != EOF)
printf("%c",file_character);

Here, the C **while** loop continues until an **EOF** marker is reached, indicating the end of the file. While the loop is active, the characters received from the disk file are directed to the monitor screen for display.

Conclusion

This section introduced you to the concept of creating a disk file from a C program. Here you also saw how to save and retrieve characters from the created disk file. These programs did some very simple tasks, but they introduce to you a very powerful feature of current C systems. In the next section you will be presented a more detailed analysis of disk file creation using C. Check your understanding of this section by trying the following section review.

7-7 Section Review

1 Describe a legal DOS file name.
2 Explain what is meant by a file pointer in C. Give an example.
3 Name the built-in C function for opening a disk file.
4 Demonstrate how the file pointer is assigned to the DOS file name.
5 State the difference between the C file opening command for writing to or reading from a file.
6 What must you always do when you are finished with an open file? Give an example.

7-8   More Disk I/O

Discussion

In the last section you were introduced to file input/output (I/O) within a C program. This section presents a more detailed description of file I/O. Here you will discover more methods of developing disk I/O. These methods will allow you to store and retrieve complex data types such as arrays and structures.

The Possibilities

Table 7–1 shows the possible conditions you can encounter when working with disk data.

Table 7–1   Disk File Conditions

Condition	Meaning
1	The disk file does not exist, and you want to create it on the disk and add some information.
2	The disk file already exists, and you want to get information from it.
3	The disk file already exists, and you want to add more information into it while preserving the old information that was already there.
4	The disk file already exists, and you want to get rid of all of the old information and add new information.

As you can see from Table 7–1, there are four possibilities when working with disk information. All of these possibilities will be presented in this section.

Besides the considerations presented in Table 7–1, in the C language there are four different ways of reading and writing data. These are listed on Table 7–2.

Table 7–2   Different Methods of Reading and Writing Data

Method	Comments
One character at a time.	Inputs and outputs one character to the disk at a time.
Read and write data and strings	Inputs and outputs a string of characters to the disk.
Mixed mode.	Used for I/O of characters, strings, floating points, and integers.
Structure or block method	Use for I/O of array elements and structures

You have already used the first two methods from Table 7–2. These allow you to perform simple I/O where you can store and retrieve characters or strings. The programs in this section will concentrate on the last two methods: mixed data types and structure or blocks of data.

File Format

You may have noted in the last section that a C program which performs disk I/O has a specific structure. This is illustrated in Figure 7–6.

Referring to Figure 7–6, you can see that a C program designed for disk I/O uses the type FILE with a corresponding file pointer. The type FILE is a predefined

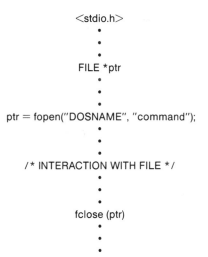

Figure 7-6   Standard C Structure for Disk I/O

structure declared in the header file $<$stdio.h$>$. This file must be included in all of your C programs designed for disk I/O. This predefined **FILE** structure helps establish the necessary link between your program and the disk operating system.

The next step in a disk I/O program is to open the file. As shown in Figure 7-6, the built-in C function **fopen("DOSNAME", "command")**; is used. As you saw before, **"DOSNAME"** is any legal DOS file name which may include an extension. (It should be noted that this may have also included an MS-DOS pathname such as **\MYFILES\MYFILE.DAT**. The purpose of the one letter string is to identify the type of file that will be opened.)

Table 7-3   C File Type Commands

String Command	Meaning
"a"	Open for appending. New data added to existing file data or a new file is created.
"r"	Open for reading. File must already exist.
"w"	Open for writing. Contents written over or new file created.
"a+"	Open for reading and appending. If file does not exist, it will be created.
"r+"	Open for both reading and writing—file must exist.
"w+"	Open for both reading and writing. Contents written over.

You have already used the "r" and "w" directives in the last section. Note from Table 7-3 how these file directives relate to the different ways you can treat a disk file.

Example File Program

Program 7–23 demonstrates several important points about C disk files. The program allows the user one of the following options for creating and reading from a simple message file:

1. Create a new file by a user supplied file name.
2. Create a new file by a user supplied file name and place a message in it.
3. Read a message from an existing file.
4. Append a message to an existing file.
5. The user is told if a given file for reading or appending does not exist.

Program 7–23

```c
#include <stdio.h>

main( )
{
 char selection[2];        /* User file selection. */
 char file_name[13];       /* Name of the disk file. */
 char user_choice[2];      /* User choice for file activity. */
 int selection_value;      /* User selection number. */
 int file_character;       /* File character to be saved. */
 FILE *file_pointer;       /* File pointer. */

  /* Display option user options: */

    printf("\n\nSelect one of the following by number:\n ");
    printf("1] Create a new file.   2] Write over an existing file.\n");
    printf("3] Add new data to an existing file.\n");
    printf("4] Get data from an existing file.\n");

  /* Get and act on user input:  */

    do
      {
        printf("Your selection => ");
        gets(user_choice);
         selection_value = atoi(user_choice);

     switch (selection_value)
       {
        case 1 :                        /* Create a new file. */
        case 2 : strcpy(selection, "w"); /* Write over existing data. */
               break;
        case 3 : strcpy(selection, "a"); /* Append data to existing file.*/
               break;
```

(continued)

Program 7-23 *(continued)*

```
        case 4 : strcpy(selection, "r"); /* Get data from an existing file. */
                break;
        default : {
                    printf("That was not one of the choices.");
                    selection_value = 0;
                }
        }  /* End of switch */

    } /* End of do */
    while(selection_value == 0);

/*  Get the file name from user:  */

    printf("\nEnter the name of the file => ");
    gets(file_name);

/*  Open the file for action:  */

  if((file_pointer = fopen(file_name, selection)) == NULL)
      {           /* File cannot be opened.  */
        printf("Can not open file %s!",file_name);
        exit(-1);
      }

/*  Write to or read from the file:  */

    switch(selection_value)
      {
        case 1 : break;                      /* No file interaction.   */
        case 2 :
        case 3 : {                           /* User writes data to file. */
                    printf("Enter string to be saved:\n");
                    while((file_character = getch( )) != '\r')
                    file_character = putc(file_character, file_pointer);
                }
                break;
        case 4 : {                           /* Get data from file. */
                    while((file_character = getc(file_pointer)) != EOF)
                    printf("%c",file_character);
                }
                break;
      }  /* End of switch */

/* Close the opened file.  */

            fclose(file_pointer);

}
```

Note: The function **getch** gets a character from the keyboard and does not echo it to the screen. The function **getche** does the same thing except it does echo the character to the screen.

Execution of the above program produces

```
Select one of the following by number:
1] Create a new file.    2] Write over an existing file.
3] Add new data to an existing file.
4] Get data from an existing file.
Your selection => 2
Enter the name of the file => MYFILE.01
Enter string to be saved: Saved by a C program.
```

The above message can be retrieved at a later date, added to or written over. Note that the program will tell the user if the desired file does not exist. With this program, any new file must first be created by the user.

Program Analysis

The file pointer is declared in the declaration part of the program:

```
FILE *file_pointer;      /* File pointer. */
```

User options are then displayed using the **printf( )** function:

```
printf("\n\nSelect one of the following by number:\n ");
printf("1] Create a new file.    2] Write over an existing file.\n");
printf("3] Add new data to an existing file.\n");
printf("4] Get data from an existing file.\n");
```

The user prompt for input becomes part of a C **do** loop. The reason for this is to repeat the prompt if the user input is not valid:

```
do
  {
   printf("Your selection => ");
```

Note that next a C **gets( )** function is used for receiving the user input. This is to avoid the difficulties of using the **scanf( )** function. The user input is then converted to a type **int** using the C **atoi( )** function that converts a string to an integer.

```
gets(user_choice);
  selection_value = atoi(user_choice);
```

A C **switch( )** function is then used to take the converted integer value and select one of five possible conditions. Note that the first two conditions will both select the "**w**" file option for writing a file:

```
switch (selection_value)
  {
```

```
case 1 :                              /* Create a new file. */
case 2 : strcpy(selection, "w"); /* Write over existing data. */
         break;
case 3 : strcpy(selection, "a"); /* Append data to existing file.*/
         break;
case 4 : strcpy(selection, "r"); /* Get data from an existing file. */
         break;
default : {
           printf("That was not one of the choices.");
           selection_value = 0;
         }
```

The C **default** condition will activate if one of the first four cases is not selected. Here the user is told that none of the possible selections were selected. Also note that in the **default** case, the variable `selection_value` is set to 0. This is done to cause a repeat of the C **do** loop and allow the user to try the selection again.

```
while(selection_value == 0);
```

Assuming the user makes a correct selection, the program will now ask the user for the name of the file. Note that again the C **gets( )** function is used:

```
printf("\nEnter the name of the file => ");
gets(file_name);
```

The next part of the program makes use of the fact that if, for any reason, a file cannot be opened, the null value (0) will be returned to the file pointer.

```
if((file_pointer = fopen(file_name, selection)) == NULL)
```

In the above statement, the C **fopen( )** function is being used to open the file. The variable `file_name` contains the name of the file entered by the user, and the variable `selection` will contain one of the file directives (r, w, or a). The file pointer `file_pointer` will be used to replace all references to the opened file. If the value of the pointer is 0 then the following compound statement will be executed:

```
{               /*  File cannot be opened.  */
  printf("Can not open file %s!",file_name);
  exit(-1);
}
```

Otherwise, the above is skipped by the program.

Next, a C **switch** is used to determine what action is to be taken with the now opened file. Note that in case 1 the user opts to simply create a new file, so there is nothing to input or output:

```
switch(selection_value)
  {
   case 1 : break;                    /* No file interaction.  */
```

In case 2 or 3, the program user has opted to save a string to the file (by writing over an existing string or appending a new string). Here, the C **while** loop is used to input the string using the C **putc** function which writes a character `file_character` to the file pointed to by `file_pointer`. The **while** loop stays active until the program user presses the -RETURN/ENTER- key producing a carriage return ($\backslash$r).

```
case 2 :
case 3 : {                          /* User writes data to file.  */
        printf("Enter string to be saved:\n");
        while((file_character = getche( )) != '\r')
        file_character = putc(file_character, file_pointer);
        }
        break;
```

In the fourth case, the user chooses to read a string from a selected file. Here, the **getc( )** function is used to get a character from the file pointed to by `file_pointer`. This process is in a **while** loop that will continue until the EOF marker is reached. The **printf( )** function is used to direct the resulting characters to the monitor screen.

```
case 4 : {                          /* Get data from file. */
        while((file_character = getc(file_pointer)) != EOF)
        printf("%c",file_character);
        }
        break;
```

The program then terminates with the necessary file closing:

```
fclose(file_pointer);
```

Mixed File Data

The previous file programs were limited to working with strings. Program 7–24 illustrates a method of working with files that allows you to input numerical as well as string data. The program illustrates a process of entering the mixed data from the parts structure program of the previous sections. This is accomplished with a new function called **fprintf**.

Program 7–24

```
#include <stdio.h>

main( )
{
 char part_name[15];     /* Type of part. */
 int quantity;           /* Number of parts left. */
 float cost_each;        /* Cost of each part. */
 FILE *file_pointer;     /* File pointer. */
```

(continued)

Program 7-24 *(continued)*

```
/* Open a file for writing: */

    file_pointer = fopen("B:PARTS.DTA", "w");

/* Get data from program user: */

  printf("Enter part type, quantity, cost each, separated by commas:\n");
  printf("Press -RETURN- to terminate input.\n");
  scanf("%s %d %f",part_name, &quantity, &cost_each);
  fprintf(file_pointer, "%s %d %f",part_name, quantity, cost_each);

/* Close the opened file.  */

    fclose(file_pointer);

}
```

Execution of the above program produces:

```
Enter part type, quantity, cost each, separated by commas:
Press -RETURN- to terminate input.
Resistor, 12, 0.05
```

Assuming that the program user carefully followed instructions, the **scanf( )** function would receive the required data and the **fprintf( )** function would store it in the opened file. The **fprintf( )** function has the form:

fprintf(FILE *stream, const char *format..argument).

This function accepts a series of arguments, and outputs the formatted stream as specified by the format string. Similar to **printf( )**.

Program Analysis

Program 7-24 has several weaknesses. It doesn't protect user input (the **scanf( )** function is used). It also doesn't let the user know if there was a problem opening the file. However, it does illustrate a simple C program that enters mixed data types into a file. The key to the program is the **fprintf( )** function that allows the user to format the input data to the disk in many different ways—in this case as a string, then as an integer, and finally as a float.

Retrieving Mixed Data

Program 7-25 demonstrates how the mixed data stored on the disk may be retrieved and sent to the monitor screen.

The key to this program is the C **fscanf( )** function. This function is similar to the **scanf( )** function except that a pointer to FILE is used as its first argument.

Program 7-25

```c
#include <stdio.h>

main( )
{
 char part_name[15];        /* Type of part. */
 int quantity;              /* Number of parts left. */
 float cost_each;           /* Cost of each part. */
 FILE *file_pointer;        /* File pointer. */

  /* Open a file for writing: */

    file_pointer = fopen("B:PARTS.DTA", "w");

  /* Get data from disk file: */

while(fscnf(file_pointer,"%s %d %f",part_name,&quantity,cost_each)!=EOF)
    printf("%s %d %f\n", part_name, quantity, cost_each);

  /* Close the opened file.  */

    fclose(file_pointer);
 }
```

Assuming that the file contained the data entered by the previous program, execution of the above program would yield:

```
Resistor 12 0.050000
```

Text vs. Binary Files

All of the file I/O you have been using up to this point has been what are called **text files**. Figure 7-7 illustrates how a text file is stored on a disk.

As shown in Figure 7-7, numbers stored on the disk in string format are stored as strings instead of as numerical values. Because of this, the storage of numerical data as strings does not use disk space efficiently. One way to increase the disk storage efficiency is to use what is called the **binary mode** of disk I/O rather than the **text mode**. The **binary file** does not store numbers as a string of characters (as is done with text files). Instead, they are stored as they are in memory—two bytes for an integer, four for floating point, and so on for the rest. The only restriction is that a file stored in binary mode must also be retrieved in binary mode, or what you get will not make sense. All that you need to do is add the letter b after the file directive. Thus **fopen("MYFILE.01", "wb")** means to open (or create) a file called **MYFILE.01** for writing in binary format. Likewise, **fopen("MYFILE.01", "rb")** means to open the file **MYFILE.01** for reading in binary format. As you may have guessed, **fopen("MYFILE.01", "ab")** means to open the file **MYFILE.01** for appending in binary format.

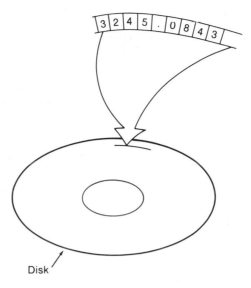

Figure 7–7   Storing Numbers in Text Format

The "**ws**" directive could have been used to indicate that a file is to be opened for writing in string format. For the previous programs, the **s** could have been added to all of the file directives resulting in a string format file. However, this is redundant, since, by default, files are opened for string format unless the directive is modified with the **b** for binary format.

Record Input

Program 7–26 shows an example of saving C data structures to a file. This is a powerful and important method of saving complex data to disk files. Observe that this program is similar to that of the mixed data type. However, the important difference is that it stores a data structure to the file. This is done using a new function called **fwrite**.

Program 7–26

```
#include <stdio.h>

typedef struct
  {
    char part_name[15];        /* Type of part. */
    int quantity;              /* Number of parts left. */
    float cost_each;           /* Cost of each part. */
  } parts_structure;
```

(continued)

Program 7–26 *(continued)*

```
main( )
{
 parts_structure parts_data;    /* Parts structure variable.  */
 FILE *file_pointer;            /* File pointer. */

 /* Open a file for writing: */

    file_pointer = fopen("B:PARTS.DTA", "wb");


 /* Get data from program user: */

    do
    {
      printf("\nName of part => ");
      gets(parts_data.part_name);
      printf("Number of parts => ");
      scanf("%d",&parts_data.quantity);
      printf("Cost per part => ");
      scanf("%f",&parts_data.cost_each);

/* Write structure to the opened file:  */

    fwrite(&parts_data, sizeof(parts_data), 1, file_pointer);

/* Prompt user for more input:  */

    printf("Add more parts (y/n)? => ");

    }   /* End of do */
    while(getche( ) == 'y');

 /* Close the opened file.  */

    fclose(file_pointer);


 }
```

Assuming that the user enters the given value, execution of the above program will yield

```
Name of part => Resistor
Number of parts => 12
Cost per part => 0.5
```

The above data will now be saved to the disk as a block of information in a binary format under the file name of **PARTS.DTA** on drive B:.

Program Analysis

The program first defines a type called `parts_structure` as a structure consisting of three members. This is the structure for the parts inventory presented in earlier sections of this chapter.

```
typedef struct
    {
       char part_name[15];        /* Type of part. */
       int quantity;              /* Number of parts left. */
       float cost_each;           /* Cost of each part. */
    } parts_structure;
```

The program then declares the required file pointer as type `FILE` and the variable `parts_data` as the type `parts_structure`. For user input, a character string is also declared.

```
main( )
{
 char numberstr[81];              /* Input for number data. */
 parts_structure parts_data;      /* Parts structure variable.  */
 FILE *file_pointer;              /* File pointer. */
```

A file on drive B: called `PARTS.DTA` is opened for writing in the binary mode (if the file does not exist, it will be created).

```
/* Open a file for writing: */

    file_pointer = fopen("B:PARTS.DTA", "wb");
```

The program user is now prompted to enter data. Note the use of the member of (.) operator. This is done inside a C **do** loop so that the user may enter more than one structure of data:

```
/* Get data from program user: */

    do
      {
        printf("\nName of part => ");
        gets(parts_data.part_name);
        printf("Number of parts => ");
        scanf("%d",&parts_data.quantity);
        printf("Cost per part => ");
        scanf("%f",&parts_data.cost_each);
```

Data is then entered into the opened file using the C **fwrite( )** function. This function has four arguments:

```
fwrite(buffer, size, n, pointer)
```

The `buffer` contains address of the data that is to be written into the file. The `size` is the size of the buffer in bytes. The `n` is the number of items of this size and `pointer` points to the file to be written to.

```
fwrite(&parts_data, sizeof(parts_data), 1, file_pointer);
```

Note that the C **sizeof( )** function is used to automatically determine the size, in bytes, of the data to go to the file.

The user is prompted for more data. The input loop will end if the character **y** is not entered.

```
printf("Add more parts (y/n)? => ");
```

```
}   /*  End of do  */
while(getche( ) == 'y');
```

The file is then closed.

```
/* Close the opened file.  */

    fclose(file_pointer);
```

Record Output

Program 7–27 illustrates the retrieving of the block of data entered by Program 7–26. Note that this program uses the same type definition of the parts structure as the previous program did.

Program 7–27

```c
#include <stdio.h>

  typedef struct
    {
      char part_name[15];        /* Type of part. */
      int quantity;              /* Number of parts left. */
      float cost_each;           /* Cost of each part. */
    } parts_structure;

main( )
{
  parts_structure parts_data;  /* Structure variable. */
  FILE *file_pointer;          /* File pointer. */

  /* Open a file for reading: */

    file_pointer = fopen("B:PARTS.DTA", "rb");

  /* Get data from file and display: */
```

(continued)

Program 7–27 *(continued)*

```
    while(fread(&parts_data, sizeof(parts_data), 1, file_pointer)==1)
      {
      printf("\nPart name => %s\n",parts_data.part_name);
      printf("Number of parts => %d\n",parts_data.quantity);
      printf("Cost of part => %f\n",parts_data.cost_each);
      }

/* Close the opened file.  */

    fclose(file_pointer);

  }
```

When executed, the program will look for the file PARTS.DTA on drive B: and open it in the binary mode for reading (the file must already exist). The data are read using the C **fread( )** function which has the form

fread(buffer, size, n, pointer)

This is very similar to the **fwrite( )** function. Again, buffer is a pointer to the data, size is the size of the data in bytes, n is the number of items of this size, and pointer points to the file to be written to.

The **fread( )** function will return the number of items actually written to the disk. Since in the input program (Program 7–26) the value of n is 1, if this value differs in the output program, then the file will no longer be read. This is accomplished by having the file read in a C **while** loop as long as this function is equal to 1.

Conclusion

This section demonstrated some key points concerning disk file I/O. Here you saw a new method of storing and retrieving different kinds of data to disk files. You also saw how to do this with blocks of data such as the structure. In the next section you will see some of the technical aspects of I/O. The case study section will present the development of a program that uses more sophisticated file techniques such as record arrays with random access to each record.

Check your understanding of this section by trying the following section review.

7–8 Section Review

1 List the four disk file I/O conditions that are possible.
2 State the different methods of reading and writing data with the C language.
3 Give the three basic C file type commands. State their meaning.
4 What is the purpose of the C **fscanf( )** function?
5 State the meaning of a block read or write function.

7-9   Program Debugging and Implementation

Discussion

This section presents some of the technical details concerning I/O using C. Here you will also be introduced to some details concerning random access files. This section will help prepare you for the case study that follows in the next section.

Inside I/O

You can think of any sequence of bytes of data as a stream of data. This stream of data can be thought of as information being sent or received in a serial form. The stream concept is illustrated in Figure 7–8.

You have actually worked with two kinds of streams, a text stream and a binary stream (text files and binary files). A text stream consists of lines of characters. Each line of characters ends with a terminating newline character ($\backslash$n). The important point about text streams (such as those stored to a disk file), is that they may not all be stored in the same way your C program stores them. This is the case with your MS-DOS where, unlike standard C, the end of line is terminated with both a newline and a carriage return ($\backslash$n$\backslash$r) character. This is not the case with a binary stream. The only requirement here is that you know the size of what was stored in the file.

In standard C, a file represents a source of data that is stored on some external media. These different types of media are illustrated in Figure 7–9.

The Buffered Stream

When performing a file I/O, an association needs to be made between the stream and the file. This is done by using a section of memory referred to as a **buffer**. Think of a buffer as a reserved section of memory that will hold data. This acts as a temporary storage place in memory. The bytes being read or written to the file are stored here. In this way, when a file is read from the disk, it is first stored in the buffer as a fixed "chunk" of data. Each of these "chunks" is the same size. What happens is that a process that reads or writes data to the disk is actually first communicating with the buffer. Hence the name **buffered stream** where a stream of data is stored in this buffer. Only when the buffer is **flushed** (data transferred from the buffer to the disk) is data actually stored to the disk from the buffer. Doing this reduces the amount of access time between the disk and program.

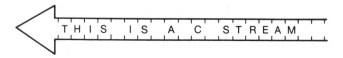

Figure 7–8   Concept of a C Stream

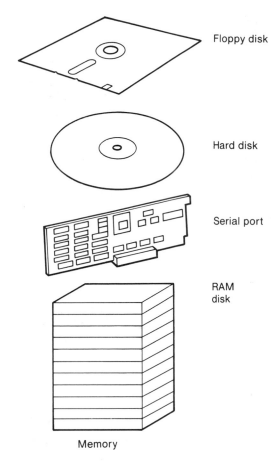

Floppy disk

Hard disk

Serial port

RAM disk

Memory

Figure 7–9   Different Types of C File Media

A buffer may not be flushed if the program is abnormally terminated. It's of interest to note that the file pointer you declare and use in your disk file programs is actually a pointer to this buffer. Thus, for stream files, the file pointer replaces all references to the file after it has been opened.

In the standard header file <**stdio.h**>, the following constants are defined:

EOF = −1

NULL = 0

BUFSIZE = 512

As you may recall, EOF is the end-of-file marker placed at the end of a disk file. NULL is assigned to a file pointer if an error has occurred and BUFSIZE is the size assigned to the I/O stream. You can redefine BUFSIZE from its assigned optimum value.

Standard Files

When you run your C program, it automatically opens three **standard files**. These are the standard input, standard output and standard error files. On your PC, these files represent the keyboard (for standard input) and the monitor (for standard output and error). This idea is illustrated in Figure 7–10.

The reason for having a standard error output is to ensure that the program always has access to your monitor. This is necessary in case you have directed your standard output somewhere else (such as to the disk). Thus if you have an error during this process, C still has access to your monitor to let you know that an error has taken place.

Redirecting Files

You can use DOS commands to **redirect files** on your disk. As an example, suppose you have the following two disk files:

READFILE.EXE $\Leftarrow$ A C program that normally gets data from the keyboard (standard input).

DATAFILE $\Leftarrow$ A character file that contains data from the keyboard—this file could have been produced when using a word processor.

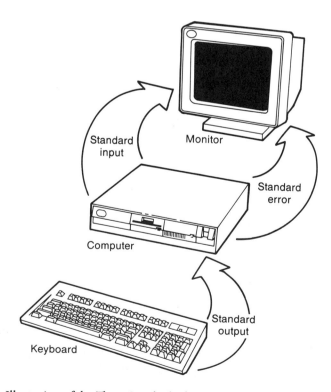

Figure 7–10   Illustration of the Three Standard Files

If you then enter the following from the DOS prompt (A>):

A> READFILE < DATAFILE

(the EXE extension is omitted), this will redirect standard input for the C READFILE to get its data from DATAFILE rather than from the keyboard.

The converse can also be done. Suppose that you have a C program called DATAMAKE.EXE that produces data (say the results of some calculations) and normally outputs the results to the monitor. Then

A> DATAMAKE > NEWDATA

will redirect the standard output so that instead of the data from the C program DATAMAKE.EXE going to the monitor, it is now going to the file called NEWDATA.

Piping Files

To **pipe** a file simply means to cause two or more separate disk files to act as if they are joined together. The DOS piping command is the |. As an example, MS-DOS contains a utility program that will do an alphabetical sort (called SORT.COM). If you want to see your disk directory displayed alphabetically sorted you would enter

A> DIR | SORT

(The extension COM is not entered.)

Filtering

Filtering is the process of modifying data in some fashion while it is in the process of being transferred from one file to another. As an example, if you have a file that contains a list of words called WORDFILE.OLD, you can copy it to another file called WORDFILE.NEW and alphabetize the words in the process. To do this you need to have the MS-DOS SORT.COM file on you disk. Then do the following:

A> SORT < WORDFILE.OLD > WRODFILE.NEW

This will cause the information in WORDFILE.OLD to be stored alphabetically in the file called WORDFILE.NEW. What you have done here is filter the data through the SORT program in the process of copying it from one file to another. A helpful aid used by many programmers is to use filtering to display their alphabetically sorted disk directory one screen at a time. To do this, both of the MS-DOS programs, SORT.COM and MORE.COM, must be accessible. From the DOS prompt, you would enter

A> DIR | SORT | MORE

I/O Levels

There are actually two levels of I/O. The one you have been working with is called the **standard I/O**. The other is called the low-level I/O. The advantage of the standard

I/O is that it requires less programming detail on your part. Its disadvantage is that you have less control over the details of the I/O process, and it is slower than low-level I/O. For most programming tasks, standard I/O will meet the I/O requirements. The standard I/O C functions actually use the low-level I/O. Low-level I/O is usually referred to as **system I/O** and is the topic of the next discussion.

A Word about File Handles

MS-DOS defines file handle 0 to be standard input (the keyboard) and file handle 1 to be standard output (the monitor screen). Turbo C defines file handle 2 as the standard error output and 3 as serial port output, with 4 as standard printer output.

Random Access Files

Up to this point, the types of files you have worked with have been **sequential access files**. This means when you work with file data, you get all of the data in one chunk from the disk. In a **random access file**, a particular data item may be accessed while ignoring the rest of the file. Program 7–28 illustrates a random access file that uses the parts inventory program. To understand this program, you must first know what is meant by a file pointer.

File Pointers

A **file pointer** is simply a pointer to a particular place in the file. What a file pointer does is to point to the byte in the file where the next access will take place with the file. As an example, every time you access a disk file, the file pointer starts at position 0, the beginning of the file. Every time you write data to the file, the file pointer ends up at the end of the file. When you do an append operation to a file, the file pointer is first set to the end of the file before new data is written to the file. The concept of a file pointer is illustrated in Figure 7–11.

There is a C function called **fseek( )** that moves the file pointer. This function will move the file pointer to the position in the file where the next file access will occur. It is used in Program 7–28 in order to access a single record from any place within the file.

Program 7–28

```
#include <stdio.h>

  typedef struct
    {
      char part_name[15];      /*  Type of part. */
      int quantity;             /* Number of parts left. */
```

(continued)

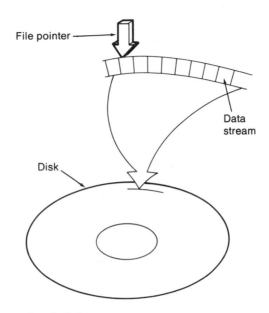

Figure 7–11   Concept of a File Pointer

Program 7–28 *(continued)*

```
     float cost_each;              /* Cost of each part.   */
   } parts_structure;

main( )
{
   parts_structure parts_data; /* Parts structure variable. */
   FILE *file_pointer;         /* File pointer. */
   int record_number;          /* Number of the record. */
   long int offset;            /* Offset of the record. */


/* Open the file for reading. */

   if((file_pointer = fopen("B:PARTS.DTA", "r")) = NULL) /* Error chk.*/
       {
           printf("Unable to open file B:PARTS.DTA");
           exit(-1);
       }
/* Get record number from user:  */

   printf("Enter the record number => ");
   scanf("%d",&record_number);

/* Compute the offset value of the selected record:  */

   offset = record_number * sizeof(parts_data);
```

(continued)

Program 7–28 *(continued)*

```
/* Go to the required file:  */

    if(fseek(file_pointer, offset, 0) != 0);  /* Error check. */
        {
            printf("Pointer moved beyond file boundary.");
            exit( );
        }

/* Read the selected file:  */

    fread(&parts_data, sizeof(parts_data),1,file_pointer);

/* Display the file data:  */

    printf("\nName of part => %s\n",part_data.part_name);
    printf("Number of parts => %d\n",part_data.quantity);
    printf("Cost of each part => %f\n",part_data.cost_each);

/* Close the opened file:  */

    fclose(file_pointer);
```

Program Analysis

The first part of the program is the same as the previous sequential access file I/O. The differences start with the variable declarations:

```
parts_structure parts_data: /* Parts structure variable */
FILE *file_pointer;         /* File pointer. */
int record_number;          /* Number of the record. */
long int offset;            /* Offset of the record.  */
```

Here, two new variables `record_number` and `offset`, are defined. The variable `record_number` will be the value of the record which the program user wishes to access. The variable `offset` will be the actual offset for the file pointer. This variable must be of type `long`. The offset is the number of bytes from a particular place in the file to start reading.

The file is first opened for reading with error checking:

```
if((file_pointer = fopen("B:PARTS.DTA", "r")) = NULL) /* Error chk.*/
    {
```

```
        printf("Unable to open file B:PARTS.DTA");
        exit(-1);
}
```

This is no different from a sequentially accessed file. The difference starts when the program user is asked for the record number. What this means is that the user must already know how many records are in the disk file and the significance of each record number. For example, the record number could correspond to a parts bin number.

```
printf("Enter the record number => ");
scanf("%d",&record_number);
```

Next, the value of **offset** must be computed. Note that this is done by multiplying the record number by the C **sizeof( )** function. The result will give the number of bytes the **offset** must have to pick up the record requested by the user.

```
offset = record_number * sizeof(parts_data);
```

Now the program uses the C **fseek( )** function to actually move the file pointer to the position of the record and starts to read the data. An error trapping statement is attached.

```
if(fseek(file_pointer, offset, 0) != 0); /* Error check. */
    {
      printf("Pointer moved beyond file boundary.");
      exit( );
    }
```

The C **fseek( )** contains three arguments. The first is the file pointer, next is the value of the offset and last is the mode. There are three values for the C **fseek( )** function mode. These are:

> 0 => Count from the beginning of the file.
> 1 => Start from the current pointer position.
> 2 => Start from the end of the file.

These values are defined as

0 => **SEEK_SET**,
1 => **SEEK_CUR**,
2 => **SEEK_END**.

The **fseek( )** function returns a non-zero value only if it fails to perform its required operation.

In the case of our program, the pointer is starting its count from the beginning of the file. It should be noted that a positive value offset moves the pointer toward the end of the file while a negative offset value moves the pointer toward the beginning of the file.

Next, the `/* Read the selected file: */`

```
fread(&parts_data, sizeof(parts_data),1,file_pointer);
```

Since the file pointer has now been moved, the C **fread( )** function is now used to do its read operation.

In the remaining part of the program, the data are displayed and the file is then closed:

```
printf("\nName of part => %s\n",part_data.part_name);
printf("Number of parts => %d\n",part_data.quantity);
printf("Cost of each part => %f\n",part_data.cost_each);
```

```
/* Close the opened file: */
```

```
fclose(file_pointer);
```

Command Line Arguments—The Basic Idea

One of the powerful features of C is its ability to allow you to use **command line arguments**. What this means is that when you execute a C program from the DOS prompt (by entering the name of the C `.EXE` file), you can enter other information at the same time that will influence the operation of the program. For example, if you have information that must be kept confidential in a program that displays a parts inventory (called `PARTS.EXE`), from the DOS prompt you could execute the program by entering

```
A> PARTS
```

and the program would execute, allowing the user to access information about any part—except its wholesale price. In order to get the information about the wholesale cost of the each part the program user would have to know a code string (say it was `cost_code_1`). This would constitute a command line and the same identical program would then be executed by

```
A> PARTS cost_code_1
```

Now when the program is executed, the program user can also get the information about the wholesale price of each part.

Developing Command Line Arguments

A command line argument is developed within function **main( )**. This is what those parentheses following **main** are for, to place the command line arguments inside them. This is illustrated in Program 7–29.

Program 7–29

```c
#include <stdio.h>

main(int argc, char *argv[])      /* Command line arguments. */
{
   int counter;       /* For counting number of arguments. */

   /* Display the number of arguments entered by the user.  */

    printf("The number of arguments you entered is: %d\n",argc);


   /* Show each of the entered arguments:  */

   for(counter = 0; counter < argc; counter++)
     printf("Argument %d is %s\n", counter, argv[counter]));

}
```

If the above program is compiled and saved under the name of **MYSTUFF.EXE**, then execution of the program from the DOS prompt with the following commands will produce the following output:

```
A> MYSTUFF ONE TWO THREE
The number of arguments you entered is: 4
Argument number 0 is MYSTUFF.EXE
Argument number 1 is ONE
Argument number 2 is TWO
Argument number 3 is THREE
```

The format for a command line argument is shown in Figure 7–12.

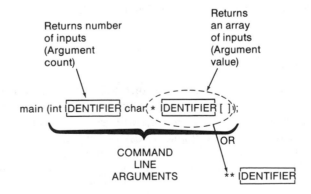

Figure 7–12 Format for Command Line Arguments

Note that the two arguments `argc` and `argv` represent the number of command line arguments and an array of pointers. Observe that the first argument (`argv[0]`) is always the pathname of the program. The identifiers `argc` (for **arg**ument count) and `argv` (for **arg**ument value) are traditional. Any legal C identifier may be used instead.

Conclusion

This section presented some of the technical details concerning C I/O. Here you saw some more of the important disk file details. Material presented here will be used in the development of the case study in the next section. Check your understanding of this section by trying the following section review.

7-9 Section Review

1   Name the two kinds of data streams used in I/O operations.
2   Describe what is meant by a buffer in I/O.
3   Name the three standard files used by a C program.
4   What are the redirection operators used by DOS? Explain what they mean.
5   State what is meant by a command line argument.

7-10   Case Study

Discussion

The case study for this chapter shows the development of a simple data base system. The system illustrates some of the major points of data base management such as saving data to a disk file, retrieving it from the file, and an example security system.

The Problem

Create a C data base program that will store data on an assortment of up to 50 parts. Each assortment is to contain the name of the part, quantity of parts, and price of each part. The program will save this data to a disk file on drive B:. The program user may randomly access this file to get information on any one of the part assortments. Only an authorized program user may input data to be stored in the disk file. This prevents the files being purged by unauthorized users.

First Step—Stating the Problem

Stating the case study in writing yields

- Purpose of the Program: Provide for the creation of an array of 50 structures. Each structure shall contain record number, name of item, quantity of items, and cost of each item. Structure array is to be stored on disk file in drive B: to be modified by authorized users only. Any user may have random access to any single record in the disk file.
- Required Input: User identification (to select personal authorized to store data in file). User selection to read file, and (if authorized) to enter new data to file. User selection for getting any stored file from the disk. Authorized user input of: Item name, Quantity, and Cost.
- Process on Input: Store arrayed data to a disk file on drive B: for authorized personnel. Random access records from same disk file for any program user.
- Required Output: Two menus, one for users authorized to store new files and one for any other program user. Prompts for program repeats. Display of file contents randomly accessed by user.

Developing the Algorithm

The steps in this data base management program are

1. Check user authorization.
2. Explain program to user.
3. Get user options.

> If authorized => Enter new data.
> Any user => View any record in the file.

4. Display output record.

This program will use a file writing function that writes arrays to the disk file and a file reading function that will allow random access to any single array stored in the file.

Program Development

Good programming practice discourages the use of global variables. Good programming practice encourages the use of block structure and top-down design. The programmer's block is developed first.

Since the Learn C environment could not be used because it will not compile a stand-alone .**EXE** file necessary for the command line arguments, a Turbo C program was used for the program instead. There are a few cases where functions that are Turbo-specific are used in the program. In these cases, the comments state −**Turbo**−.

The complete program is shown in Program 7–30.

Program 7-30

```c
#include  <stdio.h>
#include  <string.h>
#include  <conio.h>
#include  <math.h>

#define  TRUE 1
#define  FALSE 0


/*****************************************************************/
/*                Parts Inventory Program                     */
/*****************************************************************/
/*              Developed by: A. C. Programmer                */
/*****************************************************************/
/*                Date: November 15, 1992                     */
/*****************************************************************/
/*   This program will inventory up to 50 different assortments*/
/* of parts. Each part assortment contains data about the name  */
/* of the part, number of parts, and the price of each part.    */
/*   The program will store and retrieve this information to    */
/* a disk file called PARTS.DTA. The program will allow only    */
/* authorized users to enter data to the disk file. No special  */
/* authorization is required to read from the files.            */
/*   Authorization for writing to the file is obtained by the   */
/* command line argument "Parts Access 123".                    */
/*****************************************************************/
/*                Type Definitions                            */
/*-------------------------------------------------------------*/
  typedef struct
    {
    char      part[20];          /* Type of part. */
    int       quantity;          /* Number of parts left. */
    double    price;             /* Cost of each part. */
    int       record_number;     /* Number of the parts record. */
    } parts_record;

/* This type definition defines each part assortment.          */
/*****************************************************************/
/*                Function Prototype                          */
/*-------------------------------------------------------------*/
void explain_program(void);
/*   This function explains the operation of the program to the */
/*   program user.                                              */
/*-------------------------------------------------------------*/
void press_return(void);
```

(continued)

Program 7-30 *(continued)*

```
/*    This function holds the screen for the program user and     */
/* displays a prompt to press the -RETURN- key when ready to      */
/* continue.                                                      */
/*--------------------------------------------------------------*/
char read_menu(int authorization);
/*     This function displays the selection menu for the program */
/* user and returns the selected value.                           */
/*--------------------------------------------------------------*/
void enter_data( );
/*     This function allows the program user to enter new data    */
/* into the file.                                                 */
/*--------------------------------------------------------------*/
void read_data(int record_number);
/*    This function retrieves a given record number from the disk*/
/* file.                                                          */
/*--------------------------------------------------------------*/
main(int argc, char *argv[ ])
{
 char in_string[10];              /* User input string. */
 char ch;                         /* Utility character.  */
 static int repeat = TRUE;        /* Repeat flag. */
 int index;                       /* Record number. */
 int authorization;               /* User authorization. */

  /* Test for user file writing authorization. */

    if(strcmp(argv[1], "Parts Access 123") == 0)
     authorization = TRUE;
    else
     authorization = FALSE;

     explain_program( );       /* Explain program to the user. */

   while(repeat)
    {

     ch = read_menu(authorization);     /* Get user selection. */
     printf("\n\n");                     /* Give some space. */

 /* Act on the input. */

    switch(ch)
     {
     case 'R' : {
                 printf("\n\n Enter record number => ");
```

Program 7-30 *(continued)*

```c
                    gets(in_string);
                    index = atoi(in_string);
                    read_data(index);
                    }
                    break;
        case 'X' : repeat = FALSE;    /* Set flag false. */
                    break;
        case 'E' : if(authorization)
                    {
                    enter_data( );    /* Enter new data. */
                    break;
                    }
        default  : printf("That was not a correct selection:\n");
        } /* End of switch  */
    } /* End of repeat  */

    } /* End of main( ) */

/*----------------------------------------------------------------*/

void explain_program(void) /* Explains program to the user. */
{
 char *string;              /* String to hold program explanation. */

  clrscr( );    /* Clear the screen -Turbo- */

  string = "\n\n\n"
     "    This program will allow you to store and retrieve\n"
     "data on up to 50 assortments of parts. Each part\n"
     "assortment lets you enter and retrieve the name of\n"
     "the part, number of parts left and price each.\n"
     "    Entering new data is available to authorized personnel.\n";

  printf("%s",string);  /* Print the above string. */

  press_return( );       /* Holds screen for the user. */

  }   /* End of explain_program( )  */

/*----------------------------------------------------------------*/

void press_return(void)    /* Holds screen for the user. */
{
```

Program 7-30 *(continued)*

```c
    gotoxy(10,25);     /* Places courser at bottom of screen. -Turbo- */
    printf("Press -RETURN- to continue.");

    getchar( );       /* Wait for user response. */

    clrscr( );        /* Clears the screen. -Turbo- */
}   /* End of press_return( )  */

/*-------------------------------------------------------------*/

char read_menu(int authorization)     /* Display the menu.  */
{
 char ch;      /* User selection.  */

 /* Display the menu.  */

    printf("\n\n Select one of the following by letter:\n\n");
    if(authorization)
      printf(" E] Enter new data.");
      printf(" R] Read data files.\n");
      printf(" X] Exit this program.");

 /* Get user selection.  */

    ch = toupper(getche( ));

      return(ch);

}   /* End of read_menu( ) */

 /*-------------------------------------------------------------*/

void enter_data( )    /* Enter new data.  */
{
 FILE *file_ptr;                  /* File pointer. */
 parts_record parts_array[50];    /* Up to 50 assortments. */
 char in_string[10];              /* User input string. */
 int index;                       /* Record number. */
 int repeat;                      /* Continue data input flag. */
 char ch;                         /* User input response. */

   repeat = TRUE;                 /* Set continue data flag. */
   index = 1;                     /* Initialize the index. */
```

(continued)

Program 7-30 *(continued)*

```
while(repeat)
 {

  /* Get name of part:  */

      printf("\n Part name => ");
      gets(parts_array[index].part);

  /* Get number of parts left: */

      printf("Number of parts left = ");
      gets(in_string);
      parts_array[index].quantity = atoi(in_string);

  /* Get cost of each part: */

      printf("Cost of each part => ");
      gets(in_string);
      parts_array[index].price = atof(in_string);

  /*  Set record number  */

      parts_array[index].record_number = index;

  /* Ask for more data.  */

      printf("Enter more data (Y/N) => ");
      ch = toupper(getche( ));

      index++;   /* Increment the index. */
if(ch != 'Y')
  repeat = FALSE;    /* Break out of loop. */

  } /* End of while. */

/* Enter data into disk file:  */

  /* Open the file for writing.  */

  if((file_ptr = fopen("B:PARTS.DTA","wb"))==NULL)
    {
```

(continued)

Program 7-30 *(continued)*

```
                printf("Can't open parts file.");
                exit( );
                }

      /* Enter the file data.  */

   fwrite(parts_array, sizeof(parts_array[1]), index, file_ptr);

      /* Close the opened file. */

          fclose(file_ptr);

}   /* End of enter_data( ) */

 /*--------------------------------------------------------------*/

void read_data(int record_number)
{
 FILE *file_ptr;                /* The file pointer. */
 parts_record record_file;      /* Structure for reading file. */
 long int offset;               /* Offset value for random access. */

 /* Open the file for reading.  */

  if((file_ptr = fopen("B:PARTS.DTA","rb"))==NULL)
      {
       printf("Can't open parts file.");
       exit( );
      }

 /* Compute the offset. */

  offset = record_number * sizeof(record_file);

/* Move the file pointer. */

  if(fseek(file_ptr, offset, SEEK_SET) != 0)
     {
      printf("Can't find the file.");
      exit( );
     }
```

Program 7–30 *(continued)*

```
/* Retrieve the file data.   */

  fread(&record_file, sizeof(record_file), 1, file_ptr);

/* Close the opened file. */

  fclose(file_ptr);

  printf("\nRecord %d\n",record_file.record_number);
  printf("Item => %s\n",record_file.part);
  printf("Quantity => %d\n",record_file.quantity);
  printf("Cost each => %f\n",record_file.price);

}   /* End of read_data( ) */

/*-----------------------------------------------------------------*/
/*  End of program.   */
```

Program Analysis

The arguments in function **main( )** get command line input from the user. This will allow authorization for putting data into the disk files. Notice that the command line arguments use double quotes. This is done so that spaces can be placed on the input string.

In the C **switch** function of **main( )**, the **case 'E'** placed before the **default** is there so that if **authorization** is FALSE, then the message in the **default** section will be displayed. Observe the functions that are unique to Turbo C. One is **clrscr( )** which clears the screen and brings the cursor to the upper left corner of the monitor. The other function is **gotoxy(x, y)** which places the cursor at the location indicated by **x** (horizontal) and **y** (vertical) on the monitor screen. From this position, text can then be displayed.

Observe the use of the function **toupper( )**. This allows the program user to enter either upper or lowercase in response to menu prompts. The member **price** must be of type **long** because the **atof( )** function converts the user input to this type. Information in the disk file is in binary mode. This is because the **fopen( )** function for entering and receiving data uses the "b" extension in its file directive.

Conclusion

This section presented the development of a simple data base system that uses disk file access and file security. The Self-Test at the end of this chapter includes specific questions about this program.

Interactive Exercises

DIRECTIONS

These exercises require that you have access to a computer and software that supports C. They are provided here to give you valuable experience and immediate feedback on what the concepts and commands introduced in this chapter will do. They are also fun.

Exercises

1   Program 7–31 illustrates the C **enum**. Try it.

Program 7–31

```
#include <stdio.h>

  enum numbers {One, Two, Three}

  main( )
  {
  enum numbers value;

          value = One;
  }
```

2   The next program illustrates the C **enum** with constants assigned to the identifiers. Do you think you know what the output will be? Try it.

Program 7–32

```
#include <stdio.h>

  enum numbers {One = 1, Two = 2, Three = 3}

  main( )
  {
  int result;

          result = One + Two;
          printf("Result = %d",result);
  }
```

3   Program 7–33 illustrates a counting loop using the C **enum**. Predict what the output of the program will be, then try it.

Program 7-33

```c
#include <stdio.h>

enum numbers {Zero, One, Two, Three}

main( )
{
enum numbers number;

    for(number = Zero; number <= Three; number++)
      printf("Digital value of enum type number = %d\n",number);
}
```

4 Program 7-34 attempts to illustrate the smallest possible structure. See if the output you get is what you would expect.

Program 7-34

```c
#include <stdio.h>

main( )
{
   struct {int number;} value;

       value.number = 5;
       printf("Value is %d",value.number);
}
```

5 Note how the next program declares the member declaration list. Will this work? Give it a try.

Program 7-35

```c
#include <stdio.h>

main( )
{
   struct {int number1, number2, number3;} value;

       value.number1 = 1;
       value.number2 = 2;
       value.number3 = 3;

       printf("Value of number1 is %d",value.number1);
}
```

6   Program 7–36 shows a use of the C **define** to replace a `long_identifier`. This is an interesting program to try.

Program 7–36

```
#include <stdio.h>

#define S long_identifier

 main( )
 {
   struct {int number1, number2, number3;} long_identifier;

      S.number1 = 1;
      S.number2 = 2;
      S.number3 = 3;

      printf("Value of number1 is %d",S.number1);
 }
```

7   Program 7–37 illustrates the use of a structure tag. See what the output is. Did you get what you expected?

Program 7–37

```
#include <stdio.h>

   struct tag

      {
        int first;
        int second;
      }

 main( )
 {
   struct tag structure;

      structure.first = 1;

      printf("Value of structure.first is %d",structure.first);
 }
```

8   The next program shows a structure utilizing a C **typedef**. See if the output of this program is not the same as the last program.

Program 7-38

```
#include <stdio.h>

    typedef struct
       {
        int first;
        int second;
       } name;

main( )
{
  name structure;

     structure.first = 1;

     printf("Value of structure.first is %d",structure.first);
}
```

9   Do you see the differences between Program 7-38 and Program 7-39. Do you know why the outputs are different?

Program 7-39

```
#include <stdio.h>

 typedef struct
       {
        int first;
        int second;
       } name;

main( )
{
  name structure;

     structure.first = 11;

     printf("Value of structure1.first is %d",structure.first);

}
```

10   Program 7-40 illustrates an example of a C **union**. Recall that a union allows the same memory location for variables of different types. Predict first what you think this program will do, then try and check the results!

Program 7-40

```
#include <stdio.h>

 typedef union
     {
      int first;
      float second;
      } value;

 main( )
 {
   value number1;

  number1.first = 12;
   printf("Value of number1.first %d\n",number1.first);

  number1.second = 34.5;
   printf("Value of number1.first %d\n",number1.first);

 }
```

11 Program 7-41 shows a structure within a structure. Analyze the program before trying it. Then check to see if it performs as expected.

Program 7-41

```
#include <stdio.h>

 typedef struct
     {
      int first;
      } structure1;

 typedef struct
   {
    structure1 second;
   } structure2;

 main( )
 {
  structure2 new_structure;

     new_structure.second.first = 25;
     printf("The value is %d",new_structure.second.first);

 }
```

12   Program 7–42 is an expansion of the previous program. It has increased the complexity of the data structure by making an array of a structure that contains another structure as its member. Study the program and see if it behaves as you expected.

Program 7–42

```
#include <stdio.h>

typedef struct
    {
      int first;
    } structure1;

typedef struct
    {
      structure1 second;
    } structure2;

typedef structure2 struct_array[5];

main( )
{
 struct_array values;

    values[1].second.first = 54;
    printf("The value is %d",values[1].second.first);

}
```

13   Program 7–43 will place a new file on your active disk. After you try the program, look at your disk files. What new file do you see?

Program 7–43

```
#include <stdio.h>

main( )
{
 FILE *file_ptr;

    file_ptr = fopen("NEWFILE.DAT", "w");

    fclose(file_ptr);

}
```

14 The next program puts something in the new file created on your disk by Program 7–43. Make sure to execute this program before trying Program 7–44.

Program 7–44

```
#include <stdio.h>

main( )
{
 FILE *file_ptr;

    file_ptr = fopen("NEWFILE.DAT", "w");

         putc('C', file_ptr);

    fclose(file_ptr);

}
```

15 The next program gets from the file what Program 7–44 put into it. Try it and see if you get what you expected.

Program 7–45

```
#include <stdio.h>

main( )
{
 FILE *file_ptr;
 char ch;

    file_ptr = fopen("NEWFILE.DAT", "r");

         ch = getc(file_ptr);
         printf("The character is %c",ch);

    fclose(file_ptr);

}
```

Self-Test

DIRECTIONS

Answer the following questions by referring to Program 7–30 in the case study section of this chapter.

1 How many different structure members are contained in the program? Name them.
2 What type of a function is **read_menu( )**? Explain.
3 Are there any global variables used in the program? If so, name them.
4 State the purpose of the arguments in function **main( )**. What are they called?
5 In the C **switch** in function **main( )**, why was the **case 'E'** placed just before the **default**?
6 What functions are unique to Turbo C? What do they do?
7 Are there any **scanf( )** functions used for user input in the program? Explain.
8 Will the program respond to the user's selection if uppercase or lowercase letters are entered during the menu prompt? Explain.
9 Why is it necessary to declare the **price** member as a type **double** in the structure?
10 Is the information in the disk file in binary or text mode? How did you determine this?

End-of-Chapter Problems

General Concepts

Section 7–1

1 What data type in C is used to describe a discrete set of integer values?
2 Give the integer value of the first declared enumerated data type in C.
3 Does the C **enum** create a new data type? Explain.
4 Demonstrate how the C **enum** data type can be assigned an integer value.

Section 7–2

5 Does the C **typedef** create a new data type? Explain.
6 State the purpose of the C **typedef**.
7 Give the resulting new name of the data type for the following code:

```
typedef struct {
                int valuel;
        } structure;
```

Section 7–3

8 Define a C structure.
9 Give an example of a common everyday system that utilizes the concept of a C structure.
10 Illustrate the syntax of a C structure as presented in this section.
11 Explain what the C member of operator does. What symbol is used for this operator for a simple structure?

Section 7-4

12 What is the identifier that names the structure type defined in the member declaration list of the structure called?

13 Give an example of the use of a structure tag.

14 What is the purpose of the $->$ symbol in C as applied to C structures?

15 State the three operations that are allowed with structures.

Section 7-5

16 State how one or more different data types may be stored in the same memory location.

17 State the difference between declaring a C structure and a C union.

18 Can a C structure be initialized by the program in which it is written?

19 Explain what the following line of code represents:

```
variable[2].number
```

Section 7-6

20 Can C structure have a member that is another structure? Explain.

21 Can a C array of structures contain an array as one of its members? Explain.

22 Can a C array of structures be a member of another array of structures? Explain.

Section 7-7

23 Is the following a legal DOS reference to a file name?

```
B:FILER.02
```

24 Explain what the DOS file name reference of problem 23 above means.

25 In C how is a file pointer declared? Show by example.

26 When performing disk file I/O with C, what must always be done before the program is terminated?

Section 7-8

27 State the different methods of reading and writing data with the C language.

28 Give the three basic C file command strings.

29 If the command for writing to a disk file in C is given, and the file does not exist, what will happen?

Section 7-9

30 State the two kinds of data streams used by C.

31 Name the three standard files used by C.

32 What is the physical representation of the three standard files used by C?

33 Can the physical representation of a standard file be changed by software? Explain.

Program Design

In developing the following C programs, use the methods developed in the case studies chapter sections. This means to use top-down design and block structure with no global variables. The function **main( )** should do little more than call other functions. Use pointers when necessary to pass variables between functions. Be sure to include all of the documentation in your final program. This should consist of, but not be limited to, the programmer's block, function prototypes, and a description of each function as well as any formal parameters you may use.

Electronics Technology

34   Modify the case study program (Program 7–30) so that the total cost of any bin is computed and displayed. This should be an added option on the menu.

35   Modify the case study program (Program 7–30) so that a second command line authorization code is necessary in order to get any pricing data about the parts (this is in addition to the command line argument for the existing authorization code for entering new data to the disk files).

36   Expand the Case Study program (Program 7–30) so that new data may be added without purging the existing data to the disk file.

37   Change the case study program (Program 7–30) so that the program user may have the records displayed in alphabetical order according to the name of the part.

38   Modify the case study program (Program 7–30) so that there is another member of the structure. This member will contain the minimum number of parts allowed for each type before reordering is necessary. If the quantity of parts falls below this number, then every time the program is activated, the program user is automatically alerted to which parts need to be reordered.

39   Create another structure for the case study program (Program 7–30) that will contain the name of the parts manufacturer, street address, city, state and phone number. This new structure will now become a member of the existing structure.

Business Applications

40   Develop a C program that uses a structure that will allow the program user to input the following data concerning a business client: name, address, phone, and credit rating (good or bad).

Computer Science

41   Create a C program that demonstrates an arrayed structure that contains an arrayed structure element.

Drafting Technology

42   Design a C program that uses a structure to keep the following information on the status of a design project: ID Number, project name, client name, due date, project completed (yes or no).

Agriculture Technology

43   Create a C program using a structure that keeps information on five different herds of cattle, each herd with a maximum of 30 cattle. The information entered by the program user is: sex, age, weight, location, and if the sex is female, milk production in quarts/day, if male, estimated market value.

Health Technology

44   Develop a C program that uses a structure for the following information about different patients for a private practice: name, address, date of birth, sex, dates of visitation, amount owed, and medical problem.

Manufacturing Technology

45   Create a C program using a structure that will keep track of three different production schedules each of which contains the following information: production ID Number, manufactured item, customer name. The program is also to contain a structure on up to five employees within the production schedule structure. This should contain the following information: employee ID number, name, title, and hourly wages.

Business Applications

46 Modify the program in problem 40 so that the information entered by the program user may be saved to a disk file. Add the requirement of user identification for access to the file by the use of a command line argument.

Computer Science

47 Develop a C program that will allow the program user to use the computer as a simple word processor. The program must allow the user to save and retrieve text files, name each file, and delete old files.

Drafting Technology

48 Expand the program in problem 42 so the information entered by the program user is saved into a disk file.

Agriculture Technology

49 Modify the program in problem 43 in order for the information entered by the program user to be saved in a disk file. The program is to also require user identification through the use of a command line argument.

Health Technology

50 Expand the program in problem 44, so that the patient information entered by the program user is saved to a disk file. Add to this the requirement of user identification through the use of a command line argument.

Manufacturing Technology

51 Modify the program of problem 45 so the information entered by the program user will be saved to a disk file. Include with this the requirement for user identification through the use of a command line argument.

8 Color and Technical Graphics

Objectives

This chapter provides you the opportunity to learn:

1 How to produce text color.
2 How computer screens are used to represent graphical displays.
3 What is needed in your computer system to produce graphics and how to find out what graphic capabilities it has.
4 The concept of a pixel and how to use it in graphics.
5 How color is generated on the IBM and most compatibles.
6 The method of drawing lines on the graphic screen.
7 Some of the special Turbo C built-in graphics.
8 Ways of creating bar graphs for technical analysis.
9 Methods of generating different graphic fonts.
10 Methods of graphing mathematical functions.
11 Techniques of styling your program to make it easier to read, debug, and modify.

Key Terms

Monochrome
Default
Color Constants
Text Screen

Graphics Screen
Graphic Mode
Graphics Adapter
Graphics Driver

Autodetection	Font
Pixel	Bit Mapped
Mode	Stroked Font
Color Palette	Scaling
Area Fill	Coordinate Transformation
Aspect Ratio	Casting
Fill-Style	

Outline

8-1 C Text and Color   8-5 Bars and Text in Graphics
8-2 Starting Turbo C Graphics   8-6 Graphing Functions
8-3 Knowing Your Graphics System   8-7 Programming Style
8-4 Built-in Shapes   8-8 Case Study

Introduction

This chapter introduces the exciting world of color. Because of the diversity of commands, Turbo C is explained in this section, and the differences with Microsoft Quick C are explained in Appendix E.

As you will learn in this chapter, both of these C systems possess the power to give text screens a new dimension by using color.

This chapter also introduces the power of the computer graph. One of the fastest growing areas in computers is in the world of computer graphics! Understanding computer graphics will help you present information as drawings and graphs.

You will also learn the secrets of adding color to graphics. The use of different text fonts is discussed. This chapter will require patience and practice—but the personal and professional rewards of this new skill far exceed your time investment. Technical programmers who understand graphic programming skills are on the cutting edge of a new technology. Starting this chapter is your exciting first step.

8-1   C Text and Color

Discussion

This section demonstrates how to employ color in your C programs. Technical programs that use color can take advantage of a new dimension of useful information. For example, imagine a program to check the progress of an automated assembly line or to monitor a power plant. Green text could be used to indicate all systems are normal, yellow text could indicate that something requires attention, red text could mean immediate attention is needed, and blinking red text could mean possible danger.

What Your System Needs

In order to produce color using Turbo C 2.0, your computer system must meet the following three requirements:

1. Have a color monitor.
2. Have a color graphics card installed.
3. Be an IBM PC, AT, PS/2 or true compatible.

The color commands for text are defined in `conio.h`. This means you must use **#include <conio.h>** in those programs that use these features.

The Color Monitor

There are two different basic types of monitors used with your personal computer. One is called a **monochrome** monitor (meaning one color); the other is called a color monitor and is capable of producing a rich variety of different colors all at the same time.

System monitors have a coating of phosphor on the inside of the display screen. Electrical currents controlled by the computer strike this phosphor and cause it to glow. This glow emits light, the color of which depends upon the type of phosphor used to coat the inside of the display screen. It is this action that allows you to see images displayed on your monitor screen.

In a monochrome monitor, only one type of phosphor is used. Thus, depending on the type of phosphor, the screen of a monochrome monitor may be green or amber or any other single color. With this type of monitor, it is not possible to change its single color.

A color monitor has three different colors of phosphors placed on the inside of the glass face in the form of thousands of tiny triads. These triads consist of the red, green, and blue light emitting phosphors. These are the primary colors used for color mixing. With a color monitor of this type, 16 different colors may be displayed using text at the same time.

Display Capabilities

If your system can produce color, then it is also capable of producing two different sizes of text. One size is the 80 column by 25 row standard that you are used to. This is the **default** text size, meaning the size automatically used by your system when you turn it on. The second size produces larger text. It is 40 columns by 25 rows. All of these variations, including the color capabilities for text, are illustrated in Figure 8–1.

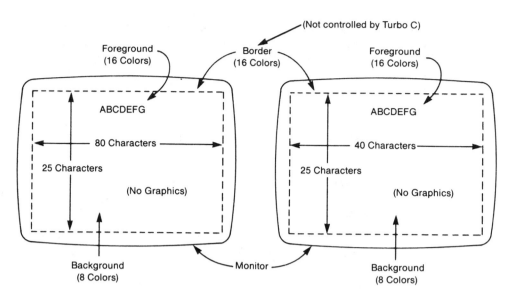

Figure 8–1   Text Display Capabilities

The **mode** Function

Turbo C has a built-in function for selecting either 40 or 80-column text. Since your system normally comes up in 80 columns, the following program will cause it to switch to 40 columns (and will only work if you have a color graphics card installed in your computer system).

Program 8–1

```
#include<stdio.h>
#include<conio.h>

  main( )
  {
    textmode(0);    /* Changes to 40 column wide text. */

      printf("This is 40 columns wide.\n");
      printf("It only works on systems with\n");
      printf("a graphics card installed.\n");
  }
```

The built-in C function is

textmode(int newmode);

Where

      Mode = An integer as described in Table 8–1.

Table 8–1   Results of Integer Values for **textmode**

Mode Value	Results
0	Produces 40×25 monochrome text. An installed graphics card is required.
1	Produces 40 × 25 color text. An installed color graphics card is required.
2	Produces 80 × 25 monochrome text. An installed color graphics card is required.
3	Produces 80 × 25 color text. An installed color graphics card is required.
7	Produces 80 × 25 monochrome text. No color graphics card installed in the system.

Note from the above table that you cannot get 40-column text without a graphics card installed in your system. The following program gives an example of the use of all variations for the built-in **textmode( )**.

Program 8–2

```
#include <stdio.h>
#include <conio.h>
 main( )
  {
     printf("This text is in the default 80 columns wide\n");
     printf("by 25 rows down.\n");
     getch( );

    textmode(0);    /* Changes to 40 × 25 Monochrome.   */
     printf("This text is now 40 columns wide\n");
     printf("but still 25 rows down.\n");
     getch( );

    textmode(1);    /* Changes to 40 X 25 Color.  */
      printf("Still a 40 column text, but on a \n");
      printf("color monitor display, it may reduce\n");
      printf("color streaking.\n");
     getch( );

    textmode(2);    /* Changes to 80 × 25 Monochrome.   */
      printf("This is the 80 column standard you are used to.\n");
      printf("it is 25 rows down.\n");
     getch( );

    textmode(3);     /* Changes to 80 × 25 Color. */
      printf("Still 80 columns but on a color monitor display, it\n");
      printf("may reduce color streaking.\n");
     getch( );

  /* textmode(7);     80 X 25 Monochrome - no graphics card required.
      printf("Still 80 columns, but for those systems that do not have\n");
      printf("a graphics card installed.\n");
      getch( ); */

 }
```

Whenever the **textmode( )** is used, the entire screen is cleared. Thus, for the above program, the screen is cleared every time the mode is changed by **textmode( )**. Note that the last **textmode(7)** has been placed in the form of a comment. The reason for this is that if you do have a color graphics card installed and try to invoke this command, you system may "hang up."

For convenience, Turbo C has defined constants that represent the numerical values of the **Textmode** procedure. These are defined as follows:

Mode Number	Mode Identifier
0	BW40
1	C40
2	W80B
3	CO80
7	MONO

Thus, the above program could have been written with **textmode(BW40)** instead of **textmode(0)**. The same is true for the other modes. Using the mode identifier rather than the mode number makes the code easier to read.

Text Color

The built-in Turbo function called **textcolor( )** will change the color of your text as shown in the following program.

Program 8–3

```c
#include <stdio.h>
#include <conio.h>

main( )
{
 int color_number;
 char color_name[10];

    for(color_number = 0; color_number <= 15; color_number++)
    {
     switch (color_number)
            {
                case 0 : strcpy(color_name, "black");
                         break;
                case 1 : strcpy(color_name, "blue");
                         break;
                case 2 : strcpy(color_name, "green");
                         break;
                case 3 : strcpy(color_name, "cyan");
                         break;
```

(continued)

Program 8–3 *(continued)*

```
              case 4 : strcpy(color_name, "red");
                       break;
              case 5 : strcpy(color_name, "magenta");
                       break;
              case 6 : strcpy(color_name, "brown");
                       break;
              case 7 : strcpy(color_name, "light gray");
                       break;
              case 8 : strcpy(color_name, "dark gray");
                       break;
              case 9 : strcpy(color_name, "light blue");
                       break;
              case 10 : strcpy(color_name, "light green");
                        break;
              case 11 : strcpy(color_name, "light cyan");
                        break;
              case 12 : strcpy(color_name, "light red");
                        break;
              case 13 : strcpy(color_name, "light magenta");
                        break;
              case 14 : strcpy(color_name, "yellow");
                        break;
              case 15 : strcpy(color_name, "white");
                        break;

     }  /* End of switch. */

   textcolor(color_number); /* This changes the color of the text.  */
    cprintf("This text is in %s\n\r",color_name);

 }  /* End of for. */

}
```

In the above program, you will not see the first **cprintf( )** function This text is in black because the background of your monitor is black. However, all the other colors will be displayed against this background.

Note that the **printf( )** function is not used when text is to be presented in color. Instead, the **cprintf( )** is required. Unlike the **printf( )** function, **cprintf( )** does not translate the (**\n**) linefeed into a carriage return/linefeed command. Thus, the **\n\r** is required in the **cprinf( )** function.

Again, for convenience, Turbo C has predefined **color constants** for the **textcolor** procedure as shown in Table 8–2.

Table 8-2 Turbo C Built-in Color Constants

Color Number	Color Constant
0	BLACK
1	BLUE
2	GREEN
3	CYAN
4	RED
5	MAGENTA
6	BROWN
7	LIGHTGRAY
8	DARKGREY
9	LIGHTBLUE
10	LIGHTGREEN
11	LIGHTCYAN
12	LIGHTRED
13	LIGHTMAGENTA
14	YELLOW
15	WHITE
+128	BLINK

Note the addition of BLINK in the above table. This command causes the text to flash on the screen. Thus textcolor (red + blink); will cause flashing red text for the next **cprintf( )** function. The following program uses the color constants in place of the color number.

Program 8-4

```
#include <stdio.h>
#include <conio.h>

main( )
{
 int color_number;
 char color_name[10];

    for(color_number = BLACK; color_number <= WHITE; color_number++)
    {
       switch (color_number)
          {
             case BLACK : strcpy(color_name, "black");
                          break;
              case BLUE : strcpy(color_name, "blue");
                          break;
             case GREEN : strcpy(color_name, "green");
                          break;
              case CYAN : strcpy(color_name, "cyan");
                          break;
```

(continued)

Program 8–4 *(continued)*

```
                    case RED : strcpy(color_name, "red");
                               break;
                case MAGENTA : strcpy(color_name, "magenta");
                               break;
                  case BROWN : strcpy(color_name, "brown");
                               break;
              case LIGHTGRAY : strcpy(color_name, "light gray");
                               break;
               case DARKGRAY : strcpy(color_name, "dark gray");
                               break;
              case LIGHTBLUE : strcpy(color_name, "light blue");
                               break;
             case LIGHTGREEN : strcpy(color_name, "light green");
                               break;
              case LIGHTCYAN : strcpy(color_name, "light cyan");
                               break;
               case LIGHTRED : strcpy(color_name, "light red");
                               break;
           case LIGHTMAGENTA : strcpy(color_name, "light magenta");
                               break;
                 case YELLOW : strcpy(color_name, "yellow");
                               break;
                  case WHITE : strcpy(color_name, "white");
                               break;

     }  /* End of switch. */

  textcolor(color_number); /* This changes the color of the text.  */
   cprintf("This text is in %s\n\r",color_name);

 }  /* End of for.  */

}
```

You can cause any text to blink, just by adding the value of 128 to the color value. The following program illustrates:

Program 8–5

```
 #include <stdio.h>
 #include <conio.h>

  main( )
  {
   textcolor(RED);
   cprintf("This text is now in red \n\r");
```

(continued)

Program 8–5 *(continued)*

```
    textcolor(4);
    cprintf("This text is also in red \n\r");

    textcolor(RED + BLINK);
    cprintf("This is blinking red text. \n\r");

    textcolor(4 + 128);
    cprintf("This is also blinking red text. \n\r");

    textcolor(RED);
    cprintf("This stopped the red text from blinking. \n\r");

    textcolor(LIGHTGREEN + 128);
    cprintf("This text is in blinking light green. \n\r");

}
```

Changing the Background

To change the background color in Turbo C, use the built-in Turbo function **textbackground(**int color**)**. This is illustrated in the following program:

Program 8–6

```
#include <stdio.h>
#include <conio.h>

main( )
{
 int color_number;
 int color_name[10];

  for(color_number = 0; color_number <= 7; color_number++)
    {
    switch(color_number)
      {
      case 0 : strcpy(color_name, "black");
              break;
      case 1 : strcpy(color_name, "blue");
              break;
      case 2 : strcpy(color_name, "green");
              break;
      case 3 : strcpy(color_name, "cyan");
              break;
```

(continued)

Program 8-6 (*continued*)

```
    case 4 : strcpy(color_name, "red");
            break;
    case 5 : strcpy(color_name, "magenta");
            break;
    case 6 : strcpy(color_name, "brown");
            break;
    case 7 : strcpy(color_name, "lightgray");
            break
    }  /* End of switch.  */

textbackground(color_number); /* This changes the color background. */

if(color_number == 7)
    textcolor(BLACK);

gotoxy(20,12);
cprintf("This background is in %s \n\r",color_name);
getch( );

}  /* End of for */

}
```

Note from the above program that there are only eight background colors possible. The **textbackground( )** function is:

textbackground(int color **);**

In the above program notice that the **textcolor( )** was changed when the background was light gray.

As with the **textcolor( )** function, Turbo C has identifier constants that represent the names of the colors. Thus **textbackground(BLACK)** has the same effect as **textbackground(0)**.

Conclusion

This section presented the important aspects of using color with text. Here you saw how to change the size of the text as well as how to invoke 16 different colors of text and 8 different background colors. Check your understanding of this section by trying the following section review.

8-1 Section Review

1   State what is needed in order to obtain text color.
2   What is needed to change the text size from 40 columns to 80 columns?
3   How many different colors can text have? Can this happen all on the same screen?

4 How many different background colors are available? What command is used to bring the screen to the desired color?

5 What choices are available to indicate the desired color? Give an example.

8-2   Starting Turbo C Graphics

Discussion

This section introduces you to the fascinating and challenging world of Turbo C graphics. In this section you will see what your system, software, and programming needs are in order to get started. You will also be introduced to the most fundamental graphics commands.

Basic Idea

If you have an installed color graphics adapter in your computer, then your monitor is capable of displaying two different kinds of screens. One screen is the **text screen** (sometimes referred to as text mode). The text screen is the one that you have been using up to this point. It does nothing more than display text—essentially characters of a preassigned shape such as the letters of the alphabet, numbers, punctuation marks, and the like. The other screen is called the **graphics screen** (sometimes referred to as the **graphic mode**). The graphics screen allows you to define your own shapes. Thus, when in graphic mode, you can create lines, graphs, charts, diagrams, pictures, and animation—essentially almost anything you can put on paper. Figure 8–2 illustrates the difference between the two modes of operation.

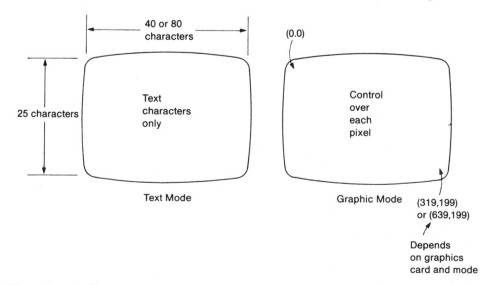

Figure 8–2   Difference between Text Mode and Graphic Mode

You can only be in one mode at a time, either in text mode or graphic mode. Your system comes up in text mode. In order to get into graphic mode, you must use special built-in Turbo C functions to get you there.

What You Need to Know

Before you can get into graphic mode with Turbo C, you must know what kind of a color **graphics adapter** your system is using. You must know this because the C program needs this information in order to use the correct built-in code. You will see how this can also be done automatically.

Turbo C 2.0 has six different programs, made especially for six different graphics adapters. This is why you need to know what kind of graphics adapter your system has—so Turbo will know which of the six different programs (called **graphics drivers**) to use. These graphic drivers all have the extension of .BGI for Borland Graphic Interface.

Fortunately, Turbo C has a built-in procedure that will allow you to find out which kind of installed graphics adapter is in your system. This is demonstrated by the following program.

Program 8-7

```
#include <stdio.h>
#include <graphics.h>       /* Necessary for graphics.  */
main( )
{
int graph_driver;                /* Specifies the graphics driver. */
int graph_mode;                  /* Specifies the graphic mode. */
char string[10];                 /* Gives the installed graphic driver. */

    /* Automatically detect installed graphics driver. */

      detectgraph(&graph_driver, &graph_mode);

      switch(graph_driver)
            {
              case 1 : strcpy(string, "CGA");
                      break;
              case 2 : strcpy(string, "MCGA");
                      break;
              case 3 : strcpy(string, "EGA");
                      break;
              case 4 : strcpy(string, "EGA64");
                      break;
              case 5 : strcpy(string, "EGAMONO");
                      break;
              case 6 : strcpy(string, "IBM8514");
                      break;
```

(continued)

Program 8-7 (*continued*)

```
           case 7 : strcpy(string, "HERCMONO");
                    break;
           case 8 : strcpy(string, "ATT400");
                    break;
           case 9 : strcpy(string, "VGA");
                    break;
           case 10 : strcpy(string, "PC3270");

    }  /* End of switch */

 if(graph_driver == -2)
  printf("\n\n No hardware detected.\n");
        else
  printf("\n\n Your installed graphics system is %s\n",string);

}
```

The above program automatically detects the kind of graphics hardware in your system. Using the program to do this is called **autodetection**. The Turbo built-in function used in the above program is:

void far detectgraph(int far *driver_address, int far *mode_address);

The C **detectgraph( )** function is a far pointer defined in the Turbo C **graphics.h** file. It takes two arguments which are addresses into which the system can place the type of adaptor and the mode numbers (you'll see what these mean shortly).

Selecting Your Graphics Driver

Once you know which installed graphics adapter you have in your system, then you must copy the appropriate graphic driver program from the original Turbo C 2.0 disks to your working disk. The Turbo C 2.0 system disks have the following graphic drivers:

> ATT.BGI = Graphics driver for AT&T 6300 graphics.
> CGA.BGI = Graphics driver for CGA and MCGA graphics.
> EGAVGA.BGI = Graphics driver for EGA and VGA
> HERC.BGI = Graphics driver for Hercules monographics.
> IBM8514.BGI = Graphics driver for IBM 8514 graphics.
> PC3270.BGI = Graphics driver for 3270PC graphics.

The Graphics Screen

Your graphics screen is divided up into **pixels**. You can think of a pixel as the smallest point possible that can be displayed on your graphics screen. Figure 8-3 presents the concept of pixels.

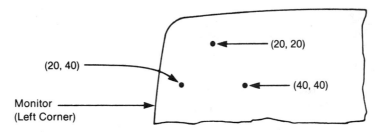

Figure 8-3   Concept of Pixels

How many pixels your graphic screen has depends on the kind of graphics adapter installed in your system. For example, CGA, MCGA, and ATT400 graphics adapters can have 320 pixels horizontally and 200 vertically. Each pixel is identified by a coordinate system of (X,Y) where X is the horizontal value of the pixel and Y is the vertical value of the pixel. The graphics screen for this hardware is shown in Figure 8-4.

Note that the top left pixel is identified as (0,0) the top right pixel (639,0), the bottom left pixel (0,199), and the bottom right is (639,199). Thus, each pixel on the graphics screen has a unique location that can be identified by the (X,Y) system.

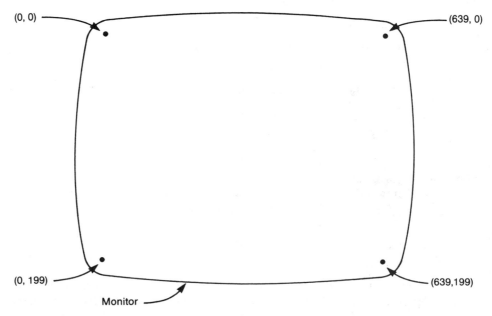

Figure 8-4   Graphics Screen for Specific Hardware

Lighting up a Pixel

The simplest graphics command is to light up a pixel on the graphics screen. To do this, there are three things your C program must do:

1. Get your system into the graphics mode with the correct graphics driver.
2. Activate the selected pixel.
3. Leave the graphics mode.

As you may suspect, Turbo C 2.0 has three built-in functions for doing the above three things. They:

1. Get you into the graphics mode:

```
void far initgraph(int far *driveradd, int far *modeadd, int far *pathadd);
```

This is a **far** function defined in **graphics.h**. The three arguments hold the addresses of the driver type, the mode address, and the address of the directory path where the graphic drivers are stored. If they are stored on the default drive, then this need not be specified.

2. Light up a pixel on the graphics screen:

```
int putpixel(int x, int y, int color);
```

Where
       x = An **int** representing the horizontal value of the pixel.
       y = An **int** representing the vertical value of the pixel.
    color = An **int** for number (or predefined color constant) indicating the color of the pixel.

3. Leave the graphics mode:

```
closegraph( );
```

This function shuts down the graphics system and restores the original screen mode before graphics was initialized.

The following program is for a system that has a CGA color graphics adapter installed.

Program 8–8

```
#include <stdio.h>
#include <graphics.h>     /* Necessary for graphics.  */

main( )
{
int graph_driver;          /* Specifies the graphics driver. */
int graph_mode;            /* Specifies the graphic mode. */

graph_driver = CGA;  /* For system with IBM color graphics adapter. */
graph_mode = 2;      /* Select a 320 X 200 screen with green, red, brown. */
```

(continued)

Program 8–8 *(continued)*

```
/* Get into graphics with selected driver and mode. */

 initgraph(&graph_driver, &graph_mode, "");

/* Plot points on the monitor screen.  */

   putpixel(160,100,0);  /* Black dot near screen center. */
   putpixel(162,100,1);  /* Green dot near center of screen. */
   putpixel(164,100,2);  /* Red dot near center of screen. */
   putpixel(166,100,3);  /* Brown dot near center of screen. */

 getchar( );      /* Hold graphics for observation. */

/* Shut down the graphics system. */

   closegraph( );

}
```

The above program places four color dots on the graphics screen. The function **putpixel(X,Y,COLOR)** puts a dot at the X and Y coordinates in the value indicated by color.

Note that the **getchar( )** function is used so that you may view the graphic. After you press the Return/Enter key, the **closegraph( )** function is initialized.

The above program uses **graph_mode := 2;**. The reason for this is explained below.

Using Color

The kind and amount of color available in graphic mode is much less than the rich color combinations you could get while in text mode. As a matter of fact, the kind and amount of color (if any) depends upon the installed graphics adapter in your system. Some graphics adapters have different **modes** of operation. As an example, the CGA graphics adapter for the above program has five different graphic modes numbered 0 through 4. Table 8–3 lists the modes available for different graphics hardware.

Table 8–3   Graphics Modes for Given Installed Hardware

Hardware	Mode # Const	Screen Size (Pixels)	Colors
CGA	0 CGA0	320 × 200	LightGreen, LightRed, Yellow
	1 CGA1	320 × 200	LightCyan, LightMagenta, White
	2 CGA2	320 × 200	Green, Red, Brown
	3 CGA3	320 × 200	Cyan, Magenta, LightGray
	4 CGAHI	640 × 200	(Color on black)

(continued)

Table 8–3 (*continued*)

Hardware	Mode # Const	Screen Size (Pixels)	Colors
MCGA	0 MCGAC0	320 × 200	LightGreen, LightRed, Yellow
	1 MCGAC1	320 × 200	LightCyan, Light Magenta, White
	2 MCGAC2	320 × 200	Green, Red, Brown
	3 MCGAC3	320 × 200	Cyan, Magenta, LightGray
	4 MCGAMED	640 × 200	(Color on black)
	5 MCGAHI	640 × 480	(Color on black)
EGA	0 EGALO	640 × 200	16 colors
	1 EGAHI	640 × 350	16 colors
EGA64	0 EGA64LO	640 × 200	16 colors
	1 EGA64HI	640 × 350	4 colors
	3 EGAMONO	640 × 350	(White on black)
HERCMONO	0 HERCMONOHI	720 × 348	(White on black)
ATT400	0 ATT400C0	320 × 200	LightGReen, LightRed, Yellow
	1 ATT400C1	320 × 200	LightCyan, LightMagenta, White
	2 ATT400C2	320 × 200	Green, Red, Brown
	3 ATT400C3	320 × 200	Cyan, Magenta, LightGray
	4 ATT400MED	640 × 200	(Color on black)
	5 ATT400HI	640 × 400	(Color on black).
VGA	0 VGALO	640 × 200	16 colors
	1 VGAMED	640 × 300	16 colors
	2 VGAHI	640 × 480	16 colors
PC3270	0 PC3270HI	720 × 350	(White on black)

As you can see from the above table, the CGA driver that will be used in the following example programs can display a maximum of only three colors at a time, from four different modes. It should be pointed out that in using the built-in Turbo C 2.0 function **initgraph( )**, Turbo has defined graphic driver constants corresponding to the number of the driver. These are listed as follows:

```
DETECT      = 0
CGA         = 1
MCGA        = 2
EGA         = 3
EGA64       = 4
EGAMONO     = 5
IBM8514     = 6
HERCMONO    = 7
ATT400      = 8
VGA         = 9
PC3270      = 10
```

Thus **initgraph(CGA,0, "")**; means initialize graphics mode using the CGA.BGI graphics driver in mode 0 (320 × 200 screen, colors light green, light red, and yellow). The CGA.BGI driver program will be on the active drive.

```
initgraph(1, 0, "");
```

means exactly the same thing. And so does:

```
initgraph(CGA, CGAC0, "");
```

Making an Autodetect Function

The following program illustrates a user defined function that will automatically detect the type of graphics hardware in your system and then automatically select the proper graphics driver. In order for this to work on any system, you must have all of the graphic drivers on the active disk or indicate where they are located with the **initgraph( )** function.

Program 8–9

```
#include <stdio.h>
#include <graphics.h>        /*  Necessary for graphics.  */

void auto_initialization(void);  /* Function to initialize to the
                                    graphics mode. */

main( )
{

    auto_initialization( );   /* Call to initialize graphics. */

    /* Graphic functions... */

    closegraph( );     /* Shut down the graphics system. */

}

void auto_initialization(void)   /* Initialize to graphics mode. */
{
int graph_driver;               /* Specifies the graphics driver. */
int graph_mode;                 /* Specifies the graphics mode. */

    graph_driver = DETECT;   /* Initializes autodetection. */
    initgraph(&graph_driver, &graph_mode, "");

}
```

The above function **auto_initialization( );** will be used with all of the graphics programs in the remainder of this chapter.

Conclusion

This section presented the necessary information you need in order to get into graphic modes using Turbo C 2.0 and IBM systems or true compatibles. Once you can get past this point, you will be able to enjoy the world of technical graphics available to you. Test your understanding of this section by trying the following section review.

8-2 Section Review

1   Name the two different kinds of screens available to you on your monitor.
2   State the difference between the graphic mode and text mode.
3   Before you can get your system into graphics operation with Turbo C 2.0, what one piece of hardware and two kinds of software will you need?
4   State how many different colors are available with color graphics.
5   Define the term pixel.

8-3   Knowing Your Graphics System

Discussion

This section will help you become familiar with your graphics system. Here you will learn about the different modes of operation for your graphics system and how to use them to your advantage in the creation of technical graphics. This section will also introduce some of the most fundamental graphic concepts and commands.

Your System's Screen Size and Colors

It's important that you know about your system's screen size and colors. If you don't, you might mistakenly try to draw lines on your graphics screen in black and never see anything against a black background. Or you might try drawing images that do not fit on your graphics screen. The size of the graphics screen and colors available (if any) depend largely on the kind of graphics hardware installed in your system.

Recall from the last section that the amount of color you can get in the graphic mode depends on the type of graphics hardware installed in your system. Table 8-3 lists the kinds of hardware and their various modes of operation. For example, from Table 8-3, a CGA (Color Graphics Adapter) has five modes of operation:

Table 8-4   Excerpt Graphics Modes for Given Installed Hardware

Hardware	Mode # Const	Screen Size (Pixels)	Colors
CGA	0 CGA0	320 × 200	LightGreen, LightRed, Yellow
	1 CGA1	320 × 200	LightCyan, LightMagenta, White
	2 CGA2	320 × 200	Green, Red, Brown
	3 CGA3	320 × 200	Cyan, Magenta, LightGray
	4 CGAHI	640 × 200	(Color on black)

The above is an excerpt from Table 8-3. It shows that for the first four modes of operation (modes 0 through 3) the screen size is 320 pixels horizontally and 200 pixels vertically. The first four modes have only three colors available at any one time. These different color modes are referred to as the **color palette** or simply the palette. Thus the palette for mode **0** of the CGA hardware is light green, light red, and yellow, while the palette for mode 1 is light cyan, light magenta, and white. The important point is that you cannot have all palettes at the same time (meaning you must be in one of the four color modes). Black is also available in all modes.

Now note that the fifth mode (mode 4) has more pixels horizontally (640) and the same number vertically (200). The background will be black and visible graphics will be a single color. Because of the makeup of the CGA adapter, this foreground color is actually set by the **setbkcolor( )** function! In this mode you can get more detail (for a CGA adapter), but you lose the ability to produce more than one color at a time.

From the above information, you can see how important it is to be able to

1. Determine the modes of your system.
2. Change the modes of your system.

First, look at how to determine the modes of your system.

Your System's Modes

The following program will determine what mode your graphics system is in as well as other useful graphic information.

Program 8-10

```
#include <stdio.h>
#include <graphics.h>

void auto_initialization(void);    /* Function to initialize to the
                                      graphic mode.    */
```

(continued)

Program 8-10 *(continued)*

```c
main( )
{
 int high_mode;        /* Highest mode range. */
 int low_mode;         /* Lowest mode range.  */
 int current_mode;     /* The current graphic mode. */
 int max_color;        /* Highest value color for current driver
                          and mode. */
 int max_horiz;        /* Maximum horizontal value. */
 int max_vert;         /* Maximum vertical value. */

    auto_initialization( );  /* Initialize graphic mode. */

 /* Determine mode range: */
 getmoderange(CGA, &low_mode, &high_mode);
 printf("\n\n CGA mode range is %d high to %d low.\n",high_mode, low_mode);

 /* Determine current mode:  */
    current_mode = getgraphmode( );
    printf(" The current graphic mode is %d.\n",current_mode);

 /* Determine maximum color value:  */
    max_color = getmaxcolor( );
    printf(" The maximum value color for this mode is %d\n",max_color);

 /* Determine maximum coordinates:  */
    max_horiz = getmaxx( );
    max_vert  = getmaxy( );
    printf(" The maximum horizontal value is %d\n",max_horiz);
    printf(" and the maximum vertical value is %d.\n",max_vert);

    getchar( );    /* Hold output for observation. */

 /* Shut down the graphics system. */
    closegraph( );

 }  /* End of main( ) */

 /*-------------------------------------------------------------------*/

void auto_initialization(void)    /* Initialize graphics mode. */
{
 int graph_driver;        /* Specifies the graphics driver. */
 int graph_mode;          /* Specifies the graphic mode.    */

 graph_driver = DETECT;  /* Initialization autodetection.  */
 initgraph(&graph_driver, &graph_mode, "");

 }  /* End of auto_initialization */
```

The results you get with the above program will depend upon graphics hardware installed in your system. If your system has an IBM CGA then the output will be

```
CGA mode range is 4 high to 0 low.
The current graphic mode is 4.
The maximum value color for this mode is 1.
The maximum horizontal value is 639
and the maximum vertical value is 199.
```

The important point about the above program is that it lets you know exactly where you are. What follows is an explanation of each of the functions used in the above program.

- Auto Initialization: This function automatically detects the type of graphics hardware and gets you into graphic mode. It will be used with every graphics program in this text. The function was explained in the last section.
- Mode Range: This function gives you the range of the modes available with your installed graphics hardware. The built-in Turbo C 2.0 function used here is

getmoderange(int graph_driver, int far *low_mode, int far *high_mode);

Where

GraphDriver = The graph driver for your system. (such as 1 or CGA)
low_mode = Returns the value of the lowest permissible mode.
hi_mode = Returns the value of the highest permissible mode.

- Current Mode: This function identifies the active mode which your system is currently in. The built-in Turbo C 2.0 function that returns the numerical value of the current mode is

int far getgraphmode(void);

Where

The value returned is the graphic mode set by **initgraph( )** or **setgraph( )**. (Refer to Table 8–3 for the significance of each graphic mode.)

- Maximum Colors: This function identifies the maximum number of colors available in the current mode. The built-in Turbo C 2.0 function that returns this value is

int far getmaxcolor(void);

Where

The value returned is the largest value valid color for the current graphics driver and mode that can be used. (In those modes where only white graphs will appear on a black background, 0 represents black and 1 represents white. In any mode, black is always 0, while the other numbers indicate colors that depend upon the installed graphics hardware and active palette.)

- Graphics Screen Size: This function gives the maximum size of the current graphics screen. There are two built-in Turbo C functions that allow you to get these values:

```
getmaxx;
getmaxy;
```

Where
> The value returned is the maximum screen coordinate for the driver and mode.

Creating Lines

In Turbo C there is a built-in function that allows you to create a line on the graphics screen. In order to use this function your system must be in graphic mode:

```
line(int X1, int Y1, int X2, int Y2);
```

Where
> X1, Y1 = Values of the starting point of the line.
> X2, Y2 = Values of the ending point of the line.

The following program causes a horizontal line to be created on the graphics screen.

Program 8–11

```c
#include <stdio.h>
#include <graphics.h>

void auto_initialization(void);     /* Function to initialize to the
                                        graphic mode. */

main( )
{
  auto_initialization( );
  line(0, 50, 299, 50);   /* Draw a line. */

  getchar( );

  closegraph( );    /* Shut down the graphics system. */

}

void auto_initialization(void)
{
  int graph_driver;       /* Specifies the graphics driver. */
  int graph_mode;         /* Specifies the graphic mode.    */

  graph_driver = DETECT;  /* Initialization autodetection.  */
  initgraph(&graph_driver, &graph_mode, "");

}  /* End of auto_initialization */
```

How far the line extends across the screen will depend upon your installed graphics hardware. If you want to ensure that the line goes all the way across the screen, do the following modification:

```
line(0,50,getmaxx( ),50);
```

Now the line will always extend across the whole graphics screen regardless of the current mode.

The following program makes a box around the graphics screen, using the built-in Turbo C functions **getmaxx( )** and **getmaxy( )**.

Program 8–12

```
#include <stdio.h>
#include <graphics.h>

void auto_initialization(void);      /* Function to initialize to the
                                         graphic mode.    */

main( )
{

    auto_initialization( );

    line(0,0,getmaxx( ),0);  /* Line across top of screen. */
    line(0,getmaxy( ),getmaxx( ),getmaxy( )); /* Line across bottom.*/
    line(0,0,0,getmaxy( )); /* Line along left side of screen. */
    line(getmaxx( ),0,getmaxx( ),getmaxy( )); /* Line along right side.*/

    getchar( );

    closegraph( );
}

void auto_initialization(void)
{
  int graph_driver;        /* Specifies the graphics driver. */
  int graph_mode;          /* Specifies the graphics mode.    */

  graph_driver = DETECT;  /* Initialization autodetection.   */
  initgraph(&graph_driver, &graph_mode, "");

}  /* End of auto_initialization  */
```

Conclusion

This section presented some very important built-in Turbo C functions for understanding your graphics system. You were also introduced to some fundamental graphic commands. In the next section, you will learn some of the powerful graphic commands of Turbo C that will allow you to create different shapes in a variety of colors and line styles. For now, check your understanding of this section by trying the following section review.

8–3 Section Review

1   State why it's important to know about your graphics system's modes and colors.
2   Explain what is meant by a color palette.
3   How many different color palettes are available for the IBM CGA color graphics card? How many colors per palette?
4   Are there any modes in which color is not available? Explain how you would find this information.
5   How could you find out how many modes your system has?

8–4   Built-in Shapes

Discussion

In this section you will learn about Turbo C built-in shapes. You will discover how to easily create different line styles, rectangles, and circles. You will also see how to fill areas of the graphics screen with different patterns and colors. This is an exciting section. With the information you learn here, you can begin to construct programs that will produce powerful technical graphics.

Changing Graphics Modes

Recall from Table 8–3 that the graphics hardware in your system may offer several different modes of operation. In the last section, you were introduced to built-in Turbo C 2.0 functions that allow you to find how many different modes your particular installed graphics system has.

For the example programs in this section, it will be assumed that you have IBM CGA graphics hardware in your system. The commands are the same no matter what acceptable system you have. For the purposes here, if your system is different, only the modes and pixel sizes may be different.

The following program demonstrates how to change graphic modes.

Program 8–13

```
#include <stdio.h>
#include <graphics.h>

void auto_initialization(void);        /* Function to initialize to the
                                          graphic mode. */

main( )
{
```

(continued)

Program 8-13 *(continued)*

```
    auto_initialization( );

/* IMPORTANT!  The following comments apply only to an IBM
        CGA color graphics adapter.  The results with
        your system's installed graphics hardware may be
        different. Refer to Table 8-3.  */

    setgraphmode(CGAC0);  /* 320 X 200 pixel screen, palette 0. */
    printf("Current color is %d\n",getcolor( ));
    line(0,0,319,199);  /* Produces a yellow line. */
    getchar( );

    setgraphmode(CGAC1);  /* 320 X 200 pixel screen, palette 1. */
    printf("Current color is %d \n",getcolor( ));
    line(0,0,319,199);  /* Produces a white line. */
    getchar( );

    setgraphmode(CGAC2);  /* 320 X 200 pixel screen, palette 2. */
    printf("Current color is %d",getcolor( ));
    line(0,0,319,199);  /* Produces a brown line. */
    getchar( );

    setgraphmode(CGAC3);  /* 320 X 200 pixel screen, palette 3. */
    printf("Current color is %d",getcolor( ));
    line(0,0,319,199);  /* Produces a light gray line. */
    getchar( );

    setgraphmode(CGAHI);  /* 640 X 200 pixel screen, white on black. */
    printf("Current color is %d",getcolor( ));
    line(0,0,639,199);  /* Produces a white line. */
    getchar( );

    closegraph( );

}  /* End of main( ) */

void auto_initialization(void)
{
 int graph_driver;       /* Specifies the graphic driver. */
 int graph_mode;         /* Specifies the graphic mode.   */

 graph_driver = DETECT;  /* Initialization autodetection.  */
 initgraph(&graph_driver, &graph_mode, "");

}  /* End of auto_initialization  */
```

The above program is commented for a system with an IBM CGA color graphics adapter. As seen in Table 8-3, this adapter has five modes of operation. Each of these modes may be referred to by a number (0 through 4) or by a built-in predefined Turbo graphic constant (CGAC0 through CGAHI). The built-in Turbo C 2.0 function that sets the graphics mode is

```
setgraphmode(int mode)
```

Where

Mode = An **int** that must be a valid mode for the installed graphics
hardware. You may use a number or one of the predefined
graphic constants.

Each time **setgraphmode( )** is activated, the graphics screen is cleared and
the graph is redrawn in the default graphics color. In the above program
setgraphmode(CGAC0) produces the same result as **setgraphmode(0)**; and
setgraphmode(CGAHI); the same results as **setgraphmode(4)**;.

Line Styles

The above program produces a diagonal line across the screen. Turbo C 2.0 allows
you to select from several different line styles. You can choose to have your lines
drawn as solid lines, dotted lines, dashed lines, or center lines. You can also select
one of two line thicknesses. This is demonstrated by the following program.

Program 8–14

```
#include <stdio.h>
#include <graphics.h>

void auto_initialization(void);      /* Function to initialize to the
                                        graphic mode.    */

main( )
{

    auto_initialization( );
    setgraphmode(CGAC0);    /* 320 X 200 pixel screen, palette 0 */

    setlinestyle(SOLID_LINE, 0, NORM_WIDTH);
    line(0,10,150,10);    /* Draws a solid line of normal width. */

    setlinestyle(DOTTED_LINE, 0, NORM_WIDTH);
    line(0,20,150,20);    /* Draws a dotted line of normal width. */

    setlinestyle(CENTER_LINE, 0, NORM_WIDTH);
    line(0,30,150,30);    /* Draws a center line of normal width. */

    setlinestyle(DASHED_LINE, 0, NORM_WIDTH);
    line(0,40,150,40);    /* Draws a dashed line of normal width. */

    setlinestyle(SOLID_LINE, 0, THICK_WIDTH);
    line(0,50,150,50);    /* Draws a solid line of thick width. */
```

(continued)

Program 8-14 *(continued)*

```
    setlinestyle(DOTTED_LINE, 0, THICK_WIDTH);
    line(0,60,150,60);    /* Draws a dotted line of thick width. */

    setlinestyle(CENTER_LINE, 0, THICK_WIDTH);
    line(0,70,150,70);    /* Draws a center line of thick width. */

    setlinestyle(DASHED_LINE, 0, THICK_WIDTH);
    line(0,80,150,80);    /* Draws a dashed line of thick width. */

    getchar( );

    closegraph( );

}  /* End of main  */

void auto_initialization(void)
{
  int graph_driver;        /* Specifies the graphics driver. */
  int graph_mode;          /* Specifies the graphic mode.    */

  graph_driver = DETECT;  /* Initialization autodetection.  */
  initgraph(&graph_driver, &graph_mode, "");

}  /* End of auto_initialization  */
```

The results of the above program are illustrated in Figure 8-5.

The built-in Turbo C 2.0 function for setting the line style is

setlinestyle(int linestyle, **unsigned** pattern, **int** thickness**);**

Where

linestyle = A type **int** that is one of the following:

Number	Constant	Result
0	SOLID_LINE	Graphic lines will be solid.
1	DOTTED_LINE	Graphic lines will be dotted
2	CENTER_LINE	Graphic lines will be center lines (long line followed by a short one).
3	DASHED_LINE	Graphic lines will be dashes.
4	USERBIT_LINE	A user defined line style.

pattern = The bit pattern for the line. For the purpose of this text, set this to 0.

thickness = A type **int** that is used to determine one of two available line thicknesses: 1 = **NORM_WIDTH** (a line of "normal" width); 2 = **THICKWIDTH** (a line thicker than the "normal" width).

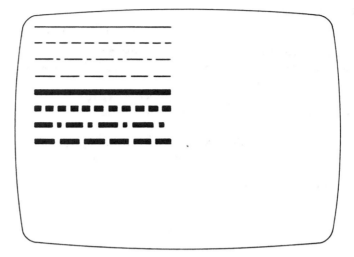

Figure 8–5   Different Graphic Line Styles

For example **setlinestyle(SOLID_LINE,0,NORM_WIDTH);** is identical to the command **setlinestyle(0,0,1);**.

Making Rectangles

Turbo C has built-in graphics procedures for making rectangles. This is a great time saving feature. As you may suspect, Turbo allows you to select the line styles for each of these rectangles. This is illustrated in the following program:

Program 8–15

```
#include <stdio.h>
#include <graphics.h>

void auto_initialization(void);      /* Function to initialize to the
                                         graphics mode. */

main( )
{

    auto_initialization( );
    setgraphmode(CGAC0);    /* 320 X 200 pixel screen, palette 0 */

    setlinestyle(SOLID_LINE, 0, NORM_WIDTH);
    rectangle(0,10,100,30); /* Draws a solid rectangle of normal width. */

    setlinestyle(DOTTED_LINE, 0, NORM_WIDTH);
    rectangle(0,40,10,60); /*Draws a dotted rectangle of normal width.*/
```

(continued)

Program 8–15 *(continued)*

```
    setlinestyle(CENTER_LINE, 0, NORM_WIDTH);
    rectangle(0,70,100,90); /*Draws a center rectangle of normal width.*/

    setlinestyle(DASHED_LINE, 0, NORM_WIDTH);
    rectangle(0,100,,100,120); /*Draws a dashed rectangle of normal width.*/

    setlinestyle(SOLID_LINE, 0, THICK_WIDTH);
    rectangle(110,10,210,30); /*Draws a solid rectangle of thick width.*/

    setlinestyle(DOTTED_LINE, 0, THICK_WIDTH);
    rectangle(110,40,210,60); /*Draws a dotted rectangle of thick width.*/

    setlinestyle(CENTER_LINE, 0, THICK_WIDTH);
    rectangle(110,70,210,90); /*Draws a center rectangle of thick width.*/

    setlinestyle(DASHED_LINE, 0, THICK_WIDTH);
    rectangle(110,100,210,120); /*Draws a dashed rectangle of thick width.*/

    getchar( );

    closegraph( );

}   /* End of main( )   */

void auto_initialization(void)
{
 int graph_driver;        /* Specifies the graphics driver. */
 int graph_mode;          /* Specifies the graphic mode.    */

 graph_driver = DETECT;  /* Initialization autodetection.   */
 initgraph(&graph_driver, &graph_mode, "");

}   /* End of auto_initialization  */
```

Figure 8-6 illustrates the results of the above program.

The built-in Turbo C function for drawing a rectangle on the graphics screen is

rectangle(int X1, int Y1, int X2, int Y2);

Where

X1, Y1 = **int** type coordinates of the top left corner of the rectangle.
X2, Y2 = **int** type coordinates of the bottom right corner of the rectangle.

As you can see from the above program, the line style of the rectangle is determined by the last **setlinestyle( )** function used in the program.

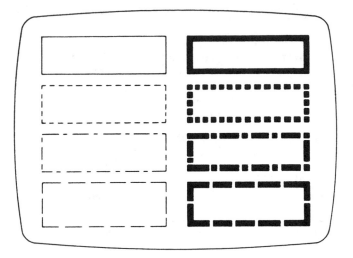

Figure 8-6   Built-in Rectangle Shapes with Different Line Styles

Creating Circles

As you may have suspected, Turbo C also has a built-in function for creating circles. However, unlike the **rectangle( )** function, the **circle( )** function is affected by the **setlinestyle( )** function in only one of two ways. The circle will always be drawn with a solid line that is one of two thicknesses. This is illustrated in the following program where the resulting circles ignore the types of lines issued by the **setlinestyle( )** functions.

Program 8-16

```
#include <stdio.h>
#include <graphics.h>

void auto_initialization(void);        /* Function to initialize to the
                                          graphic mode.    */

main( )
{

    auto_initialization( );
    setgraphmode(CGAC0);     /* 320 X 200 pixel screen, palette 0 */

    setlinestyle(SOLID_LINE, 0, NORM_WIDTH);
    circle(30,30,30);    /* Draws a solid circle of normal width. */

```

(continued)

Program 8–16 *(continued)*

```
    setlinestyle(DOTTED_LINE, 0, NORM_WIDTH);
    circle(30,60,30);    /* Draws a solid circle of normal width. */

    setlinestyle(CENTER_LINE, 0, NORM_WIDTH)
    circle(30,90,30);    /* Draws a solid circle of normal width. */

    setlinestyle(DASHED_LINE, 0, NORM_WIDTH)
    circle(30,120,30);    /* Draws a solid circle of normal width. */

    setlinestyle(SOLID_LINE, 0, THICK_WIDTH)
    circle(110,30,30);    /* Draws a solid circle of thick width. */

    setlinestyle(DOTTED_LINE, 0, THICK_WIDTH)
    circle(110,60,30);    /* Draws a solid circle of thick width. */

    setlinestyle(CENTER_LINE, 0, THICK_WIDTH)
    circle(110,90,30);    /* Draws a solid circle of thick width. */

    setlinestyle(DASHED_LINE, 0, THICK_WIDTH);
    circle(110,120,30);    /* Draws a solid circle of thick width. */

    getchar( );

    closegraph( );

  }  /* End of main( )  */

void auto_initialization(void)
{
  int graph_driver;        /* Specifies the graphics driver. */
  int graph_mode;          /* Specifies the graphic mode. */

  graph_driver = DETECT;   /* Initialization autodetection. */
  initgraph(&graph_driver, &graph_mode, "");

}  /* End of auto_initialization  */
```

Figure 8–7 shows the result of the above program. Again, note that there are only two line styles for the circle.

The built-in Turbo C function for creating circles on the graphic screen is

```
circle(int X, int Y, int radius);
```

Where
    X, Y   = The **int** coordinates of the circle.
    radius = The circle radius of type **int**.

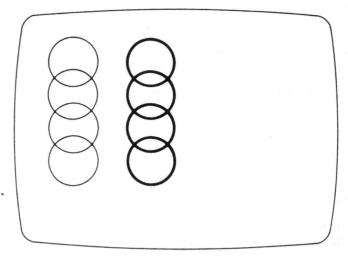

Figure 8–7   Circles Created by the **circle( )** Function

Area Fills

There may be times when you want the area within an enclosed shape such as a circle, rectangle, or any shape you may create to be of a different color from the background. Turbo C has a built-in graphics function for filling in enclosed areas thus making the area inside a shape different from its background. This is demonstrated by the following program.

Program 8–17

```
#include <stdio.h>
#include <graphics.h>

void auto_initialization(void);     /* Function to initialize to the
                                        graphic mode.    */

main( )
{

    auto_initialization( );

    setgraphmode(CGAC0);

    setlinestyle(SOLID_LINE, 0, NORM_WIDTH);
    circle(30, 30, 30);
    floodfill(30,30,3);
```

(continued)

Program 8–17 *(continued)*

```
    getchar( );
    closegraph( );

}

void auto_initialization(void)
{
  int graph_driver;        /* Specifies the graphics driver. */
  int graph_mode;          /* Specifies the graphic mode.    */

  graph_driver = DETECT;   /* Initialization autodetection.  */
  initgraph(&graph_driver, &graph_mode, "");

}   /* End of auto_initialization  */
```

The result of the above program is shown in Figure 8–8.

The built-in Turbo function for an **area fill** (filling in an enclosed area) is

floodfill(int X, int Y, int border);

Where

X, Y   = Any coordinates inside the area of the object to be filled.

border = Color of the fill (must be the color of the object border).

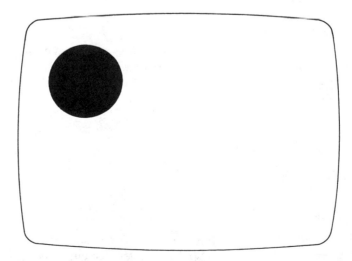

Figure 8–8   Results of Filling an Area of an Object

Aspect Ratios

The ratio of a screen's height to its width is called its **aspect ratio**. A square screen has an aspect ratio of 1.0. The aspect ratios of different color adapters are not the same. This means a circle may look like an ellipse if the aspect ratio is something different from 1.0. The CGA adapter has an aspect ratio of 1.6, while the EGA's is 1.8286.

Turbo C tries to automatically compensate for these differences to produce true circles and squares when you want them. You can get the aspect ratio of your system from the built-in Turbo C function,

`getaspectratio(int far *X_asp, int far *Y_asp);`

> Where   `X_asp` = The horizontal aspect factor.
> `Y_asp` = The vertical aspect factor.

The built-in Turbo C function

`setaspectratio(int X_asp, int Y_asp);`

will change the default aspect ratio of the graphics system.

Conclusion

This section presented the basic graphic building blocks offered by Turbo C. Here you saw how to create different line styles, rectangles that use the line styles, and circles that only use some of the line styles. You also saw how to fill areas while in the graphic mode.

In the next section, you will see how easy it is to create technical bar graphs using the built-in features of Turbo C. For now, test your understanding of this section by trying the following section review.

8-4 Section Review

1   State what effect changing a graphic mode could have on the graphics screen.
2   How many different line styles are available in Turbo C? State what they are.
3   Give the built-in Turbo C shapes introduced in this section. Which one(s), if any, have limited line styles?
4   What does a flood fill do? How do you determine what area will be filled?

8-5   Bars and Text in Graphics

Discussion

This section shows you how to create bar charts and how to put text into your graphics. Being able to manipulate text by changing its size and style (font) as well as being able to write vertically will greatly enhance your technical graphics. In addition, you will see how to create the bars used in bar charts and how to change color and built-in bar-fill patterns.

Creating Bars

Bar charts are a common method of presenting information. An example bar chart is shown in Figure 8–9.

As you can see from the figure, the basic graphic for the bar chart is a rectangle. You could construct a rectangle in Turbo graphics by simply using the `rectangle( )` function. However, there is an easier way as shown in the following program.

Program 8–18

```c
#include <stdio.h>
#include <graphics.h>

#define TRUE 1
#define FALSE 0

void auto_initialization(void);       /* Function to initialize to the
                                          graphic mode.    */


main( )
{
    auto_initialization( );
    setgraphmode(0);

    bar(20,20,70,100);
    bar3d(90,20,140,100,10,FALSE);
    bar3d(160,20,220,100,10,TRUE);

    getchar( );

    closegraph( );

}

void auto_initialization(void)
{
  int graph_driver;        /* Specifies the graphics driver. */
  int graph_mode;          /* Specifies the graphic mode.    */

  graph_driver = DETECT;  /* Initialization autodetection.   */
  initgraph(&graph_driver, &graph_mode, "");

} /* End of auto_initialization  */
```

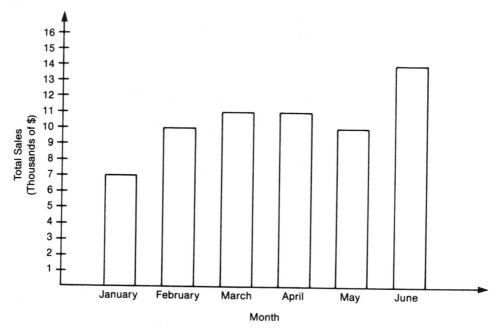

Figure 8–9   Typical Bar Chart

The graph generated by the above program is shown in Figure 8–10.

Note from the figure that three bars are generated. One is "flat," and the other two have some depth, one without a top and the last one with a top. The built-in Turbo function that produces these bars is

```
bar(int X1, int Y1, int X2, int Y2);
```

Where
  X1, Y1 = The top left point of the bar.
  X2, Y2 = The bottom right point of the bar.

```
bar3d(int X1, int Y1, int X2, int Y2, int depth, int top);
```

Where
  X1, Y1 = The top left point of the bar.
  X2, Y2 = The bottom right point of the bar.
  depth = The number of pixels deep for the three-dimensional outline.
  top = If non-zero, a top is put on the bar.

The above bars are drawn in the current color and fill-style. A discussion of fill-styles follows.

Filling in the Bars

There may be times in a bar graph display where many bars must be used to represent many different types of information on the same graph. Since the use of color is

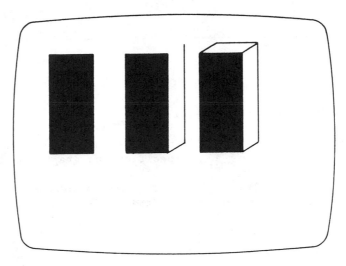

Figure 8-10   Resulting Graph for Program 8-18

limited on some graphics adapters, Turbo has a built-in function for automatically filling the area within a bar with one of 12 different **fill-styles**. One of the fill-styles is demonstrated by the following program.

Program 8-19

```
#include <stdio.h>
#include <graphics.h>
#define TRUE 1
#define FALSE 0

void auto_initialization(void);        /* Function to initialize to the
                                          graphic mode.    */

main( )
{

    auto_initialization( );

    setgraphmode(0);
    setfillstyle(LINE_FILL,2);
    bar(20,20,70,100);
    bar3d(90,20,140,100,10,FALSE);
    bar3d(160,20,220,100,10,TRUE);

    getchar( );
```

(continued)

Program 8–19 *(continued)*

```
    closegraph( );

}

void auto_initialization(void)
{
  int graph_driver;        /* Specifies the graphics driver. */
  int graph_mode;          /* Specifies the graphic mode.    */

  graph_driver = DETECT;   /* Initialization autodetection.  */
  initgraph(&graph_driver, &graph_mode, "");

}   /* End of auto_initialization  */
```

The graph generated by the above program is shown in Figure 8–11.

As you can see from the figure, each of the bars has a distinctive pattern. The commands for each of the bars are exactly the same as before, but now the pattern inside the bar is different. This is done by the built-in Turbo C function **setfillstyle( )**.

setfillstyle(int pattern, **int** color);

Where

        pattern = An **int** from 0 to 11 that sets the style of the fill pattern. (A value of 12 allows a user-defined fill pattern. Refer to the Turbo C manuals.)

        color   = An **int** that sets the color of the fill pattern.

Turbo C has defined a set of constants in the graphics library as shown in Table 8–5.

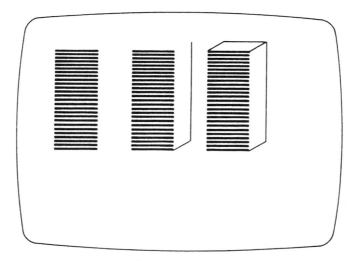

Figure 8–11   Bar Graph Generated by Program 8–19

Table 8-5 Fill Patterns Constants

Number	Constant	Bar is filled:
0	Empty_Fill	in the background color.
1	Solid_Fill	with a solid color.
2	Line_Fill	with horizontal lines.
3	LtSlash_Fill	with light slashes.
4	Slash_Fill	with thick slashes.
5	BkSlash_Fill	with thick back slashes.
6	LtBkSlash_Fill	with light back slashes.
7	Hatch_Fill	with light hatch marks.
8	XHatch_Fill	with heavy cross hatches.
9	InterLeave_Fill	with an interleaving line.
10	Wide_Dot_Fill	with widely spaced dots.
11	Close_Dot_Fill	with closely spaced dots.

The following program illustrates the use of several of the above constants and the reason for having tops on or off a three-dimensional bar.

Program 8-20

```
#include <stdio.h>
#include <graphics.h>
#define TRUE 1
#define FALSE 0

void auto_initialization(void);       /* Function to initialize to the
                                          graphic mode.     */


main( )
{

    auto_initialization( );
    setgraphmode(0);

    setfillstyle(SLASH_FILL, 3)
    bar3d(10,100,50,150,10,FALSE);

    setfillstyle(LINE_FILL, 1);
    bar3d(10,50,50,100,10,FALSE);

    setfillstyle(HATCH_FILL, 2)
    bar3d(10,10,50,50,10,TRUE);
```

(continued)

Program 8–20 *(continued)*

```
    setfillstyle(WIDE_DOT_FILL, 2);
    bar3d(50,85,90,150,10,TRUE);

    getchar( );

    closegraph( );

}

void auto_initialization(void)
{
  int graph_driver;        /* Specifies the graphics driver. */
  int graph_mode;          /* Specifies the graphic mode.    */

  graph_driver = DETECT;   /* Initialization autodetection.  */
  initgraph(&graph_driver, &graph_mode, "");

}  /* End of auto_initialization  */
```

The resulting graph from the above program is shown in Figure 8–12. Note that the two bottom stacked bars used the top as FALSE (equal to 0) and the top one had this item TRUE (equal to 1).

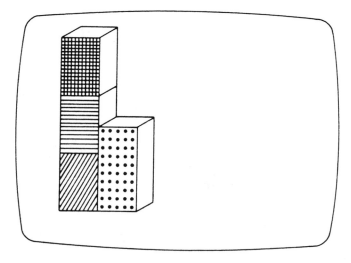

Figure 8–12   Results of Program 8–20

Text with Graphics

The use of text with graphics is an important feature of technical graphics. Turbo C provides many advanced methods for presenting text on your graphics screens. The basic text graphic command is illustrated in the following program.

Program 8–21

```
#include <stdio.h>
#include <graphics.h>

void auto_initialization(void);      /* Function to initialize to the
                                        graphic mode.    */

main( )
{

    auto_initialization( );

    outtext("This is the default font...\n");
    outtext("See where this is...");

    getchar( );

    closegraph( );

    }

void auto_initialization(void)
{
  int graph_driver;        /* Specifies the graphics driver. */
  int graph_mode;          /* Specifies the graphic mode.    */

  graph_driver = DETECT;   /* Intialization autodetection.   */
  initgraph(&graph_driver, &graph_mode, "");

}   /* End of auto_initialization  */
```

Execution of the above program results in the following text appearing, starting at the top left of the monitor:

`This is the default font...See where this is...`

Note that there is no carriage return (as you would find in a **printf( )** function). Instead, the built-in Turbo function **outtext( )** sends a string of characters to the current position of the graphics pointer. The built-in function is

outtext(char far *textstring);

The graphics pointer is the point on the graphics screen from which some action will take place. It is similar to the text cursor, only it isn't visible. The `outtext( )` function will produce a string of text from the position of the graphics pointer and leave the pointer at the end of the text string. One method of moving the text string is to use the built-in Turbo C function `outtextxy( )`.

`outtextxy(int X1, int Y1 char far *textstring);`

Where
X1, Y1 = The position on the graphics screen from which the string is to start.

Changing Text Fonts

Turbo C allows you to select different text styles (called **fonts**). Turbo C actually has two major types of graphic text styles. One is **bit mapped**; this is the default font. The other type is a **stroked font**. The difference is that characters in bit mapped fonts are composed of rectangular pixel arrays. Characters in stroked fonts are composed of line segments whose sizes and directions are defined relative to some starting point.

Turbo C allows you to change the size of graphic text. When you enlarge bit mapped fonts, the pixels are simply enlarged. Doing this makes the text look "blocky" with large stairstep effects. However, with a stroked font, text enlargement is made by making each component line segment longer; thus these fonts look more natural. The difference is illustrated in Figure 8–13.

The main advantage of bit mapped fonts is speed. Stroked fonts take longer because of their many line segments.

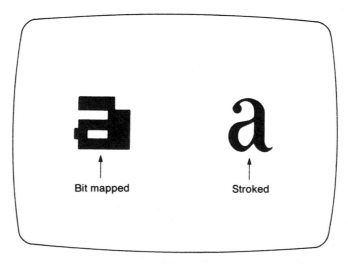

Figure 8–13   Difference between Bit-Mapped and Stroked Fonts

The built-in Turbo C function that selects the font is

`settextstyle(int font, int direction, int size);`

Where

          font = The style of the font.

   direction = The direction of the text to be displayed (0 left to right, 1 bottom to top).

        size = Size of the text (1 to 10). A size of 1 produces a character of 8×8 pixels; a size of 2 produces a character of 16×16 pixels and so on.

The different text styles supported by Turbo C are

0 = DEFAULT_FONT   **(Bit mapped)**
1 = TRIPLEX_FONT   **(All the rest are stroked)**
2 = SMALL_FONT
3 = SANSSERIF_FONT
4 = GOTHIC_FONT

The active disk must contain the following Turbo C fonts

`TRIP.CHR`   `LITT.CHR`   `SANS.CHR`   `GOTH.CHR`

These come on your Turbo C 2.0 distribution disks. Other built-in constants for the `settextstyle( )` function are

0 = HORIZ_DIR
1 = VERT_DIR

Thus, `settextstyle(TRIPLEX_FONT, HORIZ_DIR, 1);` means exactly the same thing as `settextstyle(1, 0, 1);`

The following program illustrates the use of graphic text with a bar graph.

Program 8–22

```
#include <stdio.h>
#include <graphics.h>
#define TRUE 1
#define FALSE 0

void auto_initialization(void);      /* Function to initialize to the
                                        graphic mode.    */

main( )
{

   auto_initialization( );
   setgraphmode(0);

    setfillstyle(SLASH_FILL, 3);
    bar3d(30,100,70,150,60,FALSE);
```

(continued)

Program 8-22 *(continued)*

```
      setfillstyle(LINE_FILL, 1);
      bar3d(30,50,70,100,10,FALSE);

      setfillstyle(HATCH_FILL, 2);
      bar3d(30,10,70,50,10,TRUE);

      setfillstyle(WIDE_DOT_FILL, 2);
      bar3d(80,85,120,150,10,TRUE);
/*  Set the text:  */
      settextstyle(DEFAULT_FONT, VERT_DIR, 1);
      setcolor(1);
      outtextxy(20,10, "TOM  DICK  HARRY");
      setcolor(2);
      outtextxy(145, 85, "OTHERS");

      getchar( );

      closegraph( );

      }

void auto_initialization(void)
{
  int graph_driver;        /* Specifies the graphics driver. */
  int graph_mode;          /* Specifies the graphic mode.    */

  graph_driver = DETECT;   /* initialization autodetection.  */
  initgraph(&graph_driver, &graph_mode, "");

}  /* End of auto_initialization  */
```

Figure 8–14 shows the result of the above program. Note that the vertical text displays actually *start* at the given coordinates and are drawn *down*.

The built-in Turbo C function **setcolor( )** determines the graphic drawing color:

setcolor(int color**);**

Where

color = The selected color, defined within the limits of your system's color graphics adapter and the mode you select.

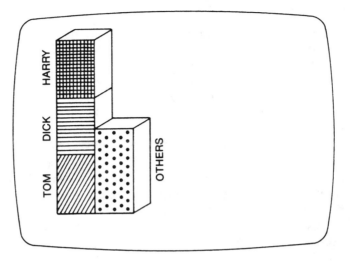

Figure 8-14   Use of Vertical Text in Graphics Mode

Conclusion

This section presented important information concerning the creation of bars used in bar graphs. Here you saw how to create three-dimensional bars and fill in their areas with different pre-made fill patterns. You also saw how to add text to graphics. The different fonts were presented as well as how to change the size of text and make it appear either vertically or horizontally.

Test your knowledge of this section by trying the following section review.

8-5 Section Review

1   Why does Turbo C have a built-in function for creating bars.
2   State what kind of bars can be displayed using built-in Turbo C functions.
3   What options are available for presenting a three-dimensional bar using Turbo C built-in functions? Why is this available?
4   Describe how you could distinguish one bar from another using built-in Turbo C functions.
5   State some of the options available to you when using text in Turbo C graphics.

8-6   Graphing Functions

Discussion

Turbo C is a powerful language system for graphing mathematical functions. Recall that a mathematical function shows the relationship between two or more quantities. This section presents the fundamentals for programming this type of technical graph. Knowing how to do this greatly enhances your technical programming skills.

Fundamental Concepts

Scaling is using numerical methods to ensure that all of the required data appears on the graphics screen and utilizes the full pixel capability. Remember that for any computer system, the number of pixels is limited. For example, in the high-resolution mode of the CGA adapter there are 640 horizontal pixels and 200 vertical pixels.

To understand how to use scaling to produce practical graphics, consider the graph of the function: $Y = X + 2$. The plot of this function is shown in Figure 8–15.

The graph shown in Figure 8–15 uses three of the four quadrants of the Cartesian coordinate system. Suppose you need to develop a computer program that will display such a graph and you want the full graphics screen to be used in this display. This means that the actual values used to plot the graph would not be the values used in the equation. Figure 8-16 illustrates this important point.

Observe from Figure 8–16, that the extreme left of the graph (the point that represents $X = -5$ and $Y = -3$) must actually have the plotted values of $X_p = 0$ and $Y_p = 199$ (the P subscript is used to denote the actual plotted values). The process used to achieve this transformation is called scaling. Also note from Figure 8-16 that the origin of the coordinate system is not in the exact center of the screen. This is done in order to make practical use of the full size of the monitor. So, not only has scaling been used, but so has **coordinate transformation**. Both processes are explained below.

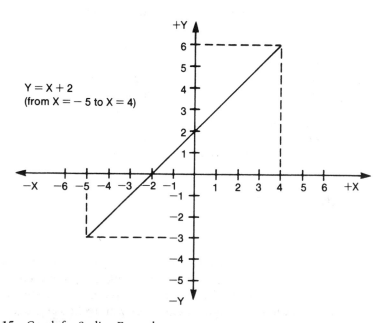

Figure 8–15   Graph for Scaling Example

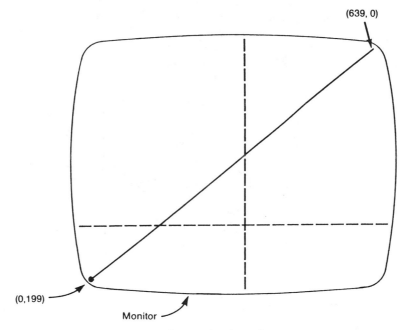

Figure 8-16   Actual Plotting Values for Graphical Display

Scaling

Observe Figure 8-17.

The horizontal scale factor can be expressed mathematically as

$$HS = P_H/(X_2-X_1)$$

Where

HS = The horizontal scale factor.
P_H = Number of pixels in the horizontal direction.
X_1 = Minimum value X.
X_2 = Maximum value of X.

The vertical scale factor can be expressed mathematically as

$$VS = P_V/(Y_2-Y_1)$$

Where

VS = The vertical scale factor.
P_V = Number of pixels in the vertical direction.
Y_1 = Minimum value of Y.
Y_2 = Maximum value of Y.

To calculate the scale factors for the graph of Figure 8-14:

For the horizontal:
$$HS = P_H/|X_2-X_1|$$
$$HS = 640/((4)-(-5))$$
$$HS = 640/(4+5) = 640/9 = 71.1$$

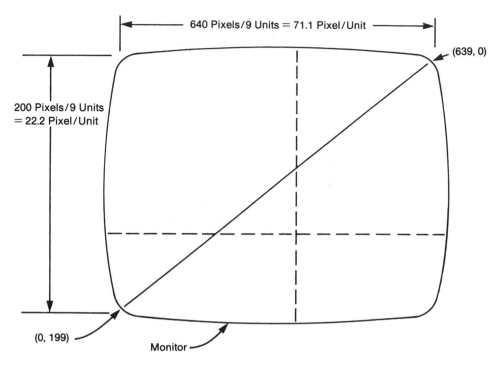

Figure 8–17   Minimum and Maximum Values of Scaling Example

For the vertical:
$$VS = P_V/|Y_2 \cdot Y_1|$$
$$VS = 200/((6)-(-3))$$
$$VS = 200/(6+3) = 200/9 = 22.22$$

What the above calculations mean is that every major division along the X axis will be 71 pixels, and every major division along the Y axis will be 22 pixels. Doing this will use the full capabilities of the graphics screen. This is shown in Figure 8–18.

Coordinate Transformation

Note from Figure 8–18, that the origin of the coordinate system used to display the example graph is not in the exact center of the graphics screen. This was done intentionally in order to display the full range of required data utilizing the full graphics screen. In order to accomplish this, coordinate transformation was required. Coordinate transformation, mentioned in the last section, is the process of using numerical methods to cause the origin of the coordinate system to appear at any desired place on the graphics screen.

The process of transforming coordinates can be expressed mathematically as follows:

For the Horizontal Transformation:
$$XT = |(HS)(X_1)|$$

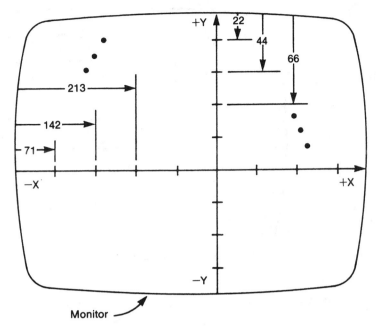

Figure 8-18 Meaning of Scaling for Example Graph

Where

XT = Horizontal transformation
HS = Horizontal scaling factor.
X_1 = Minimum value of X.

For the Vertical Transformation:

$$YT = |(VS)(Y_2)|$$

Where

YT = Vertical transformation.
VS = Vertical scaling factor.
Y_2 = Maximum value of Y.

For the above example this becomes

Horizontal Transformation:
$$XT = |(HS)(X_1)|$$
$$XT = |(71.1)(-5)| = |-355|$$
$$XT = 355.$$
Vertical Transformation:
$$YT = |(VS)(Y_2)|$$
$$YT = |(22.2)(6)| = |133|$$
$$YT = 133.$$

What the above calculations mean is the origin of the coordinate system will be at the locations of $X_G = 355$ and $Y_G = 133$. This is illustrated in Figure 8-19.

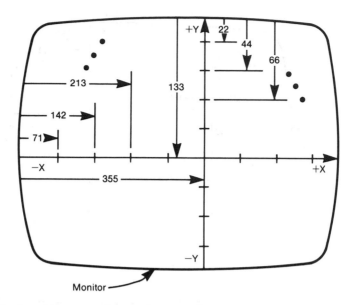

Figure 8-19   Actual Location of Origin for Example Graph

The only thing left to do now is to develop equations that can be used to display the data on the graphics screen.

Putting It Together

To display the graph of any continuous function related by two variables when scaling and coordinate transformations are used, the following equations are necessary:

Horizontal Value of Graph
$$XG = XT + (X)(HS)$$
Where

XG = Graphical X value.
XT = Horizontal transformation.
HS = Horizontal scaling factor.

Vertical Value of Graph
$$YG = YT - (Y)(VS)$$
Where

YG = Graphical Y value.
YT = Vertical transformation.
VS = Vertical scaling factor.

The following program plots the graph of the function $X = Y + 2$. The program asks the user for the maximum and minimum values of X. It then draws the transformed coordinate system and scales the plot of the graph. Note that it makes no difference what pixel size the graphic screen has as long as Turbo C recognizes it.

Program 8–23

```
#include <stdio.h>
#include <graphics.h>

void auto_initialization(void);      /* Function to initialize to the
                                        graphics mode.    */

float Y_value(float X);                    /* Function for display. */

main( )
{
 float max_X;                  /* Maximum horizontal value.   */
 float max_Y;                  /* Maximum vertical value.     */
 float min_X;                  /* Minimum horizontal value.   */
 float min_Y;                  /* Minimum vertical value.     */
 float hor_scale;             /* Horizontal scaling factor.  */
 float vert_scale;            /* Vertical scaling factor.     */
 float hor_trans;             /* Horizontal transformation.  */
 float ver_trans;             /* Vertical transformation.    */
 int hor_pixel;               /* Horizontal pixel position.  */
 int ver_pixel;               /* Vertical pixel position.     */
 float X_cord;                 /* X coordinate for graphing.  */

  auto_initialization( );      /* Get into graphics mode.   */

  printf("\n\n Maximum X = ");
  scanf("%f",&max_X);

  max_Y = Y_value(max_X);     /* Calculate maximum Y value. */

  printf(" Minimum X = ");
  scanf("%f",&min_X);

  min_Y = Y_value(min_X);     /* Calculate minimum Y value. */
/* Calculate horizontal scale.  */
     if(max_X == min_X)
       hor_scale = getmaxx( );
     else
       hor_scale = getmaxx( )/(max_X − min_X);

/* Calculate vertical scale.   */
     if(max_Y == min_Y)
       vert_scale = getmaxy( );
```

(continued)

Program 8–23 *(continued)*

```
        else
          vert_scale = getmaxy( )/(max_Y - min_Y);

  /* Calculate horizontal and vertical transformations.  */
      hor_trans = (hor_scale * min_X);
        if(hor_trans < 0)      /* Convert to a positive value. */
          hor_trans *= -1;
      ver_trans = (vert_scale * max_Y);
        if(ver_trans < 0)      /* Convert to a positive value. */
          ver_trans *= -1;

/* Display calculated results.  */
      printf("Horizontal scale = %f\n",hor_scale);
      printf("Vertical scale = %f\n",vert_scale);
      printf("Horizontal transformation = %f\n",hor_trans);
      printf("Vertical transformation = %f\n",ver_trans);
      printf("Enter X to eXit => ");

/* Draw the coordinate system.  */
      setlinestyle(SOLID_LINE, 0, THICK_WIDTH);
      line(0, (int)ver_trans, getmaxx( ), (int)ver_trans);
      line((int)hor_trans, 0, (int)hor_trans, getmaxy( ));

/* Draw the graph.  */
      X_cord = min_X;

    do
     {
      hor_pixel = (int)(hor_trans + X_cord * hor_scale);
      ver_pixel = (int)(ver_trans - Y_value(X_cord) * vert_scale);
        putpixel(hor_pixel, ver_pixel, 1);
        X_cord = X_cord + 0.1;
     } while (hor_pixel < getmaxx( ));

      while('X' != toupper(getchar( )));

      closegraph( );

  }

/*-----------------------------------------------------------*/

float Y_value(float X_value)              /* Function for display. */
{
```

(continued)

Program 8–23 *(continued)*

```
      return(X_value + 2.0);
}

/*-----------------------------------------------------------------*/

void auto_initialization(void)
{
  int graph_driver;        /* Specifies the graphics driver. */
  int graph_mode;          /* Specifies the graphic mode.    */

  graph_driver = DETECT;  /* initialization autodetection.   */
  initgraph(&graph_driver, &graph_mode, "");

}  /* End of auto_initialization  */
```

The resulting graph is shown in Figure 8–20.

Analyzing the Program.

First, the function `auto_initialization` takes place. This is done so that the Turbo system knows what kind of graphics system the computer has (if any at all). The equation to be graphed is put in the form of a function. This is done so that it is easy to change the relationship between X and Y if you wish to plot a different formula.

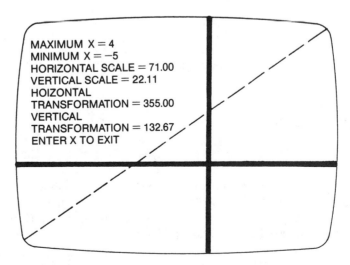

Figure 8–20   Resulting Computer-Generated Graphic

The program computes the horizontal and vertical scale as well as the horizontal and vertical transformations necessary for the graphics screen. Note that if the maximum and minimum values of X or Y are equal (this could happen if plotting a quadratic), then the scales are changed to the maximum pixel values of the given graphics driver.

As an aid in seeing the chosen values, **printf( )** statements are used to display the transformation and scaling information on the upper left portion of the screen. The graph is then constructed using the built-in Turbo C function **putpixel**. However, the **putpixel** function requires a type **int**, and all of the graph values up to this point have been of type **float**. In order to convert these numbers to type **int**, the process of **casting** is used. For example, if `hor_trans` is of type **float**, it can be converted to a type **int** by

```
(int)hor_trans;
```

This is called casting in C. You will note that this was done several times in this program.

The value of X is incremented by 0.1 each time. This can be changed depending upon the resolution you want (and time it takes to generate the graph).

Conclusion

This section introduced you to the practical requirements for displaying the graph of a formula. Here you were introduced to the concepts of scaling and coordinate transformation. These techniques will prepare you for plotting technical data that relates one variable to another.

Test your understanding of this section by trying the following section review.

8–6 Section Review

1 Define scaling as it applies to computer graphics.
2 Define coordinate transformation as it applies to computer graphics.
3 State the factors that determine the values of the scaling factors.
4 State the factors that determine the values of the coordinate transformation values.

8-7   Programming Style

Discussion

As you gain experience writing your own C programs, you will begin to develop a unique style. You will begin to standardize how you format your program commands, how long you make your declared identifiers, how and when you capitalize, and when you use the _ to break up identifiers.

There are no hard and fast rules for doing these very individualistic elements of programming. The important thing is to make your programs so they are easy to understand and modify. It's important that you see many different programming styles. Reading programming magazines is a good resource for seeing other programming styles.

This section presents some ideas concerning programming styles that you may want to consider using in your own work.

C Reserved Words

All C reserved words are given in lowercase. In this text, they are also set in **boldface** to make them stand out in the programs. It's not recommended that you boldface your C commands, but it is suggested that you distinguish them from the body of the rest of the program. A lot of published material does not do this, and you may find this more difficult to understand.

Naming Things

C will let you use very long identifiers. This can be a mixed blessing. Consider the following two program lines:

```
E = I * R; /* Compute voltage */
voltage = current * resistance;
```

The first line uses familiar single-letter symbols to represent an electronic formula (familiar at least to electronic students and professionals). The second formula spells out the relationship. If you are used to using certain formulas with specific single-letter variables (and if using these variables is a generally accepted standard), then your source code may be more readable if you stay with the single letter identifier. As an example,

```
Xt = Xl - Xc; /* Total reactance */
```

may be more recognizable to an electronics technician than

```
TotalReactance = InductiveReactance - CapacitiveReactance;
```

Your Structure

One of the main advantages of the C language is the rich variety of program structures available. Usually beginning programmers tend to shy away from user-defined scalar types, **switch** statements, and structures. However, if you just keep these in mind, they can be very useful in the construction of a C program. Consider, for example, the program used in the last section to generate the graph of a mathematical relationship between two variables. Its programming style could have been improved by using a C structure.

Program 8–24

```c
#include <stdio.h>
#include <graphics.h>

void auto_initialization(void);       /* Function to initialize to the
                                         graphic mode.    */

float Y_value(float X);               /* Function for display. */

typedef struct
  {
    float max_X;         /* Maximum horizontal value.   */
    float max_Y;         /* Maximum vertical value.     */
    float min_X;         /* Minimum horizontal value.   */
    float min_Y;         /* Minimum vertical value.     */
  } coordinate_record;

  typedef struct
    {
      float hor_scale;     /* Horizontal scaling factor.  */
      float vert_scale;    /* Vertical scaling factor.    */
      float hor_trans;     /* Horizontal transformation.  */
      float ver_trans;     /* Vertical transformation.    */
    } translation_record;

main( )
{
  coordinate_record coord_data;  /* Initial coordinate values. */
  translation_record trans_out;  /* Translated coordinate values. */
  int hor_pixel;                 /* Horizontal pixel position.  */
  int ver_pixel;                 /* Vertical pixel position.    */
  float X_cord;                  /* X coordinate for graphing.  */

  auto_initialization( );      /* Get into graphic mode.  */

  printf("\n\n Maximum X = ");
  scanf("%f",&coord_data.max_X);

/* Calculate maximum Y value. */
    coord_data.max_Y = Y_value(coord_data.max_X);

  printf(" Minimum X = ");
  scanf("%f",&coord_data.min_X);

/* Calculate minimum Y value.  */
    coord_data.min_Y = Y_value(coord_data.min_X);

  /* Calculate horizontal scale.  */
    if(coord_data.max_X == coord_data.min_X)
      trans_out.hor_scale = getmaxx( );
```

(continued)

Program 8-24 *(continued)*

```
        else
          trans_out.hor_scale = getmaxx( )/(coord_data.max_X-coord_data.min_X);

/* Calculate vertical scale.  */
        if(coord_data.max_Y == coord_data.min_Y)
         trans_out.vert_scale = getmaxy( );
        else
         trans_out.vert_scale = getmaxy( )/(coord_data.max_Y-coord_data.min_Y);

 /* Caculate horizontal and vertical transformations.  */
        trans_out.hor_trans = trans_out.hor_scale * coord_data.min_X;
        if(trans_out.hor_trans < 0)     /* Convert to a positive value. */
          trans_out.hor_trans *= -1;
        trans_out.ver_trans = trans_out.vert_scale * coord_data.max_Y;
        if(trans_out.ver_trans < 0)     /* Convert to a positive value. */
          trans_out.ver_trans *= -1;

/* Display calculated results.  */
        printf("Horizontal scale = %f\n",trans_out.hor_scale);
        printf("Vertical scale = %f\n",trans_out.vert_scale);
        printf("Horizontal transformation = %f\n",trans_out.hor_trans);
        printf("Vertical transformation = %f\n",trans_out.ver_trans);
        printf("Enter X to eXit => ");

/* Draw the coordinate system.  */
       setlinestyle(SOLID_LINE, 0, THICK_WIDTH);
       line(0, (int)trans_out.ver_trans,getmaxx( ),(int)trans_out.ver_trans);
       line((int)trans_out.hor_trans,0,(int)trans_out.hor_trans, getmaxy( ));

/* Draw the graph.  */
        X_cord = coord_data.min_X;

do
{
hor_pixel=(int)(trans_out.hor_trans + X_cord * trans_out.hor_scale);
ver_pixel=(int)(trans_out.ver_trans-Y_value(X_cord)*trans_out.vert_scale);
        putpixel(hor_pixel, ver_pixel, 1);
        X_cord = X_cord + 0.1;
    } while (hor_pixel < getmaxx( ));

      while('X' != toupper(getchar( )));

      closegraph( );

  }

/*-------------------------------------------------------------*/

float Y_value(float X_value)            /* Function for display. */
{
```

(continued)

Program 8–24 *(continued)*

```
      return(X_value + 2.0);
}

/*--------------------------------------------------------------*/

void auto_initialization(void)
{
  int graph_driver;        /* Specifies the graphics driver. */
  int graph_mode;          /* Specifies the graphic mode.    */

  graph_driver = DETECT;   /* initialization autodetection.  */
  initgraph(&graph_driver, &graph_mode, "");

}  /* End of auto_initialization  */
```

Note the saving of program code in the above program by using C **typedef struct** compared to how the program was originally done in the last section. Another important point is to observe that there are no global variables used in the program. All variables are local, thus protecting them.

Conclusion

This section presented some ideas you may want to consider to help improve your programming style. You saw some suggestions for using identifiers as well as taking advantage of other types of program structure. Check your understanding of this section by trying the following section review.

8–7 Section Review

1   State the guiding principle in the style you choose for developing C programs. For procedures and functions?
2   Should variables always be more than one or two letters long? Explain.
3   State what kind of structure you should use in a C program.

8–8   Case Study

Discussion

The case study for this chapter demonstrates the use of text color, graphics, and different text fonts. Here you will see the development of a technical program used to generate a sine wave. The sine wave has a wide range of applications to many areas of technology, especially in the area of electronics technology.

The Problem

Create a demonstration program in C that will generate a sine wave. The program user may input the values of the amplitude, number of cycles, and the phase. The program is not to use any global variables, and it must demonstrate the use of the C **typedef struct** where appropriate. The user has the option of repeating the program.

First Step—Stating the Problem

Stating the case study in writing yields

- Purpose of the program: Provide a program that will generate a sine wave. User input is the amplitude, number of cycles, and phase of the sine wave. User has program repeat option.
- Required Input: Amplitude, number of cycles, and phase of the sine wave. Is the program to be repeated.
- Process on Input: perform required calculations.
- Required Output: Sine wave which has the amplitude, number of cycles, and phase of the user input.

Developing the Algorithm

Developing the algorithm requires the following steps:

1. Explain program to user.
2. Prompt user for input:

   Amplitude
   Number of cycles
   Phase angle

3. Display the sine wave.
4. Ask for program repeat (do not repeat program explanation).

Arithmetic Functions

This program requires the use of one of the C built-in math functions. (The ANSI C standard math functions are listed in Appendix G.) The one that will be used for this program is the built-in **sin( )** function, where

```
double sin(double X);
```

where X is the angle expressed in radians.

Program Development

In keeping with the idea that this program is to be a model program with C **typedef struct**, Program 8–25 has been developed. Note that there are actually two C **typedef** structures used in the program. The program also makes extensive use of text color and displays the instructions in the 40-character-wide screen.

Observe the block structure and the fact that a whole structure of information is passed from the calling function to the called function. This is another advantage of using structures in C; many different data types may be easily passed between functions by using structures in their arguments.

Program 8-25

```
#include <stdio.h>
#include <graphics.h>
#include <conio.h>
#include <math.h>

#define PI = 3.14159
#define TRUE 1
#define FALSE 0

/*****************************************************************/
/*                 Sine Wave Generation Program                 */
/*****************************************************************/
/*                 Developed by: A. C. Student                  */
/*****************************************************************/
/*    This program will generate a sine wave.  The amplitude,   */
/*  number of cycles, and the phase may be determined by the    */
/*  program user.                                               */
/*****************************************************************/
/*                      Type Definitions                        */
/*-------------------------------------------------------------*/
 typedef struct
   {
     float degrees;     /* Number of degrees. */
     float radians;     /* Number of radians. */
     int X_value;       /* Value of X coordinate. */
     int Y_value;       /* Value of Y coordinate. */
   } wave_values;
/*  This type definition defines the values of the sine wave.  */
/*-------------------------------------------------------------*/
 typedef struct
   {
     char user_input[10]; /* User input string. */
     int amplitude;       /* Height of the sine wave. */
     int cycles;          /* Number of cycles to be displayed.  */
     int phase;           /* Phase of the sine wave. */
   } user_input;
/*  This type definition defines the values inputted by user. */
/*****************************************************************/
/*                     Function Prototypes                      */
/*-------------------------------------------------------------*/
```

(continued)

Program 8–25 *(continued)*

```c
void explain_program(void);
/*   This function explains the purpose of the program to the */
/*   program user.                                             */
/*-----------------------------------------------------------*/
void press_return(void);
/*   This function holds the current screen for the program   */
/* user until the -RETURN- key is depressed.                  */
/*-----------------------------------------------------------*/
void auto_initialization(void);
/*   This function initializes the graphics screen.           */
/*-----------------------------------------------------------*/
void sine_wave_display(user_input values);
/*   values = Structure of user input values.                 */
/*   This function displays the actual sine wave.             */
/*-----------------------------------------------------------*/
int program_repeat(void);
/*   Gives program user the option of repeating the program.  */
/*************************************************************/

main( )
{
 user_input in_values;          /* Values inputted by user. */

   explain_program( );   /* Explain program to user. */

  do
   {
 /* Get user input:  */
   printf("\n\n Amplitude => ");
   gets(in_values.user_input);
   in_values.amplitude = atoi(in_values.user_input);
   printf("\n Cycles => ");
   gets(in_values.user_input);
   in_values.cycles = atoi(in_values.user_input);
   printf("\n Phase => ");
   gets(in_values.user_input);
   in_values.phase = atoi(in_values.user_input);

   sine_wave_display(in_values);  /* Display the sine wave. */
  } while (program_repeat( ));

     closegraph( );  /* Shut down the graphics system.  */
     textmode(C80); /* Return to normal text mode. */

  }  /*  End of main( )  */

 /*-----------------------------------------------------------*/
```

(continued)

Program 8–25 *(continued)*

```
void explain_program(void)          /* Explain program to user. */
{
 char *string;          /* String to hold program explanation. */

        textmode(C40);      /* Make sure you're in text mode.  */
        textbackground(BLUE); /* Sets the color of the text background. */
        clrscr( );          /* Clear the screen.              */

        textcolor(RED);     /* Sets color of the text. */

    string = "\n\n\n"
        "  This program will display a sine wave\n\r"
        " in graphics mode.  You may input the\n\r"
        " values of the amplitude, phase and the\n\r"
        " number of cycles of the sine wave.\n\r"
        "\n"
        "  In order to keep all of the wave\n\r"
        " on your monitor screen, keep the \n\r"
        " amplitude less than 75.";

    cprintf("%s",string);      /* Print the above string. */

    press_return( );           /* Holds screen for user. */

 }   /* End of explain_program( )  */

/*-------------------------------------------------------------*/

void sine_wave_display(user_input values)
{
  wave_values display;          /* Structure for sine wave display. */

 /* Get into graphic mode:  */
   auto_initialization( );

   setgraphmode(3);
   setbkcolor(1);

 /* Draw the graph:  */
   for(display.X_value = 0; display.X_value <= 319; display.X_value++)
       {
       display.degrees = display.X_value;
       display.radians = display.X_value * PI/180;
       display.Y_value = (int)(values.amplitude*
       sin((double)(values.cycles*display.radians+values.phase)));
       putpixel(display.X_value, 75 - display.Y_value, 2);
       }

 }  /* End of sine_wave_display( )  */

/*-------------------------------------------------------------*/
```

(continued)

Program 8–25 *(continued)*

```c
void auto_initialization(void)
{
 int graph_driver;          /* Specifies the graphics driver. */
 int graph_mode;            /* Specifies the graphic mode.    */

 graph_driver = DETECT;  /* Initialization autodetection.  */
 initgraph(&graph_driver, &graph_mode, "");

} /* End of auto_initialization */

/*------------------------------------------------------------*/

void press_return(void)        /* Holds screen for user.  */
{

    gotoxy(10,25);  /* Places curser at bottom of screen. */
    textcolor(YELLOW);    /* Sets color of text. */
    cprintf("Press -RETURN- to continue.");
    getchar( );        /* Wait for user response. */

    clrscr( );        /* Clears the screen.  */
    textcolor(WHITE);    /* Returns text color to white. */

} /* End of press_return( )  */

/*------------------------------------------------------------*/

int program_repeat(void)        /* Repeat program option. */
{
 char response;                 /* User input response. */

    settextstyle(SMALL_FONT, HORIZ_DIR, 3);
    setcolor(1);
    outtextxy(120,140, "Do you want to repeat the program (Y-N)? => ");

   do
    {
     response = toupper(getche( ));
    } while ((response != 'Y') && (response != 'N'));

    if (response == 'Y') return(TRUE);
    else
     return(FALSE);

 } /* End of program_repeat( )  */

/*------------------------------------------------------------*/

 /* End of program.  */
```

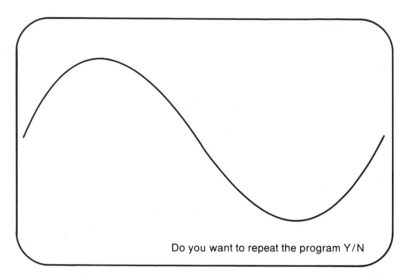

Figure 8–21   Typical Output from Case Study

Program Analysis

Observe that the string functions in **explain_program( )** are terminated with a \n\r is because the **cprintf( )** function is used. Recall that unlike the **printf( )** function, the **cprintf( )** does not give an automatic carriage return with just the \n. The **cprintf( )** function is used in order to display text color.

Observe how the C **while** is used in the function **program_repeat( )** to ensure that the only acceptable input response by the user is a Y or N (upper or lowercase). Note that the program does not use any global variables. There are two structures in the program. Their identifiers are **wave_values** and **user_input**. Notice that the user input variables are passed to the function **sine_wave_display( )** by passing the name of the structure as an argument.

Conclusion

In this section you had the opportunity to see the development of a graphics technology program. There is no Section Review here since the questions for this section are covered in the Self-Test for this chapter.

Interactive Exercises

DIRECTIONS

Because of the nature of this chapter and the differences in computer systems that display color text and graphics, this section differs from the other interactive exercise sections.

This section asks questions about the computer system(s) to which you have access. The questions are designed as a guide to help you find out important information that is useful in developing the programs in the end-of-chapter problems.

Exercises

1 Is your system capable of displaying text in color? How did you find this out?
2 How many different colors of text can your system display? Can all these different colors be displayed at the same time?
3 If your system can display text in color, can it also display different background colors? How many?
4 What kind of graphics adapter does your system have? How did you find this out?
5 For the graphics adapter in your system, how many different modes does it have? How did you find this out?
6 What .BGI file do you need on your disk in order to operate your graphics system?
7 For your graphics system, what is the maximum number of horizontal and vertical pixels available? How did you determine this?
8 What is the maximum number of drawing colors available with your system? How was this determined?
9 For your system, how many palettes does it have? How did you determine this?
10 State the maximum number of colors that can be displayed on your graphics screen at the same time. How was this determined?

Self-Test

DIRECTIONS

Answer the following questions by referring to Program 8-25 in the case study section of this chapter.

1 How many structures are used in the program? What are their identifiers?
2 Are there any global variables in the program? If so, name them.
3 How many function prototypes are in the program? Name them.
4 What are the local variables declared in function **main( )**?
5 State why the function program_repeat( ) is of type **int**.
6 In the function **explain_program( )**, why is each of the strings terminated with a **\n\r**?
7 What is the reason for using the **cprintf( )** function in function explain_program( )
8 How are the user input variables passed to the function sine_wave_display( )?
9 State the purpose of the C **while** in function program_repeat( ).

End-of-Chapter Problems

General Concepts

Section 8–1

1 State what your computer system must contain in order to produce text in color using Turbo 2.0 C.

2 How many columns of text can be displayed with an installed graphics adapter?

3 What is the maximum number of colors text may have? Can all of these text colors be displayed at the same time?

4 Explain how you would cause colored text to blink.

Section 8-2

5 Explain the difference between a text screen and a graphics screen.

6 State what your system needs, in terms of hardware and software, in order to get a Turbo C program into the graphic mode of operation.

7 Define the term pixel.

8 State which color graphics adapters supported by Turbo C can produce 16 colors in graphics.

Section 8-3

9 What built-in Turbo C function would you use to determine the modes available on your system?

10 State what Turbo C command you would use in order to determine the number of available colors for any given graphics mode of your system.

11 Explain the purpose of the built-in Turbo C functions **getmaxx( )** and **getmaxy( )**.

12 State the meaning of the term palette as used in Turbo C.

Section 8-4

13 Explain how you would change the palette on the graphics screen using Turbo C.

14 What built-in Turbo C function would you use to cause dotted lines to be created?

15 Name two ways of creating a rectangle using Turbo C.

16 When creating a rectangle in Turbo C, can you change the style of the line used to create the rectangle? Explain.

17 What are the line styles for creating a circle in Turbo C?

Section 8-5

18 There are three varieties of bars that are built into Turbo C; state what they are.

19 Explain what is meant by a Turbo C fill-style.

20 State when you would want the top of a three-dimensional bar not to appear.

21 How many different fill-styles for bars are available in Turbo C?

22 What built-in Turbo C function determines the fill-style of a bar?

Section 8-6

23 State what is meant by scaling when graphing a mathematical function.

24 Explain what is meant by coordinate transformation when graphing a mathematical function.

25 Why is coordinate transformation used?

26 How are the values of the scaling factors determined?

27 State how the coordinate transformation values are determined.

Section 8-7

28 What is meant by programming style?

29 State what is considered to be good programming style.

30 Are there any general rules to follow when developing your own variable identifiers? Explain.

31 What C structures should you keep in mind when developing programs?

Program Design

For the following programs, use the structure that is assigned to you by your instructor. Otherwise, use a structure that you prefer. Remember, your program should be easy for anyone to read and understand if it ever needs modification. Note that for these programs, you will need a system with color graphic capabilities.

Electronics Technology

32 Develop a C program that makes use of text color to display the wattage value of a resistor. The user input is the resistance of the resistor and the voltage across the resistor. The color of the answer is to be influenced by the resulting wattage values as given below:

$$
\begin{array}{rcl}
\text{microwatts} & = & \text{blue.} \\
\text{milliwatts} & = & \text{purple.} \\
1 \text{ to } 10 \text{ watts} & = & \text{green.} \\
10 \text{ to } 100 \text{ watts} & = & \text{yellow.} \\
100 \text{ to } 1\text{KW} & = & \text{red.} \\
\text{over } 1\text{KW} & = & \text{Flashing red.}
\end{array}
$$

The relationship for power is

$$P = I^2R$$

33 Create a C program that will display a graph of the relationship of the voltage across a resistor vs. the current in the resistor and the value of the resistor. User input is the value of the resistor and the range of currents to be graphed. The X axis represents the current, the Y axis the voltage. This relationship is known as Ohm's law: $E = IR$.

34 Make a C program that will display the reactance of an inductor for a given range of frequencies. The user input is the value of the inductor and the range of frequencies to be graphed. The X axis represents the frequency, the Y axis the inductive reactance. The relationship is $X_L = 2\pi FL$.

35 Develop a C program that will display the impedance of a series RLC circuit. User input is the value of the resistor, inductor, and capacitor. The X axis represents frequency, the Y axis impedance. The relationship is $Z_T = \sqrt{(R^2 + (X_L - X_C)^2)}$

36 Create a C program that will display the **resultant wave of two sine waves**. The user input is the amplitude, frequency, and relative phase of each sine wave.

Business Applications

37 Make a C program that will display a bar graph of the amount of sales, measured in dollars, for five sales persons in a given year. The X axis is to represent each sales person and the Y axis the amount of money. The user input is the amount of sales for each salesperson.

Computer Science

38 Develop a C program that will place a coordinate axis system anywhere on the graphics screen. User input is the line style and location of the coordinates.

Drafting Technology

39 Create a C program that will display a rectangle with dimension lines. User input is the size of the rectangle in pixels. The dimension lines are to contain the size of the rectangle in pixels.

Agriculture Technology

40 Develop a C program that displays the land defined by four posts. The program user may enter the coordinates of each post, and the program returns with a display of lines connecting each of the posts.

Health Technology

41   The department head needs a C program that will display the temperature readings of three different patients taken five times during the day. The X axis represents the time a temperature reading is made; the Y axis represents the temperature reading in degrees F. Use a different line style for each patient.

Manufacturing Technology

42   A machine shop requires a C program that will display a stacked bar graph of the number of different parts used in an assembly process each day (during the assembly process, some of the parts are lost or damaged). There are 5 different parts used in the process and the graph is to display the results for five days. The X axis represents the day and the Y axis the number of parts. User input is the day and the number of parts used in that day.

Business Applications

43   Expand the program in problem 37 so that the name of each sales person may be displayed vertically along the bar representing his/her sales for the time period.

Computer Science

44   The built-in Turbo C function for developing an arc is

```
arc(X,Y, StartAngle, EndAngle, Radius);
```

Where

$$X, Y = \texttt{int}$$
$$StartAngle, EndAngle = \texttt{int}$$
$$Radius = \texttt{int}$$

The function draws a circular arc from the coordinates X, Y with a radius of Radius. The arc travels from StartAngle to End Angle and is drawn in the current drawing color. Create a C program that will allow the program user to select the variables for the above arc. The program is to display the arc along with the values of the variables selected by the program user.

Drafting Technology

45   The built-in Turbo C function for creating an ellipse is

```
ellipse(X,Y,StartAngle,EndAngle,XRadius,YRadius);
```

Where

$$X,Y = \texttt{int}$$
$$StartAngle, EndAngle = \texttt{int}$$
$$XRadius, YRadius = \texttt{int}$$

The function draws an elliptical arc with X and Y as the center point. XRadius and YRadius are the horizontal and vertical axes. The ellipse travels from StartAngle to EndAngle and is drawn in the current color.

Create a C program that will allow the program user to select the variables for the above ellipse. The program is to display the ellipse along with the values of the variables selected by the program user.

Agriculture Technology

46 Expand the program in problem 40 so that the program user may enter any number of posts. The program will then display the land enclosed by the posts.

Health Technology

47 Expand the program in problem 41 so that each patient's blood pressure (taken the same number of times each day) is displayed on the same graph with the temperature. Here, use a different line color to represent the patients' blood pressure.

Manufacturing Technology

48 Create a C program that will display a bar graph of the production output of ten factory workers for a given month. The Y axis is to represent each individual factory worker and the X axis the amount of production output (assume 100 units maximum). The program user is to enter the production output for each of the ten workers along with the name of each worker. Make sure that each bar in the graph is easily distinguished from the others and that the name of the worker appears horizontally along the corresponding bar.

9 Hardware and Language Interfacing

Objectives

This chapter gives you the opportunity to learn:

1 The basic concepts of assembly language programming.
2 Microprocessor architecture as it applies to assembly language programming.
3 Methods of interfacing between C and assembly language programs.
4 The basic concepts of the IBM BIOS system.
5 Using BIOS interrupts with the C language.
6 Modifying the C library using Turbo or Microsoft C environments.
7 Developing your own library using Turbo or Microsoft C environments.
8 Some of the common utility programs used by both the Turbo and Microsoft C environments, such as the **MAKE** utility.
9 An introduction to the concepts of hardware interfacing.
10 Interfacing between data and printer ports using C.

Key Terms

Architecture
Internal Registers
Write Operation
Read Operation
General Purpose Registers

Segment Registers
Flags Register
Accumulator
Base Register
Count Register

Data Register
Stack Pointer
Stack
Base Pointer
Source Index
Destination Index
Instruction Pointer
Segmentated Memory
Segmentation Registers
Code Segment
Data Segment
Stack Segment
Extra Segment
Flag Register
Assembler
Calling Convention
Memory Models

Near Pointer
Far Pointer
Pseudo Variables
Inline Assembly
Inline Machine Code
Program Utility
MAKE Utility
Library Utilities
BIOS
DOS Calls
Interrupt
Interrupt Handler
Interrupt Service Routine
Interrupt Vector
Vector Table
Interrupt Numbers

Outline

9-1 Inside Your Computer
9-2 Assembly Language Concepts
9-3 Memory Models
9-4 Source Code to Assembly
    Language
9-5 Pseudo Variables and Inline
    Assembly

9-6 Programming Utilities
9-7 BIOS and DOS Interfacing
9-8 Program Debugging and
    Implementation
9-9 Case Study

Introduction

The last chapter of this book presents some important advanced topics. These topics include the introduction of assembly language and how to let your C programs interact with this language. You will discover how to accomplish this using either the Turbo C or Microsoft C environment.

This chapter also introduces you to the IBM BIOS. You can think of this as the software that interfaces between your program and the actual circuits that make up your computer. There are many useful tricks to be learned here. These can be used to increase your ability to control the operation of the computer using the C language.

The last topic to be presented is how you can create your own library of useful C functions that you have developed. Again, both the Turbo and Microsoft C environments are presented.

The case study demonstrates the development of a program that allows you to use your knowledge of the C language to give you a program that will allow interfacing.

This chapter is a good ending to the discussion of this powerful programming language. It is well worth the required learning effort.

9-1   Inside Your Computer

Discussion

Before you can appreciate what assembly language does and how it can interface with your C programs, you need to know some things about the inside of your computer. That is what this section will discuss. Here you will get enough of an introduction to the **architecture** of your IBM computer so that the other sections that follow will "make sense."

Basic Architecture

You don't need to know every detail about the microprocessor (μP) in your computer—there are whole books devoted to just that subject—but you do need to know something about its major features. The parts of the μP (8086/8088) you need to know about to develop a simple assembly language program are outlined in Figure 9–1.

As shown in Figure 9–1, the major parts of the μP consist of **internal registers**. You can think of a register as a place for storing bit patterns just like they can be stored in memory. The difference is that a register inside the μP can change these bit patterns by shifting, incrementing, decrementing, as well as performing arithmetic and logical operations. Recall that you were introduced to the concept of bit manipulation using C in the case study of Chapter 5.

The microprocessor interacts with the computer's memory in two fundamental ways:

1. It selects a memory location and copies a bit pattern from one of its internal registers to that memory location. This is called a **write operation**.
2. It selects a memory location and copies a bit pattern from that memory location into one of its internal registers. This is called a **read operation**.

As you can see, the internal registers in the μP can be divided up into four groups: the **general purpose registers**, the pointer and index registers, the **segment registers**, and the **flags register**. Each of these register groupings will be presented in this section.

General Purpose Registers

Figure 9–2 shows some of the details of the general purpose registers.

Each of these registers can be accessed either as four separate two-byte (16-bit) registers or as eight separate one-byte (8-bit) registers. This means they can store a 16-bit variable or an 8-bit variable.

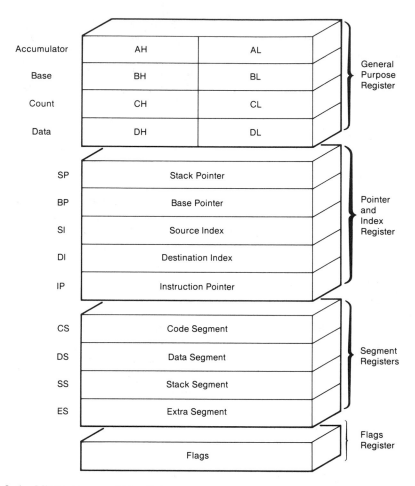

Figure 9-1   Microprocessor Major Parts

When one of these registers is accessed as a 16-bit register, it is referred to by the letter representing the first letter of the register name followed by the letter X. Thus, if an accumulator is referenced as AX, this means to treat the accumulator as a single 16-bit register. When a register is accessed as an 8-bit register, it is referred to again by using the first letter of the register name, but this time the letter is followed by either an H or an L. As you can see from Figure 9-2, each of the registers is divided into a lower half (representing the 8 least significant bits of a 16-bit number) and the upper half (representing the most significant 8 bits of a 16-bit number).

Thus, when the accumulator is to be represented as two separate 8-bit registers, the designation is either AH or AL (for a reference to the first half (AH) or last half (AL) of this register). The purpose of each of these registers is given in Table 9-1.

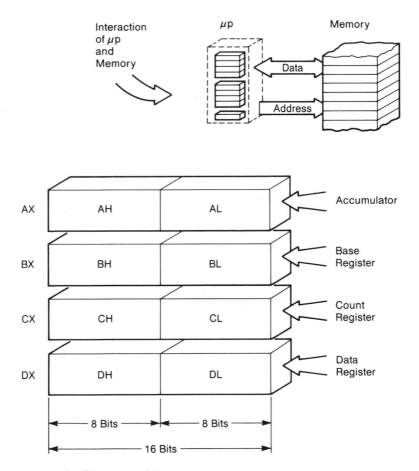

Figure 9-2 Details of the General Purpose Registers

Table 9-1 The General Purpose Registers

Register	Function
Accumulator	Stores the temporary result of an arithmetic or logical operation. Also used to interact directly with other hardware called I/O ports.
Base Register	Used to indicate specific memory locations. It is often used to store the value of a pointer; thus it will point to a memory location.
Count Register	Used as a counter for specific instructions. These include bit shift and rotate instructions or as a counter in a loop operation.
Data Register	A general register. It is usually used to hold parts of answers for multiplication and division as well as I/O port numbers when involved in I/O operations.

Pointer and Index Registers

The registers just presented are primarily used to interact with data. The next set of registers, the pointer and index registers, are primarily used to interact with memory locations used by the data. The details of these registers are shown in Figure 9–3.

As shown in Figure 9–3, all of the special purpose registers are 16-bit registers. The purpose of each of these registers is presented in Table 9–2.

Table 9–2   Special Purpose Registers

Register	Function
Stack Pointer	Used to sequentially address data in a part of memory (called the stack) that is set aside especially for this purpose.
Base Pointer	A general purpose register sometimes used to hold an address that indicates the beginning (base) of the stack.
Source Index	Usually used to address source data with the string instructions.
Destination Index	Usually used to address the destination of data required by string instructions.
Instruction Pointer	Used to select the next instruction to be executed.

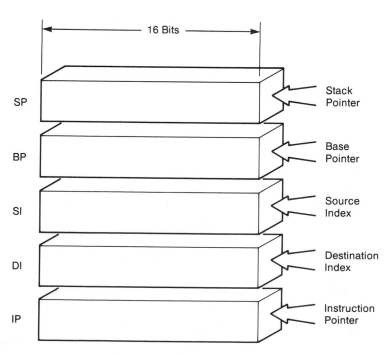

Figure 9–3   Details of the Special Purpose Registers

Figure 9–4 gives you an idea of how the stack pointer (SP) and the base pointer (BP) are used to identify the area of memory used as the **stack**.

This arrangement allows the μP to access data in memory by the order in which the data is stored. This process is referred to as **Last In First Out** or **LIFO**. What this means is that data is stored (pushed) into the stack in sequential order (one memory location after the other). This order goes toward lower memory locations, meaning the stack grows down. When data is to be copied from the stack (referred to as popping the stack), it is copied in sequential order—only this time from a lower memory location toward a higher memory location. In other words, the last piece of data to be pushed into the stack is the first piece of data to be popped from the stack. More will be said about the operation of the stack in the next section of this chapter.

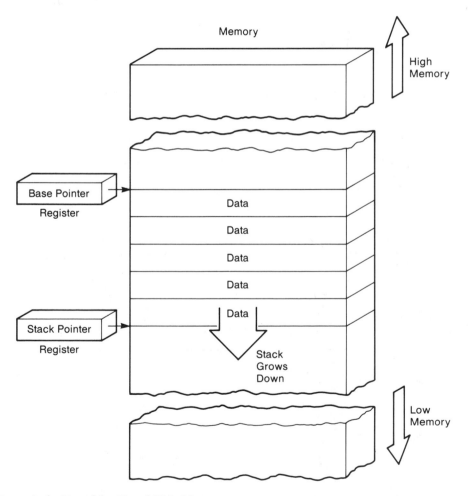

Figure 9–4   Use of the SP and BP in Memory

Memory Segmentation

The 8086 μP uses what is referred to as **segmented memory architecture**. What this means is that since the internal addressing registers of the μP are only 16 bits wide, they are capable of addressing only from $0000\ 0000\ 0000\ 0000_2 = 0$ to $1111\ 1111\ 1111\ 1111_2 = 65,535$. This represents a total of 65,536 memory locations. (This is normally stated as 64K, where a K represents 1,024. Thus $65,536/1,024 = 64K$). If this were the maximum amount of memory that the μP could access, then the capabilities of your computer would be quite limited. However, those with an IBM or compatible can address up to 1Mb (1 megabyte = 1,048,576 memory locations). This requires a binary number consisting of 20 bits ($1111\ 1111\ 1111\ 1111\ 1111_2 = 1,048,575$). To accomplish this task with only 16-bit registers, a process known as segmentation is used. This process requires **segmentation registers**. These are illustrated in Figure 9–5.

The purpose of each of the segmentation registers is outlined in Table 9–3.

Table 9–3   Segmentation Registers

Register	Function
Code Segment	Used to interact with the section of memory that contains the program code. Works with the instruction pointer (IP) register to address memory.
Data Segment	Used to interact with the data register to access data in memory.
Stack Segment	Interacts with the stack pointer (SP) to help access data in the stack.
Extra Segment	Special segment register that is normally used to help with the addressing of string instructions.

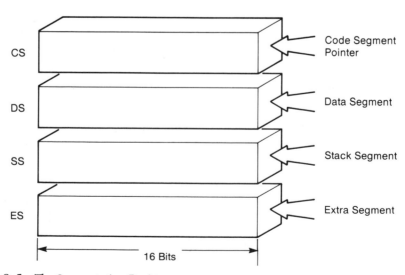

Figure 9–5   The Segmentation Registers

Calculating the Address

The actual address is determined by using two values produced by the μP. One value is called the segment address and the other is called the offset. The segment address is contained in a segment register, and the offset is taken from one of the index or pointer registers. For example, the **stack segment** register would work with the stack pointer register.

To get the actual address in memory, the bit pattern in the segmentation register is shifted left four bits and then added to the contents of the register containing the offset. Thus if the stack segment register contains $3F86_{16}$ and the offset contains $05DA_{16}$, then the absolute address would be found as shown in Figure 9–6.

Because the segment register is always shifted left four bits, its representation as a 20-bit binary number makes its last four bits zero. Thus, memory segments can only start at every 16 bytes through memory.

Figure 9–7 shows how memory is partitioned by the use of the segmentation registers.

Flag Register

The **flag register** uses individual bits to keep track of the results of many of the operations performed by the microprocessor. This register is sometimes referred to as the status register, and its contents are called the program status word. For example, one of the bits indicates if the result of the last arithmetic or logic operation was a zero. Another bit indicates if the last arithmetic operation produced a carry. The details of this register are not necessary to know for what is to follow in this chapter.

Conclusion

This section presented the foundation you will need in order to understand the basics of assembly language for your computer. This information will also be helpful in understanding how the C language itself interacts with your computer's internal workings. Check your understanding of this section by trying the following section review.

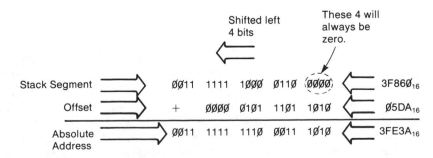

Figure 9–6 Calculation of Absolute Address

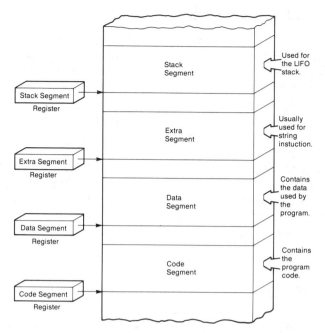

Figure 9-7 Allocation of Memory Using Segmentation

9-1 Section Review

1 Name the four major groupings of the registers in the 8086 μP.
2 Describe the difference between a register inside the μP and a memory location in the computer's memory.
3 State how the μP interacts with the computer's memory.
4 Which registers inside the μP can be treated as 8-bit or 16-bit registers?
5 Explain how a 16-bit μP can access up to 1Mb of memory.

9-2   Assembly Language Concepts

Discussion

Assembly language is about as close as you can get to having complete and absolute control over your computer without the use of a soldering iron. Assembly language programs run faster than equivalent programs written in other languages, and assembly language programs take up less memory than high-level languages trying to do the same thing. But there is a big price to pay for all of this. You must be intimately familiar with the inner workings of your computer, and you must be a very, very careful programmer. Another disadvantage of assembly language programs is that they work for only one specific kind of microprocessor family and are not very portable.

With all this, there may be times where you need the advantages of an assembly language program as a part of your C program. And, if for no other reason, as a technically-oriented person you should have an idea of what assembly language looks like and how it interacts with your C program.

Basic Idea

To use assembly language requires an intimate knowledge of the structure (architecture) of your computer and its microprocessor. When speed, control, and compactness are required, then assembly language is used.

What Assembly Language Looks Like

An assembly language is specific to a family of microprocessors. An assembly language uses mnemonics in place of the 1s and 0s of a machine language. A mnemonic is a short word used as a memory aid to indicate a process on a specific microprocessor. For example, the assembly language instruction

```
mov ah,04
```

means (for the 8088/8086 μP family) to put the number 4 into the AH register (the high byte of the accumulator).

As with any language above machine language, an assembly language program must be converted to machine language before it can be executed. This is done by a program called an **assembler**. This idea is illustrated in Figure 9–8.

The name assembly language comes from the assembler program that assembles the mnemonics into a machine code.

Types of Assembly Language Instructions

There are over 200 different instructions available for the microprocessor in your PC. Each of these instructions has a unique mnemonic to represent it. Generally speaking, these instructions can fall into one of the categories shown in Figure 9–9.

Converting C to Assembly Language Code

Both Microsoft and Turbo C can convert your C program source code into assembly language. As a matter of fact, this can be a real help in developing your own assembly language programs. First write what you want to have happen in C (or as close as you can get to what you want to have happen), then use your Microsoft or Turbo operating system to convert what you have written in C to assembly language. In order for this to be useful, you need to know some of the working details of how C uses your computer's memory. In this way, the assembly language programs generated by your C programs will make sense.

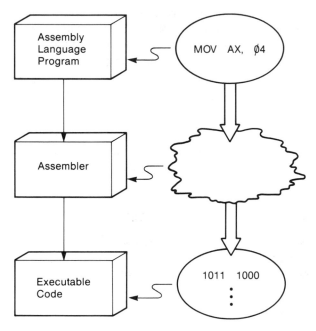

Figure 9–8 Idea of an Assembler

C at the Machine Level

When a C function uses memory, there are certain tasks that it must perform. Keep in mind that your C program, when executed, is actually causing the μP inside your system to perform specific tasks. It will cause bit patterns to be transferred between the μP's internal registers and memory as well as many other details.

A C function will use the stack to store key parts of the function. For this process, the stack pointer and base pointer are used. This process is illustrated in Figure 9–10.

The sequence shown in Figure 9–10 establishes a reference that will be used to access data located on the stack. The space just above the base pointer value contains the contents of the instruction pointer so that the program knows where to return to when the called function is completed. Note in the figure that the equivalent assembly language commands are given for each of these processes.

Saving Space for Data

A C function will save memory space for any locally declared variables in the stack. This is done by decreasing the value of the stack pointer (called going toward the top of the stack).

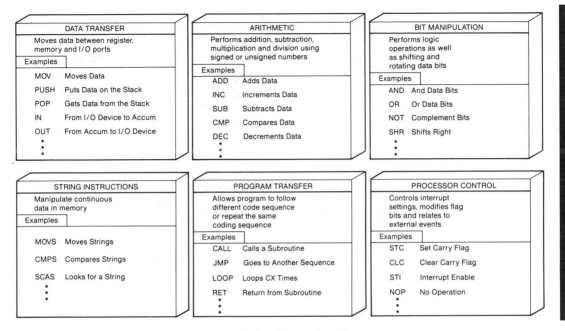

Figure 9-9   General Catalogs of μP Instructions (8086/8088 family)

This is shown in Figure 9-11.

If, for example, the called C function has two locally declared **int** type variables, then four memory locations will be preserved for their values.

Saving μP Contents

There are times when C may need the contents of some of the μP registers preserved. This is usually because the values of these registers will be needed by the calling function, and the called function may modify them in some way in order to perform its required process. Figure 9-12 shows how C would preserve the contents of the SI and DI registers on to the top of the stack.

Note that a **push** command is used to store these contents. When returning from the called function, a **pop** command will be used to set these internal registers to their original values.

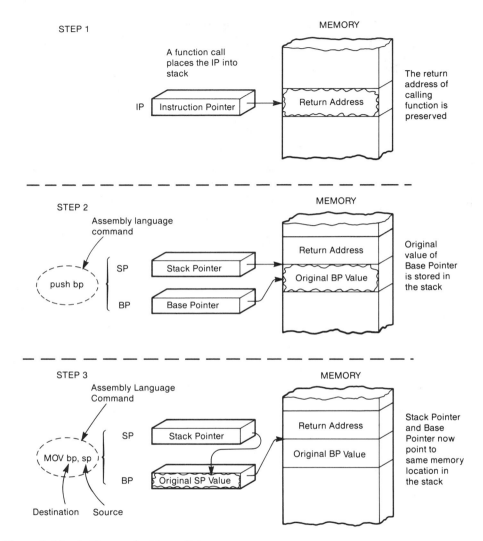

Figure 9-10   Setting up the Frame Pointer

C Arguments

So far, you have seen how a call to a C function sets up the top of the stack. The bottom of the stack is also used for function arguments. Suppose, for example, that the prototype for the called C function looks like this:

```
void function1(int val1, int val2);
```

What is known as the C **calling convention** will cause the function parameters to be pushed on the stack in a right-to-left order. This will immediately be followed

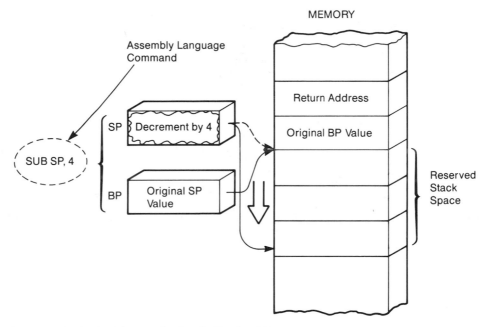

Figure 9-11   Saving Memory for Locally Declared Data

by the address the C program must return to when the called function is completed. This process is illustrated in Figure 9-13.

The complete stack of a called C function is shown in Figure 9-14.

When the called function has completed its execution, the original value of the IP will be returned to that register, and the data in the stack will no longer be used. This data is never erased, but it will be written over by the next called function. This is why data that are local to a function are available only when the function is active. This feature protects local data from being modified by another part of the program.

Storing Static Variables

Remember that a static variable has a lifetime that exceeds that of the function in which it is declared. Because of this, a variable of this type cannot be stored in the stack. Instead, variables that have a life longer than the function that called them are stored in the data segment of memory. In this manner, their values can be preserved.

Conclusion

In this section you got a basic idea of what assembly language looks like. Here you also saw how a called C function uses memory inside your computer. This

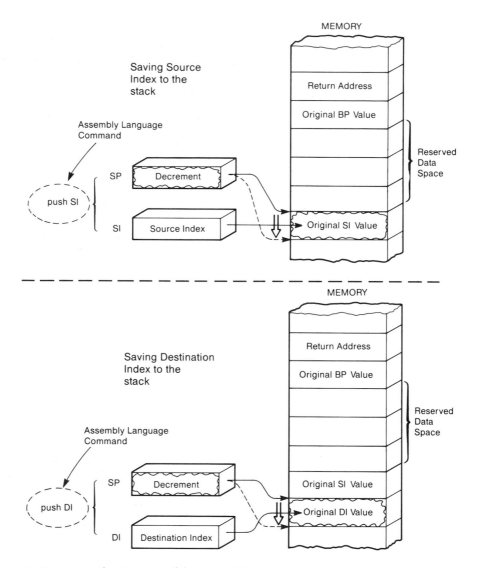

Figure 9–12 Saving the Contents of the Internal Registers

information will be needed in order to analyze assembly language programs produced from your C source code. Check your understanding of this section by trying the following section review.

9–2 Section Review

1   State three advantages of assembly language programs.
2   Give some disadvantages of assembly language programs.

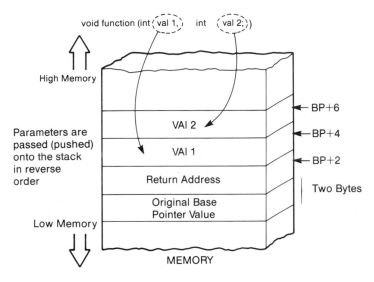

Figure 9-13   The C Calling Convention

3   What is a mnemonic?
4   When a C function is called, what is the first internal register placed on the stack?
5   Where on the stack does C store locally declared variables? Function parameters?

9-3   Memory Models

Discussion

This section shows how C handles segmented memory. The purpose of this section is to make you aware of options you have to use all of the memory available in your computer. Recall from the first section of this chapter that the way the microprocessor in your PC gets around the limitations of just 64K of memory is by using segmentation registers and offsets. This allows it to access up to 1Mb of memory. You can control how much memory your C program may access by using different kinds of **memory models**.

Basic Idea

If all of your executable code was confined to one 64K segment, then all of the segmentation registers could have a fixed value. Doing this means that only the 16-byte offset would need to be changed as different locations (addresses) in memory are accessed. As an example, if all of the data in your program was limited to a single 64K segment, then the DS register could stay fixed and only the 16-bit (2-byte) offsets would have to change. This idea is illustrated in Figure 9-15.

MEMORY

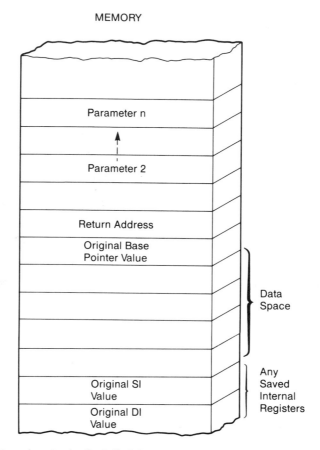

Figure 9–14   Complete Stack of a Called C Function

Confining segments to only 64K of memory has the advantage of speed since only the offset needs to be changed. This also results in more compact code since the address of your variables only needs to occupy two bytes of memory.

The disadvantage of this system is that not all of your computer's memory is usable. Confining the four available segments to 64K allows you a total of 64K $\times$ 4 = 256K of memory access—only 25% of its full 1Mb potential. For example, some C programs may require more than a 64K segment for data. This may come about with a large data base management system where it is necessary to store large amounts of data at the same time. The concept of extending data access to more than one 64K segment is illustrated in Figure 9–16.

As you can see from the above discussion, confining your segments to a maximum size of 64K requires only a 16-bit (2-byte) size pointer. To expand beyond this restriction requires a 32-bit (4-byte) size pointer—because the 16-bit segment address must be given as well as the 16-bit offset. As you will see, the 16-bit pointer is called

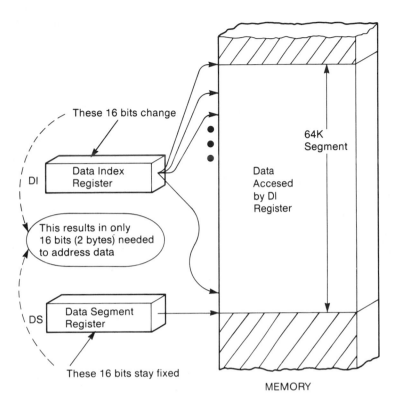

Figure 9-15   Confining Data to One 64K Segment

a **near pointer** and a 32-bit pointer is called a **far pointer**. Up to this point almost all of your C programs have used near pointers.

It's important that your C program knows what type of pointers to use. Recall from the last section how a C function uses the stack. To store the values of far pointers requires twice as much stack space as storing the values of near pointers (four bytes vs. two bytes).

Memory Models

Both Turbo and Microsoft C offer options as to how memory will be allocated by your executed C program. The memory models for Turbo C are listed in Table 9-4.

Near and Far Pointers

You can declare C types to be near or far by using the keyword **near** or **far**. As an example, if a pointer is to hold a 16-bit address it may be declared as **int near** ***ptr**;. If, on the other hand, the pointer is to hold a 32-bit address, it may be declared as **int far *ptr**;. However, it is usually best to use the default values

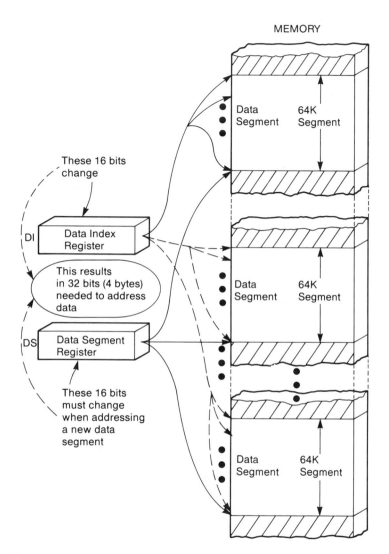

Figure 9-16 Expanding Data Access to More Than One 64K Segment

of the program models. It's important to note that the keywords **near** and **far** are not part of the ANSI C standard. They are useful only in systems (such as the 8086 family) that need to use memory segmentation. This is important if you want to create C code that can be used on other systems that may not have this requirement.

Conclusion

This section presented the concept of a memory model and the options offered to you by both Turbo and Microsoft C. Here you saw the significance of the different

Table 9-4   Turbo C Memory Models

Model Name	What It Does
TINY	This produces a total of 64K for the whole program. The reason for this is that all four segment registers are placed at the same address. This means that code, data, arrays, and the stack all share the same 64K of memory. All pointers are always near. This is a good model to use when you have a system of only 64K of memory.
SMALL	In this model, your program code has one 64K segment and your program's static data has one 64K segment. Static data does not create more memory as more data space is needed. Stack and extra segments start at the same address. Pointers are near. Use this model for most of your general purpose programming applications.
MEDIUM	Here the code segment is allowed to use up to the full 1Mb of addressable memory. The data segment is still limited to a single 64K segment. In this mode, far pointers are used for code, and near pointers are used for data. Use this model for large programs that do not keep much data in memory.
COMPACT	This model is just the opposite of the MEDIUM mode. Here, the code is limited to a single 64K segment while data is allowed to use up to the full 1Mb of addressable memory. Use this model when you have a limited amount of code with large quantities of data. Far pointers are used for data, and near pointers are used for code.
LARGE	For this model, both the MEDIUM and COMPACT models are combined. Here you have the potential of a full 1Mb for both code and data. Far pointers are used for both code and for data. This model should be used for large programs that must process large amounts of data.
HUGE	The difference between this model and the LARGE is that static data may exceed 64K. In the other models, static data is limited to 64K.

memory models and the requirements for addressing within these different models. The Turbo C or Microsoft C manuals contain much more detailed information concerning these memory models and should be consulted if you need to change pointer characteristics from the defaults offered by the different models. Check your understanding of this section by trying the following section review.

9-3 Section Review

1   State the advantage of confining all of your C code to a 64K segment.
2   Describe what is meant by a memory model.
3   State the difference between the MEDIUM and COMPACT memory model.
4   What is the difference between a far pointer and a near pointer?

Table 9–5   Microsoft C Memory Models

Model Name	What It Does
SMALL	Code and data each have their own 64K segments. In this model the total size of the program cannot exceed 128K. Near pointers are used for both code and data. This is a good model to use for most general applications.
MEDIUM	This model limits data to one 64K segment but allows multiple segments of code. Here, data pointers are near, and code pointers are far. This is a good model to use when you have large amounts of code and small amounts of data.
COMPACT	This model is just the opposite of the MEDIUM. Here code is limited to one 64K segment, but data can occupy multiple segments. Far pointers are used for data, and near pointers are used for code. Use this model when you have large amounts of data and code that can be confined to one 64K segment.
LARGE	Both code and data segments may occupy as many 64K segments as needed up to the full 1Mb of addressable memory. Both data and code pointers are far. Use this model for when you have large amounts of code and data in your executed program.

9-4   C Source Code to Assembly Language

Discussion

In this section you will discover how to convert your C source code into assembly language. Here also you will learn something about reading the resulting assembly language output. This is an important section because you get a built-in tutorial on assembly language programming. How? By writing what you want to do in C and then converting it to assembly language. This will result in a template that you can then use to modify or expand for your own programming needs.

Doing the Conversion

Your Learn C does not create stand alone programs. If you have Turbo C, you can convert your C source code into assembly language. To do this you must have the Turbo line compiler on your disk (**TCC.EXE**). This is not the Turbo C Integrated Development Environment (IDE) that has all of the pull-down menus (**TC.EXE**). The Turbo C command line (**TCC.EXE**) offers several options that are not available with the integrated development environment. Both of these systems come complete with the Turbo C 2.0 system and are explained in detail by the accompanying manuals. There isn't enough space in this chapter to cover all of the details concerning the use of the command line compiler. For the purpose of this section, you will be shown how to use it to convert a C function into assembly language source code.

Assume that you have written a short C function as shown in Program 9-1.

Program 9-1

```
/* FUNCT1.C  */

function_1(int value)
{

  return value = value + 5;

}
```

The above function could be used as a part of a C program and called from **main( )**. Assume that the above function is stored on your disk as FUNCT1.C. Once this is done, from the DOS command enter the following:

```
TCC -S -EFUNCT1.ASM FUNCT1.C
```

Essentially what this does is to evoke TCC.EXE; the -S states that it is to convert a C source code to assembly source code. The -EFUNCT1.ASM directs it to execute a new file for the assembly source code and store it on the disk with the name FUNCT1.ASM, and it will find the C source code in the file named FUNCT1.C. It's important to note that while C source code files are given an extension of .C, assembly language source code files are given an extension of .ASM.

Once you do the above, the TCC will take a few seconds while it does the job you asked of it. Then you will find a new file called FUNCT1.ASM on your disk. What it contains is shown in Program 9-2.

Program 9-2

```
     ifndef    ??version
?debug         macro
     endm
     endif
     ?debug    S "funct1.c"
_TEXT      segment byte public 'CODE'
DGROUP         group    _DATA,_BSS
     assume    cs:_TEXT,ds:DGROUP,ss:DGROUP
_TEXT          ends
_DATA          segment word public 'DATA'
d@  label      byte
d@w label      word
```

(continued)

Program 9-2 (continued)

```
_DATA          ends
_BSS           segment word public 'BSS'
b@  label      byte
b@w label      word
    ?debug     C  E99B4C6B120866756E6374312E63
_BSS           ends
_TEXT          segment    byte public 'CODE'
;    ?debug    L 5
_function_1    proc       near
    push       bp
    mov        bp,sp
;    ?debug    L 8
    mov        ax,word ptr [bp+4]
    add        ax,5
    mov        word ptr [bp+4],ax
    jmp        short @1
@1:
;    ?debug    L 10
    pop        bp
    ret
_function_1    endp
_TEXT          ends
    ?debug     C E9
_DATA          segment word public 'DATA'
s@  label      byte
_DATA          ends
_TEXT          segment    byte public 'CODE'
_TEXT          ends
    public     _function_1
    end
```

Wow! That's a lot of code for such a small C function. For the purpose of this chapter, concentrate on the assembly code. It has been made boldface in the program. The rest of the information is essential overhead for the assembler and debugger. The assembly code that does the work required by the function is shown in Program 9-3.

Program 9-3

```
push      bp
mov       bp,sp
move      ax,word ptr [bp+4]
add       ax,5
mov       word ptr [bp+4],ax
pop       bp
ret
```

Meaning of the Instructions

What happens here, is based upon an understanding of the material presented in Section 9–2. Each of the assembly language instructions and the resulting process is explained in Figure 9–17.

As you can see from the figure, the processes involved represent the actions explained in Section 9–2.

Conclusion

This section gave you an opportunity to see how you can convert C source code into assembly language source code. You also saw how to interpret the resulting assembly language code and relate it to the C function. It was not the intention of this section to explain every detail about the resulting assembly code. That is a subject beyond the scope of this book.

Test your understanding of this section by trying the following section review.

9–4 Section Review

1   Briefly describe what is needed to convert C source code to assembly language source code.
2   What is the extension normally used for an assembly language source file?
3   Explain what is meant by the following assembly language instruction **push bp**.
4   State the difference between the meaning of **bp+4** and **[bp+4]**.

9–5   Pseudo Variables and Inline Assembly

Discussion

There are other options available for getting some of the power offered by assembly language from your C programs. For example, Turbo C offers the choice of using **pseudo variables** or **inline assembly**. This section will introduce you to both of these methods.

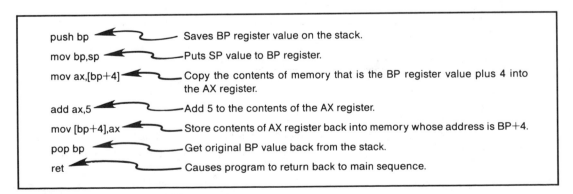

Figure 9–17 Assembly Language Process

Introduction to Pseudo Variables

In Turbo C a pseudovariable is an identifier that corresponds to one of the 8086 family internal registers. These are listed on Table 9–6.

Table 9–6   Pseudo Variables—Turbo C

Pseudo Variable	Type	μP Internal Register
_AX	unsigned int	Accumulator (16-bit) AX
_AL	unsigned char	Accumulator Lower Byte AL
_AH	unsigned char	Accumulator Upper Byte AH
_BX	unsigned int	Base Register (16-bit) BX
_BL	unsigned char	Base Lower Byte BL
_BH	unsigned char	Base Upper Byte BH
_CX	unsigned int	Count Register (16-bit) CX
_CL	unsigned char	Count Lower Byte CL
_CH	unsigned char	Count Upper Byte CH
_DX	unsigned int	Data Register (16-bit) DX
_DL	unsigned char	Data Lower Byte DL
_DH	unsigned char	Data Upper Byte DH
_SP	unsigned int	Stack Pointer SP
_BP	unsigned int	Base Pointer BP
_DI	unsigned int	Destination Index DI
_SI	unsigned int	Source Index SI
_CS	unsigned int	Code Segment CS
_DS	unsigned int	Data Segment DS
_SS	unsigned int	Stack Segment SS
_ES	unsigned int	Extra Segment ES

What these allow you to do is to access the internal registers of your μP directly from your Turbo C program. This gives you the option of loading any of these registers with a given value before calling a function or to test and see their current values.

The following program illustrates an application using this feature.

Program 9-4

```
#include <stdio.h>

main( )
{
 unsigned char lower;
 unsigned char upper;
 unsigned int total;

  _AL = 6;
  _AH = 10;
  lower = _AL;
  upper = _AH;
  total = _AX;

  printf("Value in AH register is %X\n",upper);
  printf("Value in AL register is %X\n",lower);
  printf("Value in AX register is %X\n",total);

}
```

Execution of the above program produces

```
Value in AH register is A
Value in AL register is 6
Value in AX register is A06
```

Using Pseudo Variables

Even though the Turbo C pseudo variables can be used as if they were global variables, you must keep in mind that they are interacting with the internal registers of your computer's μP, not any locations in memory. Because of this, you cannot use the address of (&) operator since the internal registers of the μP do not have an address. It's also necessary to keep in mind that your C program (or any other program) is constantly using these registers. Thus, whatever you put into one of them may not be there very long, and whatever value you read from them may not stay that way for long. The point here is to make use of the interaction with the register as soon as possible in your source code.

One note of caution. Storing values directly into the CS, BP, SP, or SS can cause your C program to crash since the contents of these registers are used by the program. Note that these pseudo variables are not ANSI C standards.

Introduction to Inline Assembly

Turbo C allows you to write assembly language code directly into your C source code. This means that you can have the power of assembly language without the hassle of first assembling a separate .ASM file and then linking it to your C code.

To use this feature, you must have a copy of the Turbo Assembler (TASM). This comes in a package separate from Turbo C and includes the Turbo Debugger. To let your C program know that you are using inline assembly, you can use the -B compliler option or put the statement

#pragma inline

in your source code.

When using inline assembly code, the following format must be observed:

asm <opcode> <operand><operands> <; or newline>

Where
        opcode = an allowable 8086 instruction.
        operand = an acceptable instruction for the corresponding opcode.

As an example, the following program is an assembly language program that will activate the speaker by actually sending on/off pulses to it.

Program 9-5

```
main( )
{
    asm    in     al,61
    asm    and    al,fc
    start:
    asm    xor    al,02
    asm    out    61,al
    asm    mov    cx,0140
    place:
    asm    loop place
    asm    jmp    start
}
```

In order to jump to a particular place in the program, Turbo C inline assembly requires that you use a label (such as **start:**) as shown in Program 9-5.

If you are familiar with assembly language programming, then you are aware of using the semicolons for the beginning of a comment

mov ah,3   ;This is a comment.

The semicolon is not allowed for this purpose when used as a part of inline assembly code in C since the semicolon there has another meaning. To use comments in inline assembly code, you must use the standard C symbols:

```
asm  mov ah,3   /* This is a comment.  */
```

Note that no semicolon is needed in the above inline assembly statement. This makes statements in inline assembly language the only C statements that do not require a semicolon. Keep in mind that inline assembly code is not an ANSI C standard.

Inline Machine Code

Turbo C offers the option of using **inline machine code**. The advantage of this is that you do not need a separate compiler, and the program may be compiled and executed from the IDE (the inline assembly code cannot—it must use the command line TCC). A machine code simply consists of the hex value of the assembly instruction. For example, the machine code for the assembly instruction **mov ah,08** is **B408**, where **B4** is the machine code for the move immediate instruction and the **08** is the value to be moved into the AH register.

The Turbo C function that does this is

```
__emit__(argument,...)
```

> Where
> > argument = generally a single-byte machine instructions.

This requires dos.h and is not an ANSI C standard.

Program 9–6 shows a use of this function. The program uses the inline machine code to place a value of 8 into the AH register and then uses the pseudo variable **_AH** to verify that the machine code did indeed do what was intended.

Program 9–6

```c
#include <stdio.h>
#include <dos.h>

main( )
{
    unsigned char value;

    __emit__(0XB4,0X08);    /* mov ah,08  */

    value = _AH;

    printf("The value in AH is %0X\n",value);

}
```

Conclusion

This section introduced two powerful features available with Turbo C. These are the pseudo variables that allowed you to directly access the internal registers of your μP, and inline assembly code where you can include assembly language instructions as a part of your C program.

Check your understanding of this section by trying the following section review.

9–5 Section Review

1   Explain what is meant by a pseudo variable in Turbo C.
2   Give an example of the use of a pseudo variable.
3   Can the address of (**&**) operator be used with a pseudo variable? Explain.
4   State what is meant by inline assembly.
5   Are pseudo variables and inline assembly accepted ANSI C standards? Explain.

9–6   Programming Utilities

Discussion

A typical C development system usually consists of many different programs. For example, one of the programs on your Learn C disk is **LC.EXE** which is the integrated development environment (IDE) that contains the built-in editor, filer, linker, and so on. In a like manner, both Turbo C and Quick C have other programs on their disks that can assist your C program development in many different ways. These ways range from searching for a word on one or more disk files to helping you create your own C libraries.

In this section, you will get an overview of some of the useful utility programs that may be available with your C system. Having a general understanding of them will help pave the way for the more detailed discussions in the systems manuals that accompany the respective C development systems.

Utility Overview

Table 9–7 lists some of the more common utility programs that may be available with a C development system.

As you can see from the above table, there is an extensive list of possible utilities available for your use in the development of a C program. This section will not replace the extensive presentations available in the manuals that accompany a professional C development system. But it will make you aware of what is available, the reason for this availability, and an overview of what the utility can do.

The MAKE Utility

The **MAKE utility** is a separate program contained on the systems disks for both Microsoft C and Turbo C. Here is the reason for having such a utility available.

Table 9-7   Common Utility Programs Used with C

Utility Type	Purpose
MAKE Utilities	A separate program that helps keep all of your source, object and .EXE files current.
File Date/Time Change (TOUCH Utilities)	A separate program that allows you to change the date and time stamp of one or more files.
Linker	A separate program used to do the linking work when source code is compiled outside of IDE.
Library Utilities	A separate program that allows you to create your own libraries that can then be accessed by your C programs.
File Search Utilities	A separate program that lets you search for a specified text in several files at once.
Conversion Utilities	A separate program that allows you to link graphic driver files and fonts directly into your C program.
Cross Reference	A separate program that allows you to get specific information concerning the contents of your object and library files.

When developing a large C program, it will usually be broken up into several different files where only one file at a time is being worked upon. Figure 9–18 illustrates this concept by showing a parts inventory program under development.

As shown in Figure 9–18, the executable program INVENT.EXE consists of the three object files, INVENT.OBJ, EXPLAIN.OBJ, and GETIT.OBJ. Further, INVENT.OBJ uses functions contained in the EXPLAIN.OBJ and GETIT.OBJ.

Each of the .OBJ files were compiled from their respective .C source code files. As an example, the GETIT.OBJ file was compiled from the GETIT.C file. In turn, the .EXE file was compiled from the respective .OBJ files. This means that INVENT.EXE consists of the code in INVENT.OBJ, EXPLAIN.OBJ, and GETIT.OBJ. All of this results in seven different disk files.

A typical file directory from such an arrangement is shown in Figure 9–19.

As you will see, it is very important to make sure that the date and time entered into your computer are correct. Note from the date/time stamps of the directory illustrated in Figure 9–19, that the .EXE file is the most recent and that the .OBJ files are more recent than their related C source code files. Of course, you should know that the one .EXE file is the only file of the seven needed in this example to execute the program. However, while the program is under development, all the other files are needed. Figure 9–20 illustrates the dependencies one file has on the other.

As you can see from Figure 9–20, changing any of the source files requires an update of other files that depend upon it. Whenever you change any of your source code files and save it to the disk, its date/time stamp will now be later than its corresponding .OBJ file and later than the .EXE file. This means that the system of files must be updated. This situation is shown in Figure 9–21.

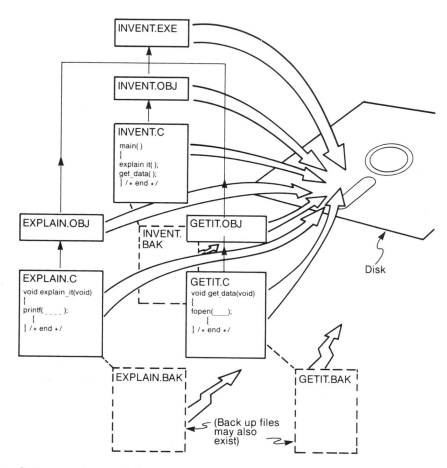

Figure 9-18 Development Files for an Inventory Program

Filename	Extension	File Size	Date Stamp	Time Stamp
INVENT	EXE	10187	9-12-92	1:15p
INVENT	OBJ	430	9-12-92	1:08p
INVENT	C	642	9-12-92	1:05p
EXPLAIN	OBJ	983	9-12-92	12:45p
EXPLAIN	C	1246	9-12-92	12:36p
GETIT	OBJ	10083	9-12-92	10:23a
GETIT	C	12436	9-12-92	10:15a

Figure 9-19 File Directory for Inventory Program

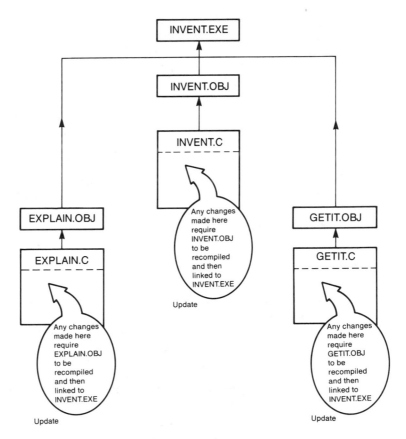

Figure 9-20 How One File Depends upon Another

INVENT	EXE	10187	9-12-92	1:15p	Does not contain most recent change
INVENT	OBJ	430	9-12-92	1:08p	
INVENT	C	642	9-12-92	1:05p	No longer current
EXPLAIN	OBJ	983	9-12-92	12:45p	
EXPLAIN	C	1287	9-12-92	3:06p	Recent update
GETIT	OBJ	10083	9-12-92	10:23a	
GETIT	C	12436	9-12-92	10:15a	

Figure 9-21 Differences in Date/Time Stamps for Recently Changed File

As you can see from Figure 9–21, the .EXE file does not include the latest changes that were made in the .C source code file. This means if the program is now executed, it will not contain the latest changes from the source code file.

Of course you could, from the IDE, compile the changed source code, updating its .OBJ file and then link all the .OBJ files to create an updated .EXE file. The resulting directory could then appear as shown in Figure 9–22.

Doing this in the IDE is fine for a small program such as the one illustrated here. However, for large programs consisting of many different source files with their resulting object files, this could become a major task. This is especially true if you are also developing your own custom .H (header) files as well as custom library files. Imagine making a change on a custom header file upon which one or more C source files depend. Making all the required updates could become a major task. This is where the MAKE utility can be very helpful. Table 9–8 lists some of the functions that can be performed by this powerful utility.

Table 9–8   Major Functions of the MAKE Utility

Environment	Capabilities
Program Development	Automatically updates executable files whenever any dependent object or source files have been modified.
Library Management	Automatically rebuilds a library any time one of the library modules is altered.
Networking	Automatically makes current a copy of a local program or file that is stored on the network whenever the master copy is updated.

There can be other features available for the MAKE utility, depending upon the development system you are using.

How MAKE Works

To use the MAKE utility, you first create a file, called a description file or MAKEFILE that contains the commands you want the MAKE utility to perform. Figure 9–23 shows the idea of the relationship between the MAKE utility, description file and your program files.

```
INVENT    EXE    10195    9-12-92    3:15p ◄─── Now updated
INVENT    OBJ      430    9-12-92    1:08p        to include
INVENT    C        642    9-12-92    1:05p        changes
EXPLAIN   OBJ      998    9-12-92    3:11p ◄─── Now current
EXPLAIN   C       1287    9-12-92    3:06  ◄─────── Changed file
GETIT     OBJ    10083    9-12-92   10:23a
GETIT     C      12436    9-12-92   10:15a
```

Figure 9–22   Resulting Directory of Updated Files

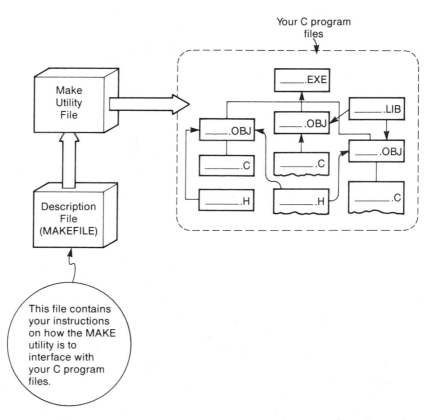

Figure 9-23 Relationship Between MAKE Description (MAKEFILE) and C Files under Development or Modification

You create a description file (**MAKEFILE**) by using a word processor (like the one in the IDE) and simply type in the instructions you want the MAKE utility to perform.

As an example, for the Turbo C MAKE utility, to give the instruction that the file **INVENT.EXE** depends upon the files **INVENT.OBJ**, **EXPLAIN.OBJ**, and **GETIT.OBJ**, you would enter

```
invent.exe: invent.obj explain.obj getit.obj
```

and simply save it on your disk giving the file the name **MAKEFILE**. Then when you're ready to use the MAKE utility, simply enter MAKE, and the MAKE utility will automatically search for a file called **MAKEFILE**, read it, and then act accordingly. What MAKE then does is to check the time and date of each of the .OBJ files with the time and date of the resultant .EXE file (and any header files you may have indicated in your **MAKEFILE**). If any of these are later than the .EXE file, then MAKE knows that the .EXE file is not current and will automatically call **TLINK**, the Turbo C linker, and relink the dependent .OBJ files.

This utility is a powerful one. The presentation here sets the foundation of information you needed to know in order to understand the specific details and features of the MAKE utility that comes with your system. It is suggested at this time that you refer to your C system manuals for specific details and then actually try a simple application of this useful utility.

Date/Time Utilities

Turbo C contains a stand-alone utility that can force a particular file to be recompiled or rebuilt. This can be done even though you may not have made any changes to the file. The Turbo C TOUCH utility will change the date and time of one or more files to the current date and time. Doing this will of course make these files newer than they were.

Library Utility

Recall that a program library is a collection of programs that have been compiled or assembled to produce a set of object modules that can be used by the C programs. Both Quick C and Turbo C provide the tools necessary to produce your own C library. In Turbo C this utility is a separate program on the system disk called TLIB.EXE and in Quick C it is called LIB.EXE.

Basically what **library utilities** can do is to make a single library from one or more .OBJ files, add or delete .OBJ files from existing libraries, get an .OBJ file from an existing library, and, depending on the system, combine libraries and list the contents of a library.

There are many advantages to creating your own C libraries. First of all, as you develop C programs in your particular area of interest, you will have certain functions that you use over and over again in more than one program. These are prime candidates for storing in a library. When linking your program, the linker automatically looks for data needed from your library. Using a library can speed up the action of the linker and also results in the use of less disk space than a collection of many different .OBJ files.

Using the Turbo C Library Utility

To create a library using the Turbo C library utility, you must have the program TLIB.EXE on your active disk drive.

The syntax of the library utility commands are

```
TLIB library-name [/C] [/E] [operations] [, listfile]
```

Where
TLIB = the command that activates the library utility.
library-name = the name you wish to give to your library.
This file will automatically have the extension .LIB added to it (or you may add it yourself). In order for your library file to function properly, the file must have the .LIB extension.

/C, /E = options for case sensitivity and creating an extended dictionary. Essentially the /C option is not normally used and you should refer to your Turbo C Reference Guide for details on the advanced /E feature.

operations = the list of operations you want the TLIB utility to perform.

listfile = an option that allows you to view the contents of the library.

To create a library named TECH.LIB that consists of modules OHMS.OBJ, REACT.OBJ and IMPED.OBJ, enter

```
TLIB TECH +OHMS +REACT +IMPED
```

The new library will then be created and stored on the active drive with the name TECH.LIB

Using the Quick C Library Utility

In the Quick C environment, you must have the file LIB.EXE on your active drive. When you enter LIB the following prompts will appear:

```
Library name:
Operations:
List file:
Output library:
```

Where

Library name = name of the library you intend to modify or create.

Operations = command symbols and object files stating what changes to make in the library.

List file = the name of a cross reference listing file. No listing file is created if you do not enter a name for this prompt.

Output library = name of the changed library to be created as output. For your first-time use of this Quick C utility it is suggested that you simply use the default values for each of the prompts.

More details for the creating of library files are contained in the respective system manuals for the Quick C and Turbo C.

File Search Utilities

Turbo C offers a powerful search utility that can search for text in several files at once. This is especially useful for large source code programs or several source code files where you must change a word and are not sure exactly where in the source code (or in which file) the word is contained.

To operate this utility, you must have the file GREP.EXE on your active drive. The command is

```
GREP [options] search-string [filespec ...]
```

Where

options = one or more single characters (as given in the Turbo C Reference Guide) that state how the files are to be searched.

`search-string` = defines the pattern `GREP` will search for.
`filespec` = specification that tells `GREP` which file(s) to search.

As an example,

```
GREP impedance *.C
```

instructs the file search utility to search all files on the active drive with the .C extension for the string `impedance`.

Conclusion

It was the intent of this section to make you aware of the many other features that may be available with your C system. Here you were introduced to some of the most common ones in two of the most popular C development systems, Quick C and Turbo C.

The details of operation for these utilities are covered in the respective Quick C and Turbo C manuals along with many suggestions and examples. You are encouraged to refer to the manuals that come with your C system. Check your understanding of this section by trying the following section review.

9-6 Section Review

1 Briefly state the purpose of a MAKE utility.
2 What does a MAKE utility look for when working to keep your files updated?
3 State the use of the Turbo C TOUCH utility.
4 What is a library utility?
5 Give the name of the Turbo C utility that will search files for a given string.

9-7   BIOS and DOS Interfacing

Discussion

This section is specifically for the IBM PC and compatibles. As you will discover, there are many different powerful programs built into the permanent memory (usually referred to as ROM for **Read Only Memory**). These permanently built-in programs are referred to as the **Basic Input/Output System** or simply **BIOS** or the IBM ROM BIOS.

You will also discover in this section that the DOS (Disk Operating System) also contains many different and powerful programs, that, once loaded into your computer's read and write memory (referred to as RAM), are also available for your use.

This section will introduce you to the advantages and disadvantages of using the built-in routines in both of these systems. Here you will also see some of the built-in C functions used for these purposes as well as a short application program for the BIOS and DOS.

BIOS and DOS

Figure 9–24 illustrates the roles that the BIOS and DOS play in your computer system.

Figure 9–24 is an important figure. As shown, both BIOS and DOS contain programs that control the Input/Output (I/O) of your computer. Most application software programs (including your C operating system) will use the I/O routines already existing in DOS rather than having to write these routines themselves. (That's why you must load DOS into your computer before you can use commercial applications programs). Also, as shown in the figure, you can have your C program use either the BIOS or the DOS for performing I/O operations. Table 9–9 lists the merits of each system.

Some BIOS Functions

Both Turbo C and Quick C support BIOS interfacing. Some of the more common I/O functions are listed in Table 9–10.

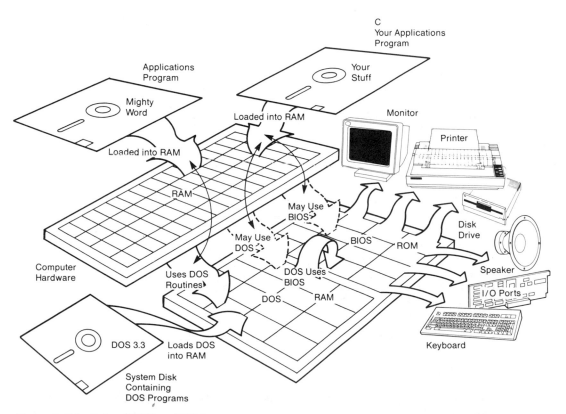

Figure 9–24   Role of BIOS and DOS

Table 9–9   BIOS and DOS I/O

I/O Type	Comments
BIOS	Faster and more direct than DOS. However, because these routines are contained in a specific computer machine architecture, they may not be configured in exactly the same way on a system made by a different manufacturer. Thus, any C program that uses BIOS directly may not be portable to these other systems.
DOS	Not as fast as accessing BIOS directly and not as versatile. However, the advantage is that your program will be more portable, because DOS can adjust for some hardware differences.

Table 9–10   Common BIOS Functions in C—Uses `<bios.h>`

NOTE: These are not ANSI C compatible and will work only on IBM PC's and true compatibles

Function		Description
Turbo C	Quick C	
`bioscom`	`_bios_serialcom`	Performs serial I/O.
`biosdsk`	`_bios_disk`	Provides many different disk access functions. Generally, this function controls the movement of the disk read/write head on a specified drive.
`biosequip`	`_bios_equipment`	This function determines the hardware and peripherals currently used by your system.
`bioskey`	`bios_keybrd`	Accesses the keyboard directly through the BIOS keyboard routines. Generally, this function checks to see if the keyboard is ready to be read, reads it, then gets the character.
`biosmemory`	`_bios_memory`	Returns the total amount of memory available on your system.
`biosprint`	`_bios_printer`	Performs a variety of printer functions, such as checking if the printer is out of paper.

DOS Calls

Both Turbo C and Quick C provide functions that interface directly to the DOS routines. Some of these are listed in Table 9–11.

Table 9–11 lists just some of the C DOS functions. There are about 40 different DOS function calls used by Turbo C and Quick C.

BIOS Application

Program 9–7 illustrates a BIOS application. This program uses the Turbo C **biosequip( )** function that returns specific information about your IBM PC (or true compatible).

Table 9-11 Common DOS Functions in C—Uses <**dos.h**>

NOTE: These functions are not ANSI C standards and require the presence of DOS (3.0 or higher for some of the C DOS functions).

Function		Description
Turbo C	Quick C	
inport		Reads a word from a specified port. (Quick C has a similar function called **inp**—it uses the <conio.h> header file.)
outport		Outputs a word to a specified hardware port. (Quick C has a similar function called **outp**—it uses the <conio.h> header file.)
peek		Returns a word at a memory location specified by an offset and segment value.
poke		Stores an integer value at a memory location specified by an offset and segment value.
keep	**_dos_keep**	Used to install a terminate-and-stay resident program in memory.
getdate	**_dos_getdate**	Gets the system date.

This function returns an integer whose value depends upon the equipment connected to your system. To understand the meaning of the returned number, it must be converted to a 16-bit binary number. As a binary number, each bit or group of bits indicates a hardware detail about your system. Figure 9–25 shows the significance of each bit for this function.

Program 9–7 illustrates the **biosequip( )** function in action.

Program 9-7

```
#include <stdio.h>
#include <bios.h>          /* Must use the BIOS header file.  */

main( )
{

   printf("The biosequip number for this system is %0X",biosequip( ));

}
```

Note that the program causes the decimal output to be expressed in hexadecimal notation. This is done so that the resulting number can easily be converted to binary.

A typical output for Program 9–7 could be:

406D

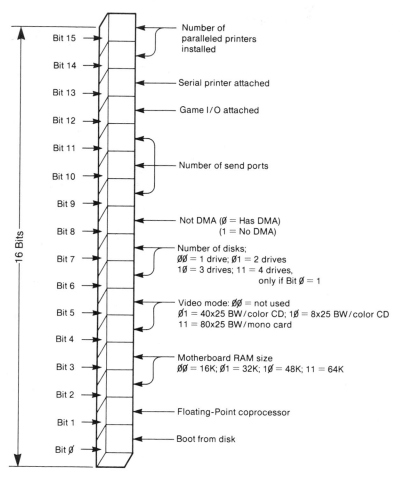

Figure 9-25   IBM PC Values for **biosequip( )** Function (Turbo C)

When the above is converted to a binary number, you have:

```
0100 0000 0110 1101
```

As you can see, since bits 15 and 14 are 01, this system has one parallel printer installed. With bit 13 at 0, there is no serial printer attached, and bit 12 at 0 indicates that there is no game I/O attached. Bits 6 and 7 are 01 indicating that there are two drives attached (a 00 indicates one drive). Bits 2 and 3 are 11 indicating that the motherboard RAM size is 64K.

DOS Call Application

Program 9-8 illustrates an application using a Quick C **DOS call**. The function used here is **_dos_getdate(** date **)**, and it has the following structure defined in the **dos.h** header file:

```
struct dosdate_t
    {
        unsigned char day;              /* Value 1 - 31   */
        unsigned char month;            /* Value 1 - 12   */
        unsigned int year;              /* Value 1980 - 2099  */
        unsigned char dayofweek;        /* Sunday = 0, through 6 */
    } *date;
```

Program 9–8

```
#include <stdio.h>
#include <dos.h>

main( )
{
  struct dosdate_t date;

  /* Get the current system date:  */

   _dos_getdate(&date);

  /* Print out the results:  */

  printf("The month is => %d\n",date.month);
  printf("The day is => %d\n",date.day);
  printf("The year is => %d\n",date.year);
  printf("Day of the week is => %d\n",date.dayofweek);

}
```

Conclusion

This section introduced you to direct hardware BIOS or DOS interfacing using Turbo C or Quick C. Here you saw the similarities and differences of both interfacing methods.

You also saw some of the most common functions used for this type of interfacing. These include a BIOS and a DOS interfacing application. Check your understanding of this section by trying the following section review.

9–7 Section Review

1   Briefly state the differences between BIOS and DOS routines.
2   What are the advantages of using BIOS routines? The disadvantage?
3   What is the advantage of using DOS routines? The disadvantage?
4   State a BIOS function and briefly describe its purpose.
5   State a DOS function and briefly describe its purpose.

9-8 Program Debugging and Implementation: Interrupts

Discussion

This section contains information concerning your system's hardware and how you can use software to directly interface with it. Here you will see how the BIOS and DOS functions you were introduced to in the previous section actually make use of your computer's memory. Armed with this knowledge you will be able to develop programs that can take full advantage of programs built into your system.

Basic Idea

An **interrupt** is a special control signal that causes your computer to temporarily stop its main program. Once this happens, it will then go to another program called the **interrupt handler** (sometimes called the **interrupt service routine**). Your computer will then take instructions from the interrupt handler. Once the requirements of this routine are completed, the computer will then return control back to where it left off in the main program. This process is called servicing the interrupt. In the process of servicing the interrupt, the μP saves the contents of some of its internal registers into the stack. Doing this allows these internal registers to be used by the service routine. When completing the service routine, the μP restores these registers from the stack. This allows the μP to pick up where it left off when the interrupt occurred. This concept is illustrated in Figure 9–26.

What an Interrupt Does

An interrupt provides access to the ROM BIOS and DOS routines. For example, the BIOS and DOS interfacing functions presented in the last section are really interrupts. When the 8086 μP is used, it reserves the first 1024 bytes to store a series of 4-byte addresses known as **interrupt vectors**. These are really far pointers that contain the offset and the segment address of the interrupt handler. Some of the more common interrupts will be presented in this section. For a complete list of all the interrupt vectors and their corresponding handlers refer to the *IBM Personal Computer Technical Reference Manual*.

Since the first 1024 bytes of memory hold a series of 4-byte addresses, this means that there are $1024/4 = 256$ different interrupt vectors that can be stored in this section of memory. However, depending on how current your system is, only about 31 of these vector locations are actually used. The rest are available for your use (or use of your software, such as you C development system).

To see what is contained in part of this memory location, Program 9–9 uses the Turbo C **peek( )** function to get the contents.

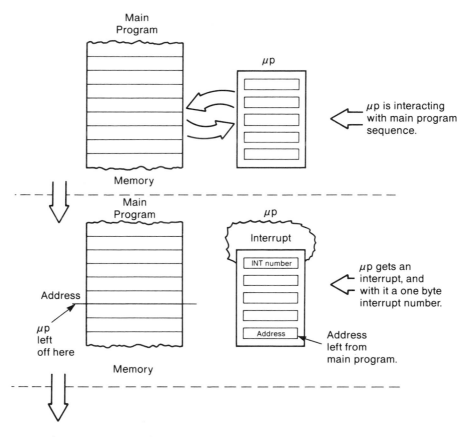

Figure 9–26   Basic Concept of Interrupt Servicing

Program 9–9

```
#include <stdio.h>
#include <conio.h>

main( )
{
 unsigned int offset;
 unsigned char value;
 int count;
 unsigned char address;
```

(continued)

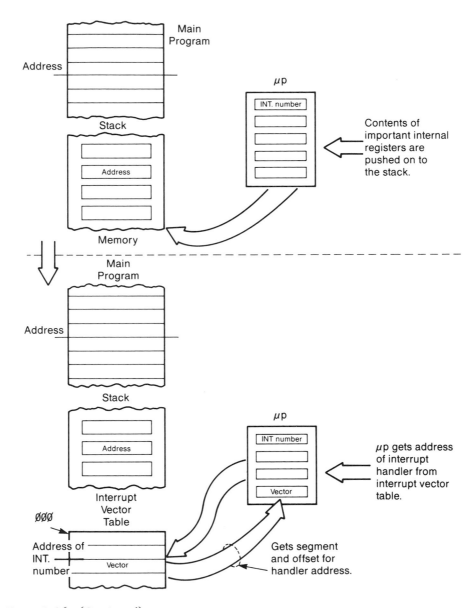

Figure 9-26 (Continued)

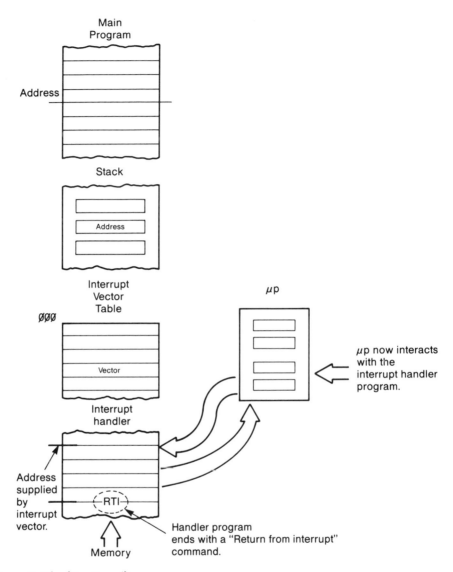

Figure 9-26 (Continued)

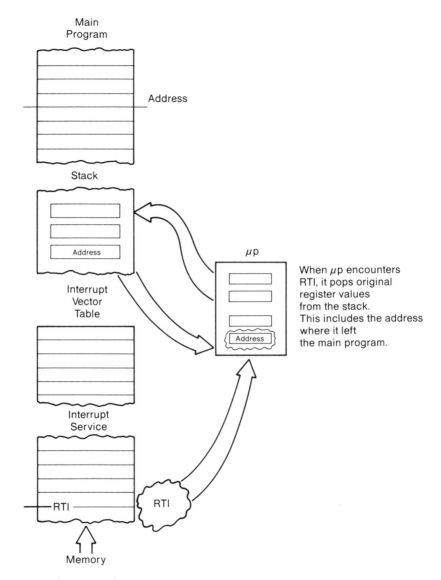

When μp encounters RTI, it pops original register values from the stack. This includes the address where it left the main program.

Figure 9–26 (Continued)

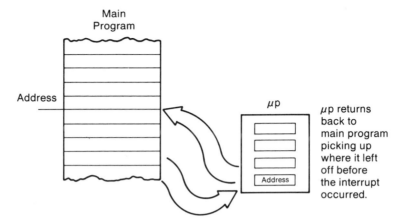

Figure 9-26 (Continued)

Program 9-9 *(continued)*

```
clrscr( );
printf("\n\n");

for(count = 0; count <= 240; count += 16)
{
  address = count;
  printf("\n 0:%2X => ",address);

    for(offset = count; offset < count + 16; offset++)
    {

    value = peek(0,offset);
    printf("%2X ",value);

    }  /* End of for */

}  /* End of for */
}
```

The output of the above program, along with an explanation, is shown in Figure 9-27.

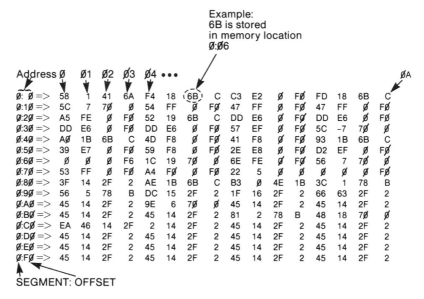

Figure 9–27   IBM PC Memory Dump of Interrupt Vector Table

Interpreting a Vector Table

A **vector table** is a sequence of memory locations that contains the memory locations (addresses) of other programs. Each vector contains 4 bytes. As an example, the Turbo C function for interfacing with the ROM BIOS that does the keyboard work is **bioskey( )**. This function does little more than actually call interrupt number 16_{16}. An **interrupt number** is an integer that indicates to the μP the memory location of the interrupt vector. This effectively makes it a pointer to a pointer. All interrupt vectors use 4 bytes of memory to store the segment and offset address of the interrupt handler. Thus, to find the address of this interrupt vector, multiply the value of the interrupt number by 4. (Be sure to first convert the interrupt number to decimal.) As an example, to get the address of the interrupt vector for the keyboard interrupt (interrupt 16_{16})

$$16_{16} = 22_{10} \times 4 = 88_{10} = 58_{16}$$

This means that the address for this ROM BIOS service routine is contained in memory location 58_{16}. Looking at Figure 9–27, from the memory dump, you can see that the line containing this address is

```
0:50 => 39 E7 0 F0 59 F8 0 F0 2E E8 0 F0 D2 EF 0 F0
```

Counting over to get the 4-byte address for interrupt vector #16, you have

```
2E E8 0 F0
```

To interpret the meaning of this address, you must take into account two things about how the IBM PC stores an address in memory:

1. The least significant byte is stored first.
2. The offset part of an address is stored before the segment part of the address.

Figure 9–28 shows how to construct the actual address for the keyboard BIOS interrupt handler.

As shown from Figure 9–28, the absolute address of the keyboard service routine is at FE82E$_{16}$.

Typical Interrupts

As you know, interrupts can be used to get ROM BIOS or DOS service routines. Table 9–12 lists some of the most commonly used BIOS and DOS interrupts.

Program 9–10 shows an application of a simple ROM BIOS interrupt application. The C program uses the Turbo C function int86( ). This is a general 8086 μP software interrupt used with <dos.h>. The format of the function is

int86(int number, &inregs, &outregs);

Where

number = The interrupt number.
inregs = Register values sent to service routine.
outregs = Register values returned from service routine.

The implementation of this program requires two structures, each of which defines a set of internal registers for the 8086 μP. The first structure defines this set as 16-bit words while the second defines them as 8-bit words. Memory is allocated for use by either structure through a C union. Note that each of the data types within the registers is declared as **unsigned int** (two bytes) in the first structure and as **unsigned char** (one byte) in the second structure. This uses interrupt number 12. As you can see from Table 9–12, this is a BIOS interrupt for checking the size of your system's memory.

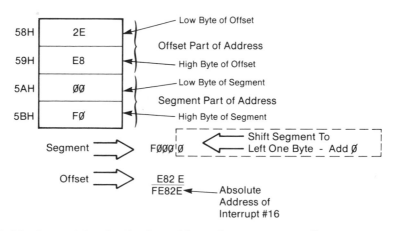

Figure 9–28 Determining the Absolute Address of an Interrupt Handler

Table 9-12 Common Interrupt Numbers—IBM PC

BIOS	Function
5	PRINT SCREEN
10	VIDEO I/O
11	EQUIPMENT CHECK
12	MEMORY SIZE CHECK
13	DISKETTE I/O
14	COMMUNICATIONS PORT I/O
15	CASSETTE I/O
16	KEYBOARD I/O
17	PRINTER I/O
18	ROM BASIC
19	BOOTSTRAP START-UP
1A	TIME OF DAY
1B	KEYBOARD BREAK (user setable)
1C	TIMER TICK (user setable)

DOS	Function
20	PROGRAM TERMINATE
21	FUNCTION REQUEST
22	TERMINATE ADDRESS—This is where control transfers to when a program terminates.
23	CTRL-BREAK EXIT ADDRESS— Done by a Ctrl-Break.
24	CRITICAL ERROR HANDLER
25	ABSOLUTE DISK READ
26	ABSOLUTE DISK WRITE
27	TERMINATE BUT STAY RESIDENT—Leaves program in memory.

Program 9-10

```
#include <stdio.h>
#include <dos.h>

#define int_number 0X12     /* The BIOS interrupt number  */

typedef struct       /* Registers as 16-bit words  */
    {
    unsigned int ax;
    unsigned int bx;
    unsigned int cx;
    unsigned int dx;
    unsigned int si;
    unsigned int di;
```

(continued)

Program 9–10 *(continued)*

```
      unsigned int flags;
      } registers_16;

  typedef struct        /* Registers as 8-bit words  */
      {
        unsigned char al;
        unsigned char ah;
        unsigned char bl;
        unsigned char bh;
        unsigned char cl;
        unsigned char ch;
        unsigned char dl;
        unsigned char dh;
      } registers_8;

  typedef union         /*  Registers as one or two bytes.  */
      {
        registers_16 reg_16;
        registers_8 reg_8;
      } registers_16_or_8;

main( )
{
registers_16_or_8 registers;   /* registers to be of type union.  */
unsigned int size;

/* Call the interrupt */
    int86(int_number, &registers, &registers);

/* Get value from AX register. */
   size = registers.reg_16.ax;

   printf("The memory size of your system is %d K.",size);

}  /* End of main( )  */
```

Creating Interrupt Handlers

Many of the interrupt vectors in your IBM PC are unassigned. This means that you (or a software vendor) can use this space to write your own interrupts. You do this by placing a C far pointer into one of the unused interrupt vector locations. With this, whenever a predefined event takes place, such as a system clock time, keypress, or some other action, these handlers can then be used to pass certain instructions to the main program.

Program 9-11 shows how to develop an interrupt handler function, install it and then test it. This program uses the Turbo C type **interrupt** for a function. It produces a short beep to the hardware output port of your PC's speaker. It's not terribly exciting, but is a pretty safe way of illustrating the development of an interrupt handler. New functions are explained after the program.

Program 9-11

```
#include <dos.h>

/*********************** Function prototypes: *****************/

void interrupt handle(unsigned bp, unsigned di, unsigned si,
                      unsigned ds, unsigned es, unsigned dx,
                      unsigned cx, unsigned bx, unsigned ax);
/* This function is the actual interrupt handler.  Note that  */
/* it is of type interrupt and contains necessary registers   */
/* in its argument.                                           */

void install(void interrupt (*faddr)( ), int vector);
/* This is the function that actually installs the address    */
/* of the interrupt handler in the location given by vector.  */

void testhandle(unsigned char bcount, int vector);
/* This function simply tests the installed routine.          */

/*************************************************************/

/*--------------Start of main program----------------------*/

main( )
{

    install(handle,10);
    testhandle(20,10);

}   /* End of main( )  */

/*----------------------------------------------------------*/

void testhandle(unsigned char bcount, int vector)
{
  _AH = bcount;
  geninterrupt(vector);

  }   /* End of testhandle( )  */

/*----------------------------------------------------------*/
```

(continued)

Program 9-11 *(continued)*

```c
void interrupt handle(unsigned bp, unsigned di, unsigned si,
                      unsigned ds, unsigned es, unsigned dx,
                      unsigned cx, unsigned bx, unsigned ax)
{
 int count_1, count_2;
 char originalbits, bits;
 unsigned char bcount = ax >> 8;

 /* Get the current control port setting */
     bits = originalbits = inport(0X61);

 for(count_1 = 0; count_1<= bcount; count_1++)
  {
   /* Turn off the speaker for a short time */
   outport(0X61, bits & 0XFC);
    for(count_2 = 0; count_2 <= 75; count_2++);   /* Empty statement. */

  /* Now turn it on for some more time */
   outport(0X61, bits | 2);
    for(count_2 = 0; count_2 <= 75; count_2++);   /* Empty statement. */

 }  /* End of for */

 /* Restore the control port setting. */
  outport(0X61, originalbits);

}  /* End of handle( )  */

/*------------------------------------------------------------*/

void install(void interrupt (*faddr)( ), int vector)
{
  setvect(vector, faddr);

}  /* End of install( )  */
```

The above program uses several of Turbo C's special functions for developing interrupt handlers and doing I/O. The first is

geninterrupt(int interrupt_number**);**

This function initializes a software trap for the interrupt.

inport(int port_number**);**

This function reads a word from a hardware port signified by port_number.

outport(int port_number **int** value**);**

This function outputs a word to a hardware port signified by `port_number`.

`setvect(vector, faddr);`

Sets the value of the interrupt vector.

Conclusion

This section has given you an introduction to interrupts. Here you saw what an interrupt is, how it is stored in your computer, and how it can be used. You also saw the development of a program that shows you how to develop an interrupt handler. Check your understanding of this section by trying the following section review.

9-8 Section Review

1 Briefly describe what is meant by an interrupt.
2 Explain the meaning of an interrupt vector.
3 State what is meant by an interrupt handler.
4 Explain how an address is stored in the IBM PC.
5 Where in the memory of the IBM PC are the interrupt vectors stored?

9-9   Case Study

Discussion

The case study for this chapter demonstrates hardware interfacing between your computer and an external device. The device in this case study is an IBM printer. This case study really shows the development of a function more then a complete program. However, this is an important development to follow because it illustrates the steps involved in analyzing an actual output port from a hardware standpoint and then implementing control over this port using the C language.

The Problem

Create a demonstration program that contains a function that will allow the user to access the parallel printer port and control the printer. The program cannot use any built-in interrupts or other preprogrammed printer control routines. The purpose of this program is to illustrate the control of a hardware port on the IBM PC using only the `inport( )` and `outport( )` functions or their equivalent. Essentially the user will be able to enter a line of text that will be echoed to the monitor screen and then outputted to the printer when the Return key is pressed. The program may terminate with a Ctrl-C.

First Step—Stating the Problem

Stating the problem in writing yields

- Purpose of the program: Provide a program that will allow the program user to enter a line of text, see it on the monitor, and have it sent to the printer when the Return key is pressed. The C program is not to use any built-in printer functions or interrupts; only the **inport( )** and **outport( )** functions or their equivalents are allowed.
- Required Input: A string of characters from the keyboard.
- Process on Input: Program is not to use built-in interrupts or preprogrammed printer routines. Only **inport( )** and **outport( )** (or equivalent) may be used.
- Required Output: The character string from the keyboard to the monitor and then, when the Return key is pressed, to the printer.

Developing the Algorithm

The algorithm for this program involves repetition of the following steps until the user enters Ctrl-C:

1. Initialize the printer.
2. Check printer status.
3. Get character from user.
4. Echo character to screen.
5. Store character in printer buffer using **outport( )**.
6. If user enters Return, print the buffer.

Program Development

The development of this program requires a knowledge of the 25-pin connector at the rear of your IBM PC that interfaces this computer with an IBM printer (or compatible). Therefore, the first phase of the development for this program will be a presentation of the technical hardware material concerning this interface connection. Since the **outport( )** and **inport( )** commands require the use of a nonresident program, Turbo C was chosen as the language for this program.

The Parallel Printer Port

When an IBM PC is connected to a parallel printer, this can be done with a 25-pin connector. Figure 9–29 illustrates the pin numbers and the purpose of the key pins used for sending and receiving signals between the computer and the printer.

As you can see from the figure, there are actually three ports associated with the computer printer interface. The data port is used to send the printable characters to the printer. The status port is used to get information from the printer so the computer will know when and if to send information to it. And the last port, the control port, is used to send control information to the printer.

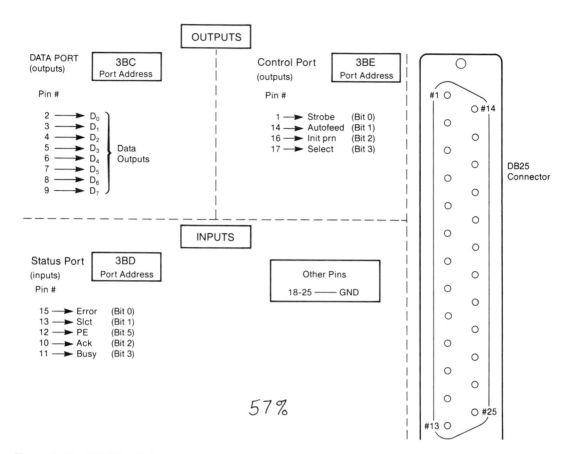

Figure 9-29 DB25 Parallel Printer Connections

Defining Data

The first step in the development of this program was to write the **#defines**. These are based on the hardware requirements of the computer printer interface:

```
#define DATA 0X3BC      /* Printer data port location.   */
#define line_feed 10
#define carriage_return 13

#define CONTROL 0X3BE   /* Printer control port location. */
#define strobe_hi 1     /* Bit 0 of control port set high */
#define init_select_prn 0XC /* Initialize and select printer */

#define STATUS 0X3BD    /* Printer status port location. */
#define ack 0X40    /* Bit 6 */
#define busy 0X80   /* Bit 7 */
```

Note: 0X means hex number follows.

Note that these are grouped so that the three separate ports are evident. A descriptive identifier is used for each port (such as DATA for the data port located at $3BC_{16}$). Just below each port identification, any important values that will be needed for the program are defined. The values for all of the control characters are found in the printer manual. Thus, under the DATA definition, the line feed data character is 10, and line_feed is defined as that value. This is also true for the value of 13 which represents a carriage return. Note that the same approach is taken with the other two data ports that have corresponding **#define** statements for their associated signals for the use of the program.

The Completed Program

Program 9–12 is the completed program.

Program 9–12

```
#include <stdio.h>
#include <dos.h>
#include <conio.h>
#define TRUE 1
#define DATA 0X3BC     /* Printer data port location. */
#define line_feed 10
#define carriage_return 13

#define CONTROL 0X3BE  /* Printer control port location. */
#define strobe_hi 1    /* Bit 0 of control port set high */
#define init_select_prn 0XC /* Initialize and select printer */

#define STATUS 0X3BD   /* Printer status port location. */
#define ack 0X40    /* Bit 6 */
#define busy 0X80   /* Bit 7 */

main( )
{
 int character;          /* Character to be printed. */

   clrscr( );   /* Clear the screen. */

  /* Strobe and activate printer. */
     outport(CONTROL, strobe_hi);
     outport(CONTROL, init_select_prn);

  do
   {
     do
     {
```

(continued)

Program 9-12 *(continued)*

```
        /* Check printer status:  */
        if((inport(STATUS) & busy));    /* If port is not busy */
            character = getchar( );
            outport(DATA, character);       /* Output a character  */
            if(!(inport(STATUS) & ack));    /* Wait for acknowledge */
        /* Output to printer buffer. */
            outport(CONTROL, strobe_hi);
            outport(CONTROL, init_select_prn);
    } while(character != '\n');

    /* Print the contents of the print buffer. */
            outport(DATA, carriage_return);
            outport(CONTROL, strobe_hi);
            outport(CONTROL, init_select_prn);

            outport(DATA, line_feed);
            outport(CONTROL, strobe_hi);
            outport(CONTROL, init_select_prn);

    } while(TRUE);    /* Repeat forever.  */

}    /* End of main( )  */
```

Program Analysis

The program contains two C **do** loops. The outer loop allows the user to keep outputting characters to the screen one line at a time, and then causes them to appear on the printer. The inner C **do** loop keeps getting characters from the keyboard, displaying them to the screen and placing them into the printer buffer. The printer buffer will not be printed until a carriage return is initiated by the program user. When this happens, the program goes into the outer loop where the contents of the printer buffer are then printed.

Note that the inner loop contains two C **if** functions. The first one checks to make sure that the printer is ready before proceeding on with the next instruction. The second C **if** function checks to make sure that the printer acknowledges receipt of the character sent to it before going on with the rest of the program.

Conclusion

This section demonstrated the use of C to interface an actual I/O port to an external device using **outport( )** and **inport( )** functions. You should not consider this to be a model printer control program. Rather consider it to be a foundation upon which you can build.

The Self-Test questions at the end of this chapter address specific questions about this program.

Interactive Exercises

DIRECTIONS

These exercises require that you have access to a computer and software that supports C. They are provided here to give you valuable experience and immediate feedback on what the concepts and commands introduced in this chapter will do. They are also fun.

Exercises

1 Program 9–13 illustrates an interaction with Turbo C pseudo variables. Try it.

Program 9–13

```
#include <stdio.h>

main( )
{
 unsigned char lo;
 unsigned char hi;
 unsigned int hi_lo;

   _BL = 0XA;
   _BH = 0XB;
   lo = _BL;
   hi = _BH;
   hi_lo = _BX;

   printf("Both together are %X",hi_lo);

}
```

2 The next program presents an example of the Turbo C inline machine code. Try the program and see what it does. Did you experience any difficulty with your system?

Program 9–14

```
#include <stdio.h>
#include <dos.h>

main( )
{
    __emit__(0XB2,0X01);      /* mov dl,1  */
    __emit__(0XB4,0X02);      /* mov ah,2  */
    __emit__(0XCD,0X21);      /* int 21 */
    __emit__(0XCD,0X20);      /* int 20 */

}
```

3   Here is a program in Turbo C that returns an equipment code number. Can you interpret the meaning of the resultant number for your system?

Program 9–15

```
#include <stdio.h>
#include <bios.h>

main( )
{
  printf("Your equipment code number is %X",biosequip( ));

}
```

4   You should know ahead of time what to expect from the output of Program 9–16. What did you get?

Program 9–16

```
#include <stdio.h>
#include <dos.h>

main( )
{
 printf("At location 0:50 is %X.",peek(0,0X50));
}
```

Self-Test

DIRECTIONS

Answer the following questions by referring to Program 9–12 in the case study section of this chapter.

1   State the meaning of **#define** DATA 0X3BC in the program.
2   What does DB25 mean in terms of your computer connections?
3   How many ports does the DB25 parallel printer connector have?
4   State the name and port number of the above ports.
5   Give the purpose of each of the above ports.
6   In the program, explain the meaning of the statement

   (inport(STATUS**) & busy)**

7   Explain how the acknowledge bit is tested in the program.
8   What does the program statement **outport(**CONTROL, strobe_hi**)** actually do?

End-of-Chapter Problems

General Concepts

Section 9–1

1 Name the general purpose registers contained in the 8086/8088 μP. State how they may be used.
2 Describe what is meant by the stack.
3 State what is meant by a LIFO stack.
4 Explain what is meant by a **push** and a **pop**.
5 Describe the process of segmentation.

Section 9–2

6 State some of the advantages and disadvantages of assembly language programs.
7 What is the short word used as a memory aid in assembly language programming called?
8 State how the stack grows as used by your C program and an IBM PC.
9 Where does C place locally declared variables on the stack? Function parameters?

Section 9–3

10 What is meant by a memory model in C?
11 State the differences between the MEDIUM and the COMPACT memory models.
12 Give the advantage of confining your C code to just one 64K segment.
13 What is the difference between a far pointer and a near pointer?

Section 9–4

14 Is it possible to convert C source code to assembly language source code? Give an example.
15 Conventionally, what kind of file is given the extension `.ASM`?
16 State the difference between the assembly language instructions: `[bp+4]` and `bp+4`.

Section 9–5

17 Explain what is meant by a pseudo variable. Name a C development system that uses them.
18 Is there any problem with the statement **&_AH** in Turbo C? Explain.
19 What is it called when you can include assembly language source code within your C source code?
20 Is a pseudo variable an accepted ANSI C standard?

Section 9–6

21 What is a utility as used in a C development system?
22 Which utility can help you keep your source, object and `.exe` files current?
23 Explain the purpose of a library utility.
24 In Turbo C, what is the purpose of the `GREP.EXE` utility?

Section 9–7

25 What does BIOS mean?
26 What does BIOS consists of?
27 Compare the advantages and disadvantages of using BIOS routines compared to using DOS routines.

Section 9–8

28 What is an interrupt?
29 State the difference between an interrupt vector and an interrupt handler.

30 What is the relationship between an interrupt handler and an interrupt service routine?

31 Explain how an address is stored by the 8086 μP family.

Program Design

For the following programs, use the structure that is assigned to you by your instructor. Otherwise, use a structure that you prefer. Remember, the program should be easy for anyone to read and understand (especially you if it ever needs modification). Note that for these programs, you will need a C system and IBM PC or true compatible. Because the advanced programming topics in this chapter concern the details of hardware and software, the following questions include primarily the areas of electronics technology and computer science.

Electronics Technology

32 Figure 9-30 shows a hardware interface between the DB25 printer connector and an electrical circuit that operates relay drivers. You should refer to Figure 9-29 for information on the pins. Note from the figure that there are two output ports and one input port.
Develop a C program that will allow the program user to activate any one of the output relays.

33 Modify the program in 32 to allow the program user to activate any one or more of the relays. Use computer graphics to actually show the condition of each of the relays.

34 Create a C program that will test the condition of each input for the hardware interface of Figure 9-30. The program should contain computer graphics that will actually show the condition of each of the switches.

35 Using the hardware interface circuit of Figure 9-30, develop a C program that will allow the user to have any one input pin determine the condition of one of the output pins.

36 Create a C program using computer graphics that will graphically show the condition of all the outputs and all the inputs for the hardware interface circuit of Figure 9-30. The user should be able to change the condition of any output.

Computer Science

37 Using your C system, develop a small assembly language program and link it to a C program.

38 Expand the requirements of problem 37 above by having at least two values passed to the assembly language program. The assembly language program will then use these values to complete a simple process of your choice.

39 If your C system has inline assembly capabilities, redo the requirement of problem 38.

40 If your C system has inline machine language capabilities, develop a simple assembly language program using inline machine language.

41 Select an interrupt vector in your computer system. Then write a C program that will do a memory dump of the entire interrupt handler for the selected vector.

All Technologies

42 Develop a MAKE utility file that will assist you in the development of any of the above programming assignments. The programming assignment should have at least three separate files.

43 Using the library utility that comes with your C system, create a library of three of your most used C functions that are original.

44 Create a C program that uses the functions placed in your library from problem 43 above.

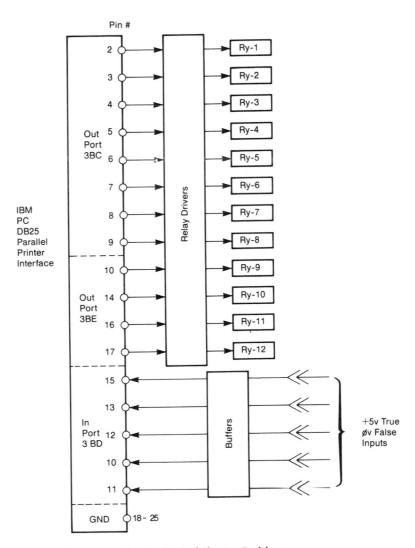

Figure 9–30 Hardware Interface for End-of-Chapter Problems

Appendices

Introduction

The word **DOS** actually stands for Disk Operating System. Among other things, it is a program or list of instructions contained on a disk that tells your computer how to operate the disk. IBM DOS is made by IBM and MS-DOS is Microsoft's version.

The information here is just enough to get you started. When you complete this section, you will know how to boot your system, initialize a disk, and copy a program from one disk to another—just scratching the surface of everything DOS can do. After you understand how to do what is shown here, you should refer to the DOS manual that comes with the computer you are using.

What to Do

Your computer system may have one of the following setups.

1. A single disk drive
2. Two disk drives.
3. A single disk drive and a hard disk drive.

The disk drives (the colon : is part of the drive name) are called:

A: Drive on the left (or top).
B: Drive on the right (or bottom).
C: Hard disk drive (if present).

DOS is contained on a floppy disk sold with the computer. It will be referred to here as the **system disk**. In most school situations there will be a copy of this disk.

Care of the Disk

1. When not using the disk, keep it inside its protective jacket.
2. Never touch an exposed surface of the disk (or allow anything else to come in contact with it).
3. Do not bend the disk or place it in a position where it could be bent.
4. Do not use sharp objects (such as pencils or ball point pens) on the disk or on any of its labels. If you must write on it, use a felt-tip pen.
5. Do not expose the disk to direct sunlight.
6. Do not expose the disk to any magnetic fields (watch out for all metal objects such as scissors, staple removers, or screw drivers—they could be magnetized).
7. Do not expose the disk to extreme heat (such as in the glove compartment of your car) or extreme cold.

Starting the System

1. Be sure the computer is OFF.
2. Insert the System Disk in drive A: as shown in Figure A–1.
3. Gently close the load lever. If it does not close, it means the disk is not correctly inserted.
4. Locate the power switch on your system. Refer to Figure A–2.
5. Your next step depends upon the system you are using. Refer to Table A–1 on page 617.

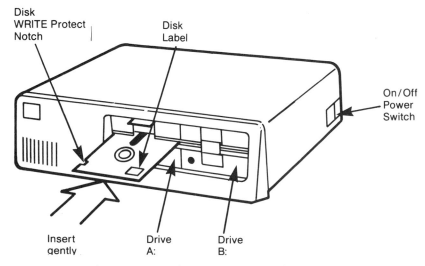

Figure A–1   Inserting the System Disk (pickup from PASCAL)

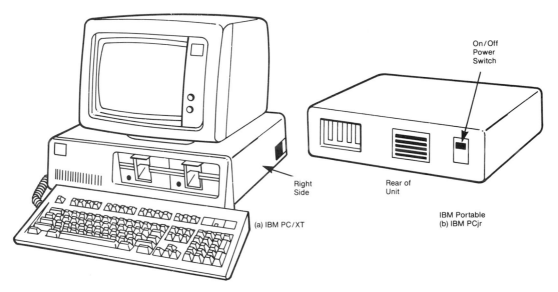

Figure A–2   Location of Power Switch

6. Turn the system on. Nothing will happen for what may seem like a long time, perhaps up to a minute, depending on how much memory the computer has. Once things begin to happen, the red light located on drive A: (where you put the System Disk) will come on. This means the disk is being used by the computer. (It is now in the process of reading the DOS from the disk and loading it into its memory.)

7. Look at Figure A–3 and make sure you know where the major control keys are on your keyboard.

8. The process of your computer reading DOS into its memory is called **booting**. It comes from "picking yourself up by your bootstraps"—essentially, this is what the computer is doing. It is getting instructions on how to operate a floppy disk from a floppy disk.

Table A–1   What Needs Turning On

System	What to Do
IBM PC/XT with a monochrome monitor	Turn on the single power switch that provides power to both the computer and the monitor.
IBM Portable PC	Turn on the single power switch that powers everything.
IBM PCjr	Turn on two switches—one for the computer, one for the monitor.
If you are using an IBM color monitor	Turn on two switches—one for the computer, one for the color monitor.

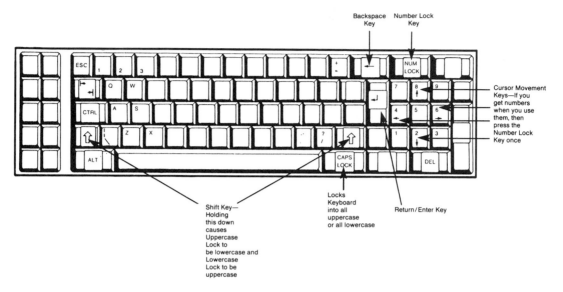

Figure A-3 IBM PC Keyboard

9. Once your computer is booted, the red light on the disk drive will turn off and you will be asked to enter the date on the screen. Do this as shown on the screen (**Month-Day-Year**), for example, **9-1-90**, then press the Return/Enter key.

10. Next the screen will ask for an input for the time. Enter this information as shown on the screen (**Hour:Minutes**), for example, **13:15** (means 1:15 PM)

11. You will then see the famous Dos prompt

    A>

    This means the computer is waiting for a DOS instruction from you; it also means that drive A: is the active (default) drive.

12. Your computer has now been booted (it contains DOS). When you remove the system disk, DOS is still inside your computer memory. It will stay there until you turn the computer off. Once the computer is turned off (on purpose or accidentally) the DOS is lost and you must start the process over again.

Preparing a New Disk

1. The first step in using a new disk is to format it—to get the disk ready to put programs on it. If your disk has been used before, formatting it will erase everything on it. Make sure there are no programs you'll need later on the used disk, because they won't be there after formatting.

2. Make sure your system is booted:

   From a one-drive system, enter FORMAT A:, remove the System Disk from drive A: and place your disk in drive A:.

   For a two-drive system, leave the System Disk in drive A: and put your disk into drive B:.

   and press Return/Enter. (Note that it makes no difference whether you use uppercase or lowercase letters. You could have entered `format b:`.)

3. Follow the instructions on the screen.

4. When the formatting process begins, the red light of the disk drive with your disk in it will light up, indicating that it is in use. Your disk is now being prepared to be used by the computer system; this process takes almost a minute.

5. When the light turns off, the computer will ask if you want to format another disk. In any case, your disk is now ready to have programs saved onto it.

Copying Programs to Your Disk (Two-Drive System)

1. Make sure your system is booted and your disk is formatted.

2. Put the disk you want to copy programs from in drive A:. This will be called the **source disk**.

3. Put the formatted disk you want to copy programs to in drive B:. This will be called the **destination disk**.

4. To see the programs that are on the source disk, enter

   `DIR/W`

   and press the Return/Enter key.

5. The names of the programs on the source disk will appear on the screen. Note that there may be two parts to the name of each program.

   `NAME.EXT`

   The `NAME` of the program will usually have letters and/or digits; the `EXT` (after the period) is optional. The `NAME` can have a maximum of eight places, the `EXT` a maximum of three places.

6. To copy a selected program from the source disk to the destination disk, enter

   `COPY SOURCE.EXT B: SOURCE.EXT`

   where `SOURCE.EXT` is the name of the program on the source disk you want to copy. For example, if the name of the program you want to copy is LC.EXE, you would enter

   `COPY LC.EXE B: LC.EXE`

   and press the Return/Enter key.

7. The red light on drive A: will come on, indicating that the selected program is being placed into the computer's memory. The red light on drive B: will then come on to indicate that the selected program is being copied to your disk. The screen will then tell you if the copy was successful. (Note that some programs are copy-protected; the manufacturer does not want copies made of the program and you will not be able to copy it to your disk.)

Appendix B—Learn C Programming Environment

About This Appendix

This appendix summarizes the meaning of the different screens and menus you will encounter if you are using the Microsoft Learn C diskette option with this text. A complete analysis of these features is covered in detail in the Microsoft *Quick C Compiler Programmer's Guide* that is available with the Microsoft Quick C system.

The Main Learn C Screen

Figure B-1 shows the Main Learn C screen. The important parts of the screen are called out on the figure.

The Learn C Menus

There are eight different Learn C menus. Their names are shown on the menu bar of Figure B-1. An overview of each of these menus is given in Table B-1.

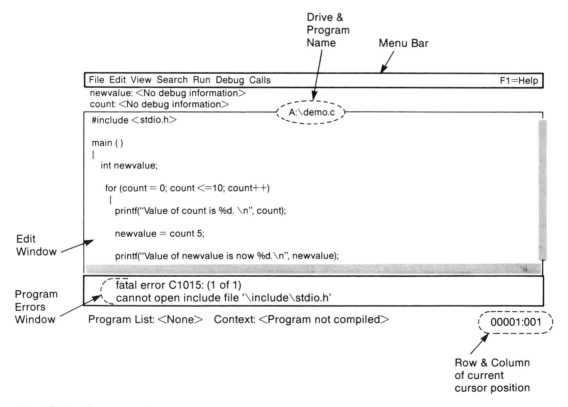

Figure B-1   The Learn C Main Screen

Table B–1   Overview of the Learn C Menus

Menu	Purpose
File	Allows you to work with files, get to DOS, and exit your Learn C program.
Edit	Lets you enter and modify your program.
View	Lets you change how your Learn C screen looks.
Search	Allows you to find any part of your program.
Run	Compiles and runs your Learn C program. Also lets you select some compile and run options.
Debug	Gives you control over the Learn C built-in debugger.
Calls	Places the cursor on specific functions used by your program.
Help	A powerful feature that uses the built-in Learn C help system. You are encouraged to make good use of this feature.

Getting into the Learn C Menus

Learn C offers you several ways to get into any of the above menus. One way of getting the File menus is to use Alt-F. As a matter of fact, that is one way of getting any of the Learn C menus:

While holding down the Alt key, press the key that is the same as the first letter of that menu.

Once you are into any of the Learn C menus, you can use the left or right arrow keys to move from one menu to the other. To get the Help menu, you must have your LEARN C SUPPORT disk in drive A: and then press the F1 key. To leave any of the Learn C menus, simply press the Esc key.

Using the Learn C Menus

The different Learn C menus are shown in Figure B–2.

The *CALLS* menus does not contain commands. It shows a hierarchy of functions that have been called by other functions. It allows you to move the cursor to the point where any of the displayed functions were called. As you can see from Figure B–2, some of the menu commands use short cut keys for most often used commands. What this means, is you don't need to get to the menu to actually compile and run your Learn C program. You can simply hold down the Shift key (the ones on the bottom left and right of your keyboard with the large up-pointing arrows) and press the F5 key at the same time.

To use any feature of a menu, you can use the up or down arrow keys to move the highlight bar over a menu feature. Once the desired feature is highlighted, you can then press the Enter key. Another way of getting any menu selection is to press the key that is the same as the highlighted letter on the menu selection. For example, to exit the Learn C environment from the File menu, you can simply press x (this is the highlighted letter in the eXit function of the File menu).

You will note that in some of the menus there are some options which are displayed in a very light type face. This means that either the feature is not available with the Learn C system or that you are not in a mode that will yet allow use of

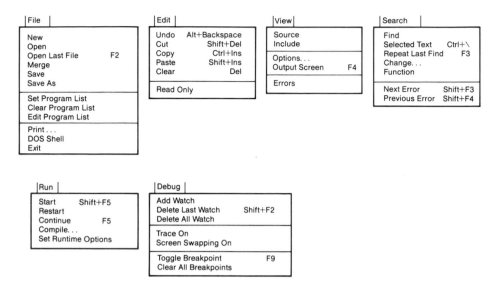

Figure B–2   The Different Learn C Menus

that feature. For example, the Edit menu does not have any of its first five options available until there is something to edit.

Working with the Menus

As long as you do not have an important program on the main Learn C screen, and your Learn C SUPPORT disk has been write protected, there is no harm you can cause by experimenting with the various menu features. Remember, use the Esc key to get out of what you are doing if you get stuck. The Tab key will move the cursor in some of the other Learn C screens.

Errors

The shortcut key for next error is Shift+F3. For the previous error it is Shift+F4.

Appendix C—Using the Learn C Help Menu

About This Appendix

This appendix presents one of the most powerful learning aids in the Learn C system— The Learn C Help menu. You can use this as an easy reference while you are programming a C program. The Learn C Help menu contains thousands of useful items about the Learn C environment as well as the C language itself. This feature can save you hours of time looking things up in a book. With the computer screen you can easily and quickly get to any C keyword and have it explained (along with an example that stays on your screen so you can reference it while you are programming). Once you get familiar with using it, you will always reference this amazing feature.

Getting Help

To use the Learn C Help menu, you must have your LEARN C SUPPORT disk in drive A: and be in the Learn C environment. To get general information (stuff that you generally need to know while programming), simply press the F1 key. Your screen will now appear as shown in Figure C-1.

The screen in Figure C-1 presents general information about the editing and debugging features of the Learn C system. Note the four boxes at the bottom of the screen (`Next`, `Previous`, `Keywords`, `Cancel`). The `Next` box is highlighted, so if you just press the Enter key, the next general information screens will appear. There are eight general information screens in all. They are:

- Editing
- Special Characters in Regular Expressions
- C Operator Precedence
- Regular ASCII Chart
- Extended ASCII Chart
- C Escape Sequences
- Formatting With `printf( )`
- C Data Types Range of Values

To get to a previous general information screen, use the Tab key to highlight the next box at the bottom of the screen (the `Previous` box) and press Enter. To leave the general information screens, you only need to press the Esc key.

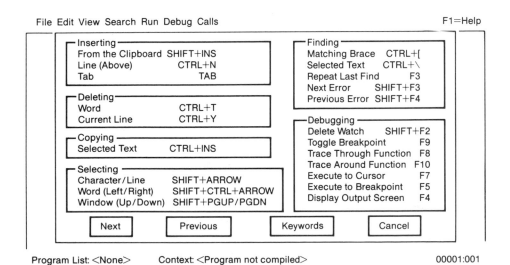

Figure C-1 The First of Eight General Help Screens

Getting Help on a Specific Topic

The general information screens are very useful to help you jog your memory about a feature of the C language or the Learn C environment. However, an even more powerful feature is yet available—getting help on a specific topic. To do this, you only need to hold down the Shift key while pressing F1. When you do this, a menu will appear as shown in Figure C-2.

There are 21 separate topics on this menu. You can select any one by using the up or down arrow keys to highlight your selection. Once your selection is made, press the Enter key. As an example, if you are in the middle of creating a graphics program, and you need the keyword for setting the drawing color, you would use the arrow keys to highlight `Graphics` and then press Enter. Your screen would now appear as shown in Figure C-3.

This menu contains all of the keywords used in Learn C for performing graphics functions. To find out about the keyword `_setcolor`, again use the arrow key to highlight the word and then press Enter. Your screen will now appear as shown in Figure C-4.

Note from Figure C-4 what has happened. You have been returned back to the main Learn C screen (where your program is), and the top of the screen now contains information about the command in question. This feature makes it easy to copy what is needed into your program.

Automatic Help

As if the above wasn't enough to help you, Learn C lets you place your cursor over any part of your C program and then get instant help on that keyword or function. All you need do is place the cursor anywhere on the word (even the space after

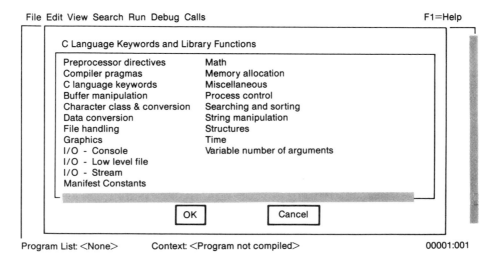

Figure C-2   Specific Topic Help Menu

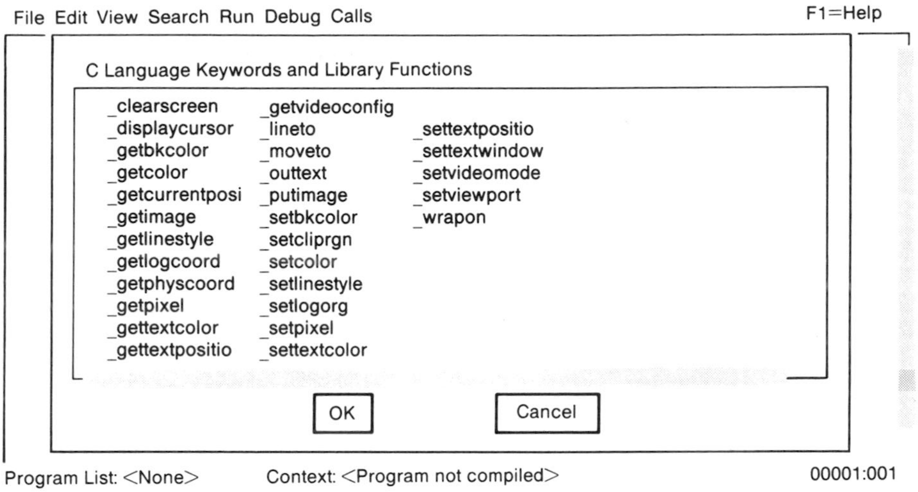

File Edit View Search Run Debug Calls   F1=Help

C Language Keywords and Library Functions

_clearscreen	_getvideoconfig	
_displaycursor	_lineto	_settextpositio
_getbkcolor	_moveto	_settextwindow
_getcolor	_outtext	_setvideomode
_getcurrentposi	_putimage	_setviewport
_getimage	_setbkcolor	_wrapon
_getlinestyle	_setcliprgn	
_getlogcoord	_setcolor	
_getphyscoord	_setlinestyle	
_getpixel	_setlogorg	
_gettextcolor	_setpixel	
_gettextpositio	_settextcolor	

OK   Cancel

Program List: <None>   Context: <Program not compiled>   00001:001

Figure C-3   The Learn C Graphics Keywords

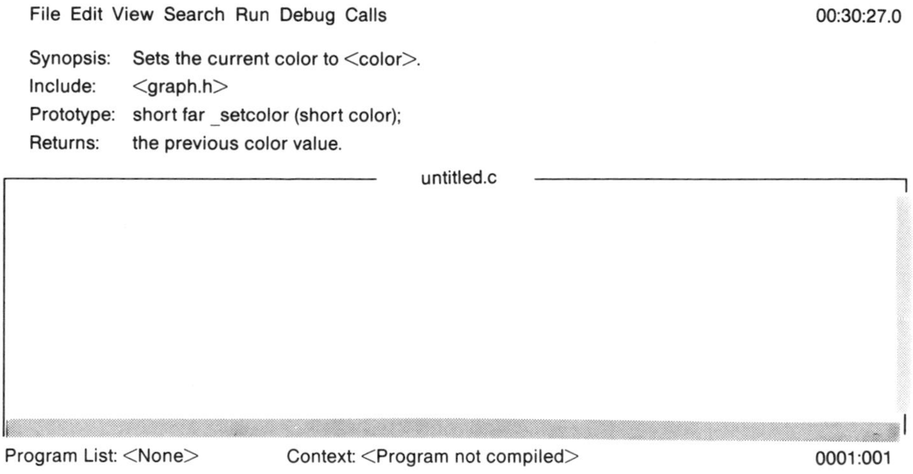

File Edit View Search Run Debug Calls   00:30:27.0

Synopsis: Sets the current color to <color>.
Include: <graph.h>
Prototype: short far _setcolor (short color);
Returns: the previous color value.

untitled.c

Program List: <None>   Context: <Program not compiled>   0001:001

Figure C-4   The Learn C Environment for Selected Function Example

the last letter will work) and then press Shift+F1 (holding down the shift key and pressing F1).

For example, suppose you want to make sure you have used the correct **#include** header file for the **printf( )** functions. Placing the cursor on **printf** and pressing Shift+F1 will give you the screen shown in Figure C-5.

Note from Figure C-5 that information concerning the **printf( )** function automatically appears at the top of your editing screen. Thus, a quick check of your program shows that you did use the correct header file for the **printf( )** function.

File Edit View Search Run Debug Calls   F1=Help

Synopsis:	Formats and prints characters and values to the standard output stream (stdout). (See the initial help screen for printf/scanf formatting characters).
Include:	<stdio.h>
Prototype:	int printf(const char *format[, argument]. . .);
Returns:	the number of characters printed.

```
                                          B:\myfirst.c

  #include <stdio.h>

  main ( )
  {
    printf("Hello world.");
  }




```

Program List: <None>   Context: <Program not compiled>   00006:002

Figure C-5   Getting Automatic Help with Learn C

What could be easier? This powerful LEARN C help feature will save you hours of looking things up in reference books or having to refer back to this text—which also helps keep your work area uncluttered.

Appendix D—The Learn C Debugger

About This Appendix

This appendix presents an introduction to the Learn C program debugger. One of the features of the debugger allows you to step thorough your C program line by line and observe the resulting values of variables. Being able to do this is a powerful learning and program troubleshooting feature.

Getting Started

Get into the Learn C environment and make sure the LEARN C SUPPORT disk is in drive A:. Enter the program shown in Figure D-1.

When the above program is compiled and executed, it will simply cause the value of the variable **number** to go from 0 to 10 in steps of 1. This program will be used to illustrate the ability of the Debug feature to go through a program step by step and show you the value of any variable you want to watch.

Using the Add Watch

You will be using the Debug feature to watch the value of the variable **number** in the program you just entered. You will be able to see its value change as you

File Edit View Search Run Debug Calls   F1=Help

```
┌──────────────────────────── B:\example.c ─────────────────────────────┐
│                                                                        │
│  main ( )                                                              │
│  {                                                                     │
│    int number;                                                         │
│                                                                        │
│     for number = 0; number <= 10; number++)                           │
│       {                                                                │
│        number = number;                                                │
│       }                                                                │
│    }                                                                   │
│                                                                        │
│                                                                        │
│                                                                        │
│                                                                        │
│                                                                        │
└────────────────────────────────────────────────────────────────────────┘
```

Program List: <None> Context: <Program not running>   00011:001

Figure D–1   C Program for Debug Example

single step through the loop in the program. This is a powerful debugging feature that will let you watch the values of many such variables while you single step through your program.

Move the cursor to any letter within the variable number. This variable appears six times in the program. It makes no difference which one you choose since they are all the same variable. You are doing this because this is the variable whose value you want to watch.

Once you have done this, get into the Debug menu (Alt-D). The pull-down Debug menu will now appear as shown in Figure D–2.

Note that the menu offers seven options. The highlight bar is currently over the Add Watch option. This is the one you want, so press the Enter key. The screen will now appear as shown in Figure D–3.

As you can see from the figure, the variable number (the one you selected) appears inside the dialogue box. You could just as well have typed it in, but using the cursor within your program is perhaps easier and less prone to error. The OK box is highlighted (if you want to highlight the Cancel box use the Tab key), so press Enter. Your Learn C main screen now looks like Figure D–4. Note the subtle change in the top portion right under the top main menu line. You see number: < >.

This addition is called the watch window, and it displays the variables whose values you will be watching as you step through your C program.

Single Stepping through the Program

To single step through your program, you do not use the ordinary compile and run sequence. The reason for this is that the C environment must imbed special code characters within your program (invisible to you), so that you can single step through the program. To allow this to happen, simply press the F8 key. When you do, your program will compile as normal, but now, instead of running, it will simply

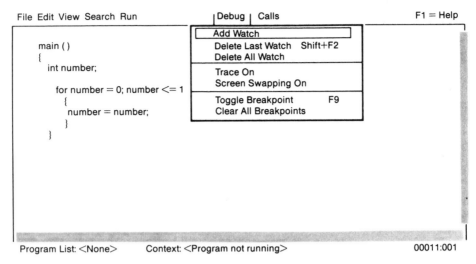

File Edit View Search Run   Debug | Calls   F1 = Help

```
    main ( )
    {
      int number;

        for number = 0; number <= 1
        {
          number = number;
        }
    }
```

| Add Watch |
| Delete Last Watch   Shift+F2 |
| Delete All Watch |
| Trace On |
| Screen Swapping On |
| Toggle Breakpoint   F9 |
| Clear All Breakpoints |

Program List: <None>   Context: <Program not running>   00011:001

Figure D-2   Debug Menu

File Edit View Search Run Debug Calls   F1=Help

B:\example.c

```
main ( )
{
  int number;

    for(numb
    {
      numbe
    }
}
```

Enter expression to add Watch Window

number

OK   Cancel

ProgramList: <None>   Context: <Program not running>   00011:001

Figure D-3   The Debug Dialogue Box

highlight the first opening brace { under **main( )**, indicating that the program has stopped there and is waiting for you to again press the F8 key.

Keep tapping the F8 key, and as you do your program will get into its loop and you can watch the value of **number** change in the watch window. Continue doing this until the program completes the conditions of the loop.

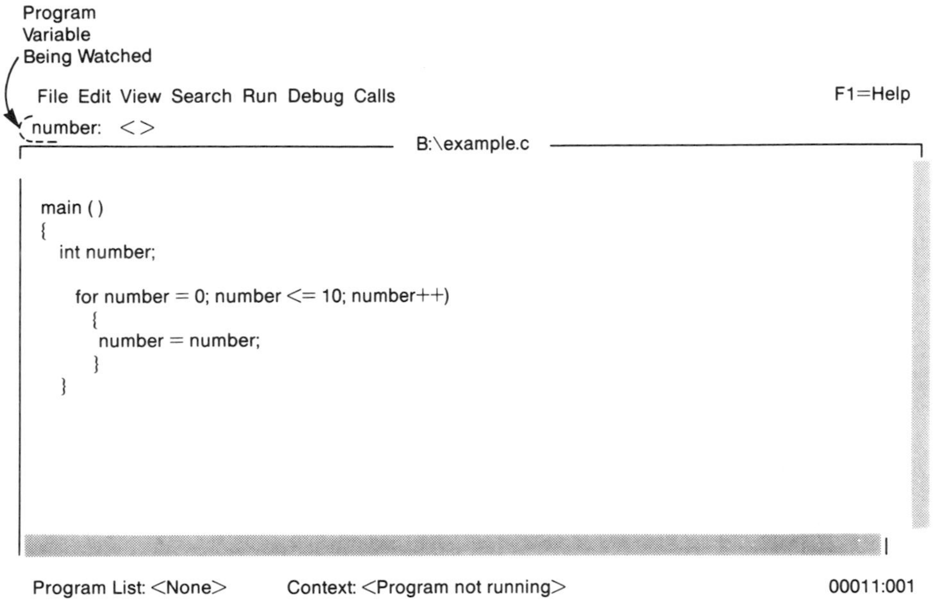

Figure D-4   Watch Window Added to Main Screen

What You Have Accomplished

You have just completed one of the most powerful features of the LEARN C debugger. The debugger also contains more advanced programming features that are presented in the Microsoft *Quick C Compiler Programmer's Guide.*

Appendix E—Learn C Color and Graphics

About This Appendix

This appendix presents the information you need to produce C programs developed in the Learn C environment that will give you color text and graphics. Chapter 8 of this text presented color and technical graphics in the Turbo C environment. The graphical concepts and technical information about the IBM PC presented in that chapter apply equally well for the material in this appendix. The main purpose of this section is to illustrate the programming commands that are unique to the Learn C environments. Armed with this information, you can then modify the programs presented in Chapter 8 to fit the Learn C programming environment.

Learn C Text Color

Learn C allows you to produce 16 colors of text and 8 background colors. You also have the option of having the text displayed in 80 or 40 columns. The commands are for an IBM PC or true compatible with a color graphics capability.

In order to actually display text in color, Learn C requires the use of a buffer to store the string to be displayed. This is illustrated in Program E-1. The program produces text on the screen in the 40-column mode in all 16 colors. Each time a key is pressed, the background color of the screen changes and the 16-color text is again displayed. This process continues until all eight background colors have been presented. An explanation of each new command follows the program.

Program E-1

```
#include <stdio.h>
#include <graph.h>

main( )
{
char buffer [255];
int text_color;
int back_color;

_setvideomode(_TEXTC40); /* 40 column text mode. */

/* Start outer loop to change the background color. */
 for(back_color = 0; back_color <= 7; back_color++)
 {
  _setbkcolor(back_color);  /* Sets the background color. */

/* Start inner loop to change the text color. */
  for(text_color = 0; text_color <= 15; text_color++)
   {
    _settextcolor(text_color);   /* Set the text color. */

    /* The method used by LEARN C to present text in color. */
    sprintf(buffer, "This is color number %d.\n",text_color);
    _outtext(buffer);

  }    /* End of the inner loop.  */

  getch( );   /* Wait for a keypress. */

  _clearscreen(_GCLEARSCREEN):  /* Clear the screen to current
                                    background color. */

}    /* End of the outer loop. */

}    /* End of main( ). */
```

Command Analysis

The Learn C function

`_setvideomode(MODE);`

sets the video mode of the system. There are 16 video modes available in the Learn C environment. These are listed in Table E-1 along with the keyword used for the MODE configuration.

Table E-1   Learn C Mode Constants for Video Modes

Mode Keyword	Adapter	Text/Pixel Size	Colors	Comments
_DEFAULTMODE	The hardware default mode.			
_TEXTBW40	CGA	Text 40 × 25	16	Shades of mono.
_TEXTCO40	CGA	Text 40 × 25	16	Colors
_TEXTBW80	CGA	Text 80 × 25	16	Shades of mono.
_TEXTCO80	CGA	Text 80 × 25	16	Colors
_MRES4COLOR	CGA	Pixel 320 × 200	4	Colors & Bkgd.
_MRESNOCOLOR	CGA	Pixel 320 × 200	4	Shades of mono.
_HRESBW	CGA	Pixel 640 × 200	2	Color & Bkgd.
_TEXTMONO	Monochrome	Text 80 × 25	1	Monochrome
_MRES16COLOR	EGA	Pixel 320 × 200	16	Colors
_HIRES16COLOR	EGA	Pixel 640 × 200	16	Colors
_ERESNOCOLOR	EGA	Pixel 640 × 350	1	Monochrome
_ERESCOLOR	EGA	Pixel 640 × 350	64	EGA
_VRES2COLOR	VGA	Pixel 640 × 480	2	Color & Bkgd.
_VERS16COLOR	VGA	Pixel 640 × 480	16	Colors
_MRES256COLOR	VGA	Pixel 320 × 200	256	Colors

The next new Learn C function changes the background color:

`_setbkcolor(color)`

The following function changes the text color:

`_settextcolor(color)`

> Where
> color = The color number of the background. (Refer to Table E-2.)

Table E–2   Color Numbers for Learn C

Color	Number	Color	Number
Black	0	Dark Gray	8
Blue	1	Light Blue	9
Green	2	Light Green	10
Cyan	3	Light Cyan	11
Red	4	Light Red	12
Magenta	5	Light Magenta	13
Brown	6	Yellow	14
White	7	Bright White	15

The **sprintf( )** function has the form:

```
sprintf(char *buffer, const char *format);
```

Where

    buffer = The storage location used by the function for storing a series of characters and values.

    format = Ordinary characters that have the same format as the **printf( )** function.

This function formats and stores a series of characters and values in **buffer**. Arguments are processed according to format specifiers. Essentially the format has the same form as the **printf( )** function.

To output text in color with Learn C use

```
_outtext(text);
```

Provides the null-terminated string that **text** points to. Does not provide any formatting. (As in **printf**.)

The Learn C function that clears the screen is

```
_clearscreen(area)
```

    Where

        area = One of the following:

                _GCLEARSCREEN — Clears the entire screen and sets it to the current color.

                _GVIEWPORT — Clears the current viewport and sets it to the current color.

                _GWINDOW — Clears the current text window and sets it to the current color.

Learn C Graphics

Program E–2 illustrates the most basic example of graphics in Learn C. The program sets the graphics mode for a CGA adapter, sets the color, and then places a single dot in the center of the graphics screen.

Program E-2

```
#include <stdio.h>
#include <graph.h>

main( )
{

  _setvideomode(_MRES4COLOR);

  _selectpalette(0);

  _setcolor(1);

  _setpixel(160,100);

    getch( );

  _setvideomode(_DEFAULTMODE);

}
```

Command Analysis

You have already seen the **_setvideomode( )** in the previous discussion. The new graphics command is

_selectpalette(number);

This function works only with the _MRES4COLOR and _MRESNOCOLOR video modes. It sets one of four palettes as shown in Table E-3.

Table E-3   Learn C Color Palettes

Palette Number	Color Value		
	1	2	3
0	Green	Red	Brown
1	Cyan	Magenta	Light Gray
2	Light Green	Light Red	Yellow
3	Light Cyan	Light Magenta	White

As shown in Table E-3, four palette color values (plus the background color—which makes the graph invisible) can be displayed at a time in the _MRESCOLOR video mode. Once the palette number has been selected, then the particular graphics color is determined by

_setcolor(color);

Where
color = The color number corresponding to Table E–2.

This command sets the current graphics drawing color.

The next new command puts a pixel at the graphics location determined by the value of X and Y.

`_setpixel(X,Y);`

Program E–2 places a dot in the center of the graphics screen.

Other Color Commands

If you have the Learn C system, you can use the Help menu (as described in Appendix C) to gain access to the other graphic commands used by Learn C. If you are using Microsoft Quick C, then you also have access to the same Help menu. It is assumed that you will be doing the exercises in Chapter 8, substituting the other color commands. If you have completed Chapter 8, you can see that the Turbo C graphics include a greater variety of programming options. Hence the reason for using Turbo C as the programming environment for the color and graphics chapter.

Appendix F—ASCII Character Set

American Standard Code for Information Interchange

Character	Code	Character	Code	Character	Code
blank	32	@	64	'	96
!	33	A	65	a	97
"	34	B	66	b	98
#	35	C	67	c	99
$	36	D	68	d	100
%	37	E	69	e	101
&	38	F	70	f	102
'	39	G	71	g	103
(	40	H	72	h	104
)	41	I	73	i	105
*	42	J	74	j	106
+	43	K	75	k	107
,	44	L	76	l	108
—	45	M	77	m	109
.	46	N	78	n	110
/	47	O	79	o	111
0	48	P	80	p	112
1	49	Q	81	q	113
2	50	R	82	r	114
3	51	S	83	s	115
4	52	T	84	t	116

Character	Code	Character	Code	Character	Code
5	53	U	85	u	117
6	54	V	86	v	118
7	55	W	87	w	119
8	56	X	88	x	120
9	57	Y	89	y	121
:	58	Z	90	z	122
;	59	[	91	{	123
<	60	\	92	|	124
=	61	]	93	}	125
>	62	↑	94	~	126
?	63	—	95	DEL	127

Appendix G—ANSI C Standard Math Functions

This appendix lists the ANSI C library routines for mathematical functions. They are listed alphabetically in Table G–1.

Table G–1   ANSI C Standard Math Functions

Function	Include File and Description:	Example

abs( )	Returns the absolute value of the integer argument n.
	`#include <stdlib.h>`
	`int abs(int n);`
		v = **abs**(−3);
		returns a value of
		v = 3

acos( )	Computes the arc cosine of the argument whose value is between 1 and −1.
	`#include <math.h>`
	`double acos (double x);`
		angle = **acos**(0.5);
		returns a value of
		angle = pi/3

asin( )	Computes the arc sin of the argument whose value is between 1 and −1.
	`#include <math.h>`
	`double asin(double x);`
		angle = **asin**(0.707);
		returns a value which is close to:
		angle = pi/4

atan( )	Computes the arc tan of the argument whose value is between − $\pi/2$ and $\pi/2$ radians.
	`#include <math.h>`
	`double atan(double x);`
		angle = **atan**(1.0);
		returns a value of
		angle = pi/4

atan2( )	Computes the arc tan of the ratio of two arguments. This function can use the sign of the results to determine in which quadrant of a Cartesian coordinate the angle is contained.
	`#include <math.h>`
	`double tan2(double x, double y);`
		angle = **atan2**(1, 2);
		returns the tangent whose angle is 1/2 in radians.

Table G–1   *(continued)*

Function	Include File and Description:	Example

`ceil( )`	Returns the ceiling of a double argument. This is the smallest integral value that is equal to or just exceeds the value of the argument. Useful in rounding a double value up to the next integer value.
	`#include <math.h>`
	`double ceil(double x);`
		ceiling = **ceil**(3.2);
		returns a value of
		ceiling = 4.0

`cos( )`	Returns the cosine of the argument in radians.
	`#include <math.h>`
	`double cos(double x);`
		ang_cos = **cos**(0);
		returns the value
		ang_cos = 1

`cosh( )`	Returns the hyperbolic cosine of the argument. If the value is larger than the system can handle a range error will occur.
	`#include <math.h>`
	`double cosh(double x);`
		value = **cosh**(x);
		returns the value of the hyperbolic cosine of x.

`div( )`	Returns the result of integer division in the form of a structure which has a resulting integer quotient and remainder. This is defined in the header file as:
	`typedef struct`
	`{`
	`    int quot;   /* Quotient */`
	`    int rem;    /* Remainder */`
	`} div_t`
	`#include <stdlib.h>`
	`div_t div(int number, int denom);`
		value = **div**(14, 3);
		returns the value of
		value.quote = 4
		value.rem = 2

`exp( )`	Returns the exponential e* (base e, where e = 2.7182818) of argument.
	`#include <math.h>`
	`double exp(double x);`
		value = **exp**(1);
		returns the value of
		value = 2.7182818...

`fabs( )`	Returns a type double that is the absolute value of the argument.
	`#include <math.h>`
	`double fabs(double x);`
		value = **fabs**(−3.94);
		returns the value of
		value = 3.94

`floor( )`	Returns the largest integral value that is less than or equal to the argument.
	`#include <math.h>`
	`double floor (double x);`
		value = **floor**(8.25);
		returns the value of
		value = 8.0

`fmod( )`	Returns the value of the floating point remainder resulting from the value of the quotient of the arguments while ensuring that the value returned is the largest possible integral value.
	`#include <math.h>`
	`double fmod(double x, double y);`
		value = **fmod**(x, y);
		returns the value:
		n = floor(x/y)
		value = n − n*y;

Table G-1   *(continued)*

Function	Include File and Description:	Example

`frexp( )`	Returns the mantissa m and the integer exponent n of the floating point argument.
	`#include <math.h>`
	`double frexp(double x, int *expptr);`
		man = **frexp**(x,&exp);
		returns the value
		man = mantissa
		exp = exponent.

`labs( )`	Returns the absolute value of a long integer argument.
	`#include <stdlib.h>`
	`long labs(long n);`
		value = **labs**(−63490L);
		returns the value of
		value = 63490

`ldexp( )`	Returns the floating point value of a double value argument with and exponent of the base two.
	`#include <math.h>`
	`double ldexp(double x, int exp);`
		value = **ldexp**(x,exp);
		returns the value
		value = x * 2^{exp}

`ldiv( )`	Returns the result of long integer division in the form of a structure that has a resulting integer quotient and remainder. This is defined in the header file as:
	`typedef struct`
	`{`
	`    long quot;   /* Quotient */`
	`    long rem;    /* Remainder */`
	`} ldiv_t`
	`#include <stdlib.h>`
	`ldiv_t div(long number, long demon);`
		value = **ldiv**(63463L,63460L);
		returns the value of
		value.quot = 1
		value.rem = 3

`log( )`	Returns the natural logarithm of the double valued argument.
	`#include <math.h>`
	`double log(double x);`
		value = **log**(3);
		returns the value of
		value = 1.098612

`log10( )`	Returns the logarithm base 10 of the double valued argument.
	`#include <math.h>`
	`double log(double x);`
		value = **log10**(3);
		returns the value of
		value = 0.47712

`modf( )`	Returns the fractional and integral parts of a floating point argument.
	`#include <math.h>`
	`double modf(double x, double *intptr);`
		value = **modf**(x,&int_pt);
		returns floating point value in location given by int_pt.

`pow( )`	Returns the value of the arguments x and y where the relationship is x^y.
	`#include <math.h>`
	`double pow(double x, double y);`
		value = **pow**(2,3);
		returns: value = 8

`rand( )`	Returns a pseudorandom integer between 0 and RAND_MAX as defined in the header file.
	`#include <stdlib.h>`
	`int rand(void);`
		value = **rand**( );
		returns a random integer to value.

Table G–1 *(continued)*

Function	Include File and Description:	Example

`sin( )`	Returns the sine of a double radian valued argument.
	`#include <math.h>`
	`double sin(double x);`
		value = sin(x);
		returns the sin of x.

`sinh( )`	Returns the hyperbolic sine of a double valued argument.
	`#include <math.h>`
	`double sinh(double x);`
		value = sinh(x);
		returns the hyperbolic sine of
		x

`sqrt( )`	Returns the square root of a positive valued integer argument.
	`#include <math.h>`
	`double sqrt(double x);`
		value = sqrt(9.0);
		returns:
		value = 3

`srand( )`	Sets the seed for the random number generator function determined by the value of the unsigned argument.
	`#include <stdlib.h>`
	`void srand(unsigned n);`
		srand(n);

`tan( )`	Returns the tangent of the argument expressed in radians.
	`#include <math.h>`
	`double tan(double x);`
		value = tan(x);
		returns the tangent of x.

`tanh( )`	Returns the hyperbolic tangent tangent of a double argument.
	`#include <math.h>`
	`double tanh(double x);`
		value = tanh(x);
		returns the hyperbolic tangent
		of x.

Answers

Answers to Section Reviews—Chapter 1

Section 1-1

1. The purpose of an editor is to allow you to enter source code. It is needed to achieve this purpose.
2. The reason for using a compiler is so the computer will have access to a program that it understands.
3. A compiler converts the source code into a code that the computer understands.

Section 1-2

1. Three advantages of the C programming language are: portability, computer control, and flexibility. Other advantages are listed in Table 1-1.
2. Portability means that a program you write on one computer will operate on another computer system with few if any changes.
3. All C programs must start with the keyword **main( )**.
4. An example of a C comment is:

   ```
   /* This is a comment. */
   ```

5. Braces indicate the beginning and the end of the program instructions.

Section 1-3

1. Program structure refers to the appearance of the source code. Good program structure will make the source code easy to read, modify, and debug.
2. It isn't necessary to give a structure to a C program for it to compile without errors. Good program structure is to make things easier for the programmer, not the computer.

3. Block structure makes the program easier to read and understand. An example would be the structure given to a business letter.

4. The reason for structured programming is to make the program easy to understand, modify, and debug.

5. The programmer's block is used to explain all of the important points about the program. It contains:
   a. Program name
   b. Developer and date
   c. Description of the program
   d. Explanation of variables
   e. Explanation of constants

Section 1–4

1. A set of characters is used to write a C program. These characters consist of the uppercase and lowercase letters of the English alphabet, the ten decimal digits of the Arabic number system, and the underscore (_). Whitespace characters are used to separate the items in a C program.

2. In C, a token is the most basic element recognized by the compiler.

3. The major data types used by C are numbers, characters and strings.

4. In C a keyword is a predefined token that has a special meaning to the C compiler.

5. The data type that handles the largest number is the `double`.

Section 1–5

1. The purpose of the `printf( )` function is to write information to standard output.

2. Characters have single quotes, and strings have double quotes.

3. The format specifier in a `printf( )` function specifies how it is to convert, print and format its arguments.

4. An argument is the actual values within the parentheses of a function.

5. You must have as many arguments as format specifiers. Extra arguments are ignored.

Section 1–6

1. A C function is an independent collection of declarations and statements.

2. An identifier is the name you give to key parts of your program.

3. All identifiers must start with a letter of the alphabet (uppercase or lowercase) or the underscore (_). The remainder of the identifier may use any arrangement of letters (uppercase or lowercase), digits (0 through 9) and the underscore—and that's it—no other characters are allowed (this means spaces are not allowed in identifiers). Do not use reserved words for your own identifiers.

4. The first 32 characters of an identifier are recognized by C.

Section 1–7

1. You can think of a variable as a specific memory location set aside for a specific kind of data and given a name for easy reference.

2. In C, you must declare all variables before using them. To declare a variable, you must declare the type and identifier of the variable.
3. Three fundamental C type specifiers are: **char**, **int**, and **float**.
4. Initializing a variable means to combine its declaration with an assignment operator.
5. To prevent a new variable from coming into your program as a result of a typing error.

Section 1-8

1. The common arithmetic operators used in C are: + => addition, − => subtraction, * => multiplication, / => division, % => remainder.
2. In integer division, C will truncate the remainder. The significance of this is that a division such as 3/5 will be evaluated to 0.
3. In C, 3 − 2 = **result**; is not allowed. You cannot have an assignment to an expression.
4. Precedence of operation means the order in which arithmetic operations are performed.
5. An example of a compound assignment is: **value** −= 5;.

Section 1-9

1. The escape sequence in the **printf( )** function causes an escape from the normal interpretation of a string.
2. When used in a **printf( )** function, the backslash character is called the escape character.
3. Three escape sequences used in the **printf( )** function are: \n => newline, \t => tab, and \b => backspace.
4. A field width specifier as used by the **printf( )** function determines the minimum number of spaces to the left of the decimal point and the maximum number of spaces to the right of the decimal point.

Section 1-10

1. The **scanf( )** function allows your program to get user information from the keyboard.
2. The **scanf( )** function is told what variable identifier to use by including the identifier in its argument preceded by an ampersand (**&**) with no space (such as **&variable**)
3. In most systems the **scanf( )** function produces a carriage return to a new line automatically.
4. Floating point values may be entered as whole numbers or using E notation.
5. To have values outputted to the screen in E notation, simply change the format specifier in the **printf( )** function from **%f** to **%e**.

Section 1-11

1. There are three error messages in LEARN C; they are: fatal error messages, compilation error messages and warning messages.
2. A fatal error message will terminate the compilation process.

3. Case sensitivity means that the C compiler makes a distinction between uppercase and lowercase letters. This means if you use identifiers with capital letters and declare them as such, you must consistently use that exact capital letter configuration for the remainder of your program.

4. It is good practice to look for semicolons when you do not understand the reason for the error message because a missing semicolon usually distorts the meaning of the program to the point where the compiler cannot conclude that the error was a missing semicolon. This results in another kind of error message depending on what follows the missing semicolon.

5. A nested comment is placing one comment inside another. It is not legal, and will prevent the program from executing. Some compilers will allow you to select this option, but it is not recommended.

Section 1–12

1. The first step in the development of a program is to state the requirements in writing.

2. The items that should be included in the program problem statement are the purpose of the program, required input (source), the process on the input and the required output (destination).

3. The first step in the actual coding of the program is to outline it using nothing more than comments.

4. The process used to develop the final program consists of coding each section of the program separately, then compiling and executing it. Any program bugs are thus removed from the program a section at a time.

Answers to Self-Test—Chapter 1

1. This program may not compile and execute on your system because an **#include <stdio.h>** directive may be required. If this is the case, be sure to use standard format and place it before anything else in the source code, starting on the far left part of the screen.

2. The program solves for the circuit current. The program user must input the values of the circuit voltage and the circuit resistance. The program will then display the value of the resulting current. This was determined by reading the comments at the beginning of the program.

3. There are three variables in the program. They are all of type **float**. This was determined by reading the /* Declaration block. */.

4. The **puts( )** function was used to explain the program to the user because it provides an automatic return to a new line. No other formatting commands were needed other than to output a series of strings to the monitor screen.

5. The program user may input the values of the voltage and resistance as a whole number, as a number with decimal fraction or by E notation. This was determined by observing the **scanf( )** function. It uses the **%f** (float) for input.

6. The output will be displayed using E notation. This was determined by observing the **printf( )** function. It uses the **%e** (E notation) for presenting output values from the program variables.

Answers to Odd End-of-Chapter Problems—Chapter 1

Section 1-1

1. The program used to enter C source code is called an *editor.*

Section 1-2

3. All C programs must start with **main( )**.
5. The opening { and closing } indicate the beginning and end of program instructions in C.

Section 1-3

7. The purpose of a programmer's block is to present all of the important information about the program.
9. No, it isn't necessary to have a programmer's block for a C program to compile.

Section 1-4

11. The underscore (_).

Section 1-5

13. Characters are represented by single quotes, a string by double quotes.
15. The name given to the actual values within the parentheses of a function is *arguments.*

Section 1-6

17. The name given to an independent collection of declarations and statements in C is a *function.*
19. The first 32 characters of an identifier are recognized by C.

Section 1-7

21. The three fundamental C type specifiers are **char, int,** and **float.**
23. The requirement that all variables must be declared prevents a new variable from coming into your program as a result of a typing error.

Section 1-8

25. The unique characteristic of integer division in C is that the answer will be *truncated.*
27. The C statement: **result *= 5;** means **result = result * 5;**

Section 1-9

29. The *escape sequence* in a **printf( )** function causes a vast departure from the normal interpretation of a string.
31. The *field width specifier* determines the number of spaces to the right of the decimal point.

Section 1-10

33. As described in this chapter, the purpose of the **scanf( )** function is to get input from the program user.
35. The program user may enter in whole numbers (no decimal part), numbers with decimal fractions, and in E notation.

Section 1-11

37. A fatal error message immediately terminates the compiling process.
39. No, compiler error messages do not always identify the problem in your program. It depends upon the type of problem in your program. A misplaced or omitted semicolon can really throw a compiler off.

Section 1-12

41. The first step in the coding of a program is to enter comments that divide the program into major sections.

Answers to Section Reviews—Chapter 2

Section 2-1

1. The C compiler does not require that a C program be structured. Structuring is done to make it easier for the programmer and others who will be responsible for it to read, debug and modify.
2. Block structure means the program will be constructed so that there are several groups of instructions rather than one continuous listing of instructions.
3. Each program block should begin with a remark that explains what the block will do.
4. Program blocks may be separated with spaces and comments that form lines (/*-------------*/) across the program code.
5. The body of a program block is highlighted by indenting the program lines from the left margin.
6. There are three types of blocks: action, branch and loop.
7. The completeness theorem states: Any program logic, no matter how complex, can be resolved into action blocks, loop blocks and branch blocks.

Section 2-2

1. A C function is a specific part of a C program that is designed to return a value.
2. A type **void** does not return any value. An example would be a function that does not return a value.
3. A function prototype declares the function type, name and parameters at the beginning of the C program.
4. The purpose of a function prototype is to let the compiler know what to expect in your program. Doing this assures that the proper amount and type of memory will be allocated to each function before it is actually used.

5. Calling a function means invoking it into action. This is done by using the function identifier in the calling function (as was done in **main( )** in this section).

6. The **exit( )** function was used in **main( )** in order to ensure that all of the necessary computer "house-cleaning" was done. This allows you to now go and use another program after your C program has terminated. The value used in its argument is **0**. By convention, this means a successful termination.

Section 2-3

1. The parameter of a function identifies the type of variable that will be passed to the function.

2. A formal parameter is the identifier that is used to identify the argument type. An actual parameter is the identifier that contains the value to be passed. These must be the same by being of the same data type. They may be different in that the formal and actual arguments do not have to have the same identifiers.

3. Passing values between functions means having a value obtained in one function influence the operation or value of a called function.

4. The only difference between the coding of a function prototype and the head of a function declaration is that the prototype must end with a required semicolon and the declaration must not have the ending semicolon.

5. One method of passing a value back to the calling function is by using the C command **return( )** where the value to be returned is placed in the function argument.

Section 2-4

1. Yes, a function may pass more than one value to a called function. The number of values to be passed must all be identified by formal parameters.

2. Yes, a function may call more than one function. The requirement here is that the function has been defined.

3. No, it makes no difference in what order called functions are defined within the program.

4. The meaning of a called function calling another function is that any function may call another function regardless of how that function was activated. This means that a function may call itself as well.

5. Recursion means a function calling itself.

Section 2-5

1. A preprocessor directive is a special instruction to the preprocessor that causes an action before the program is compiled and executed.

2. The **#include** directive instructs the preprocessor to substitute one set of tokens with another set of tokens.

3. A macro is a preprocessor instruction.

4. Constants are usually defined in C by using preprocessor commands. Conventionally constants defined in this manner are all UPPERCASE.

5. Yes, parameters may be used with **#define**. An example would be: **#define cube(x) x*x*x**.

Section 2-6

1. The advantage of creating your own header files is that you may build a library of data specific to your technology area. This will save you replicating the same code over and over again.
2. As presented in this section, the information contained in your header files consists of a series of **#define** statements.
3. The extension given to a C header file is .h.
4. You call your header files into your C program by the statement: **#include** "**file.h**" where file is the legal DOS name of your header file.

Section 2-7

1. The main goal in the development of the case study for this section was to increase readability and understanding of what the program will do while still preserving the fundamental characteristics of the C language.
2. The following information should be included in the program user's block:

   I. Program Information
     A. Name of program
     B. Name of programmer
     C. Date of development
       1. Optional date of last modification
         a. Name of person who did last modification
  II. Program Explanation
     A. What the program will do
     B. What is required for input
     C. What process will be performed
     D. What will be the results
       1. State all units for variables
 III. Describe All Functions
     A. Use function prototypes
     B. Explain purpose of each prototype

3. Using **#define** statements can result in the saving of much program code. More importantly, it allows for the eventual development of header files where these same statements may be used in future programs.
4. The last thing that is usually asked of the program user in a typical technology program is if the program is to be repeated.

Answers to Self-Test—Chapter 2

1. Four functions are defined in the program. They are: **main( )**, **explain_program( )**, **get_values( )** and **calculate_and_display( )**.
2. The total number of functions used in the program is seven. The extra three are **exit( )**, **printf** and **scanf( )**.

3. The identifiers used for formal arguments in the program are: `f`, `l`, `r`, and `v`.

4. The identifiers used for actual arguments in the program are: `resistor`, `inductor`, `frequency` and `voltage_s`.

5. The `get_values` function passes values to the `calculate_and_display` function.

6. Values are passed from one function to the other by using as actual arguments the identifiers of the calling function as arguments of the same type and number. An example from the program is:

`calculate_and_display(frequency, inductor, resistor, voltage_s);`

7. There are eight variable identifiers used in the program. They are: `f`, `l`, `r`, and `v`, `resistor`, `inductor`, `frequency` and `voltage_s`.

8. The minimum change to make to display the value of the inductive reactance would be in the function of `calculate_and_display`. For inductive reactance.

9. Yes the program will accept E notation because the input variables are of type `float`.

Answers to Odd End-of-Chapter Problems—Chapter 2

Section 2-1

1. Block structure is a method of breaking the program into distinct groups of code so the program is easier to read and understand.

3. The loop block has the ability to go back and repeat a part of the program.

Section 2-2

5. The part of a C program that gives the compiler specific information about the functions that will be defined in the program is called the function prototype.

7. A function is called by using the function identifier within another function along with any actual arguments.

Section 2-3

9. The type assigned to a function when no value is to be returned by it is type `void`.

11. The C statement `return( )` is used to return a value from the called function to the calling function.

Section 2-4

13. Yes, a function may call more than one other function.

15. Yes, a called function may call another function.

Section 2-5

17. The preprocessor directive presented in this chapter is **#define**.
19. An example of using parameters with the **#define** is

   #define square(x) x*x

Section 2-6

21. You invoke your own header file by using:

   #include "myfile.h"

   assuming your header file is named **myfile.h**.

Answers to Section Reviews—Chapter 3

Section 3-1

1. Relational operators are symbols that indicate the relationship between two quantities.
2. The relational operators used in C are:
   > greater than
   < less than
   >= greater than or equal
   <= less than or equal
   == equal to
   != not equal to
3. The two conditions allowed for relational operators are TRUE or FALSE.
4. The statement that a relational operation in C returns a value means that a value of 1 is returned if the relation is TRUE and a value of 0 is returned if the relation is FALSE.
5. The difference between the C operation symbol = and == is that the single equals sign (=) is an assignment operator. It assigns the value on the right side of this operator to the memory location of the variable on the left side. The double equals sign (==) tests to see if the value of the data on the right is equal to the value of the data on the left. No assignment or transfer of information takes place.

Section 3-2

1. An open branch offers an option to the program that may or may not be executed depending upon a given condition. In either case, program execution will always go forward to the next statement.
2. The **if** statement has the form

   if (expression) statement

   This means that if **expression** is TRUE then **statement** will be executed. If **expression** is FALSE then **statement** will not be executed.

3. A compound statement consists of more than one statement and is enclosed by braces { }.
4. An **if** statement may call another function. This is no different from the use of a compound statement as a part of the **if**.

Section 3-3

1. A closed branch is a branch that forces the program to take one of two alternatives.
2. The difference between the **if** and the **if...else** is that the **if** represents an open branch while an **if...else** represents a closed branch.
3. Yes, compound statements may be used with the **if...else**. The requirement is that the statements must be enclosed in brackets { }.
4. Function calls may be used with the **if...else**. This is done by placing the function call for each statement in the **if...else**.
5. The **if...else if...else** statement in C will essentially give an option among three choices. The first two choices will depend upon the condition of the first two expressions. If both expressions are FALSE then the third condition (the one following the last **else**) will be executed.

Section 3-4

1. The logical AND operator compares the TRUE/FALSE condition of two expressions. If both are TRUE then the final results are TRUE; otherwise the results are FALSE. The **&&** symbol is used for the logical AND in C.
2. The logical OR operator compares the TRUE/FALSE condition of two expressions. If both are FALSE then the final results are FALSE; otherwise the results are TRUE. The || symbol is used for the logical OR in C.
3. The operation of the logical NOT causes the opposite to take place. Thus NOT TRUE is FALSE, and NOT FALSE is TRUE.
4. An example of combining a relational and logical operation is: $(3 == 3) \&\& (5 < 10)$.
5. C evaluates the AND expression by first evaluating the expression on the left. If this is FALSE, the expression on the right is not evaluated. C evaluates the OR by first evaluating the expression on the left as before. However, this time, if the expression is TRUE then the expression on the right is not evaluated.

Section 3-5

1. Mixing data types means performing operations on data of different types.
2. The type **int** is upgraded to **float**, and the resulting addition is a type **float**.
3. The rule for mixed data types in C is that the data types are upgraded according to the highest ranking type in the expression.
4. A cast in C forces a change in data type.
5. An lvalue expression is an expression that refers to a memory location.

Section 3-6

1. The purpose of the **switch** statement in C is to allow a selection from several different options.
2. The other keywords that must be used with the **switch** are **case** and **break**. An optional keyword **default** may also be used.
3. The purpose of the keyword **default** in the **switch** statement is to activate a statement if there was no match in any of the **switch** selectors.
4. Function calls may be made from a **switch**. All that is necessary is that the function definition be available to the program.

Section 3-7

1. Omission of the **break** in a C **switch** causes execution of the statement following the selected **case** label.
2. To make sure that the **default** is always executed, omit all **breaks** in the C **switch**.
3. The conditional operator consists of three expressions. If the first expression is TRUE, then evaluation of the second expression takes place; otherwise, evaluation of the third expression takes place.
4. Any value other than zero makes the conditional operator TRUE. A value of zero makes the conditional operator FALSE.

Section 3-8

1. Conditional compilation means that part of your C program may or may not be compiled depending upon certain conditions within your program.
2. A compilation directive is a command to the compiler.
3. The purpose of the **#ifdef** directive is to cause compilation of a defined section of C source code under certain conditions.
4. Conditional compilation directives are normally used in large C programs.

Section 3-9

1. The first step in the development of a program is to state the purpose of the program in writing.
2. The purpose of a troubleshooting flow chart is to aid the technician in servicing a particular system.
3. A program stub is a part of the program that is intentionally left incomplete with just enough coding to ensure that the program flow will activate it at the correct time.
4. For the program developed for this case study, the user input represents the actual measurements and observations of the hypothetical robot.

Answers to Self-Test—Chapter 3

1. There are six functions in the program. They are: **main( )**, **explain_program( )**, **arm( )**, **power_unit( )**, **light_check( )** and **arm_drive_disconnect( )**.

2. No, there are no open branches.
3. Yes, the closed branch is contained in the function `arm_drive_disconnect`.
4. The function that uses the C **switch** is `light_check`.
5. The meaning of

   `measurement = (measurement > 30)? 30 : measurement;`

   is that if the value of `measurement` is greater than 30 it will be set equal to 30. Otherwise, its value will not be changed.
6. There are two, `measurement` and `light_status`.
7. There is only one function that returns a value, it is `arm( )`.
8. The value returned is the voltage measurement inputted by the program user.
9. The variable `light_status` is not of type **float** because a C **switch** cannot use a type **float**.

Answers to Odd End-of-Chapter Problems—Chapter 3

Section 3-1

1. The six relational operators presented in this chapter with their meanings are:

>	Greater than
>=	Equal to or greater than
<	Less than
<=	Equal to or less than
==	Equal to
! =	Not equal to

3. The = sign in C means assignment; it does not mean equal.

Section 3-2

5. The concept of an open branch is that the program always goes forward to new information and that an option that may or may not be taken exists.
7. A compound statement in C consists of two or more statements enclosed by an opening { and a closing }.

Section 3-3

9. In a closed branch, program flow always goes forward and there are now two options, one of which must be taken.
11. A compound **if...else** in C uses the **if...else if** statements and makes a selection from more than two alternatives.

Section 3-4

13. A logical operation is any operation that will produce one of two values, usually designated by TRUE or FALSE.

15. The logical OR operation in C compares two statements. If both are FALSE, then the result of the OR operation will be FALSE. For any other combination of the two statements, the result will be TRUE. The symbol for the logical OR operation in C is ||.

Section 3-5

17. Yes, it is legal in C to add a type **int** to a type **char**. This is called mixing data types.
19. A cast in C forces a change in data type.

Section 3-6

21. The C **switch** is used when there is a selection from several different options.
23. The keyword **break** in a C **switch** is used to terminate the selected statement.

Section 3-7

25. The purpose of the keyword **default** in a C **switch** is to identify the part of the C switch that will be executed if no match is made with any of the **cases**.

Section 3-8

27. A command to the compiler is called a compiler directive.
29. One conditional compiler directive is **#ifdef**.

Section 3-9

31. The property of the C language that allows the information in a troubleshooting flow chart to be developed into an interactive computer program are the decision-making statements of the language.

Answers to Section Reviews—Chapter 4

Section 4-1

1. The three major parts of the C **for** loop are

   - The value at which the loop starts.
   - The condition under which the loop is to continue.
   - The changes that are to take place for each loop.

2. Yes, you can have more than one statement in a C **for** loop. You must enclose the statements between opening and closing braces { }.
3. The meaning of ++**Y** is to increment Y before an operation on the variable.
4. The comma operator allows you to have two sequential C statements.

Section 4-2

1. The construction of the C **while** loop is

```
while(expression)
    statement
```

2. A **while** loop will be repeated as long as expression is TRUE (not zero).
3. The loop condition is tested first, before statement execution.
4. A good use of a **while** loop is in programming conditions when you do not know how many times the loop is to be repeated.

Section 4-3

1. The construction of the C **do while** loop is

```
do
    statement
while(expression);
```

2. The **do while** loop will be repeated as long as expression is TRUE.
3. The loop condition is tested in a C **do while** loop after statement is executed.
4. Generally, the **while** loop is preferred over the **do while** because it is considered good programming practice to test the condition before executing rather than after.

Section 4-4

1. A nested loop is having one program loop inside the other.
2. The structure used with nesting loops should make it clear where each loop begins and where it ends. This can be done by indenting the body of each loop and using comments to make it clear where each loop ends.
3. All three of the C loop types may be nested.
4. The C **do** loop will always be activated at least once. This may cause problems if it is used as a part of a nested loop because every time its outer loop is activated, the C **do** will also become active at least once, no matter what the condition of its loop counter.

Section 4-5

1. A run time error is an error that will take place during program execution.
2. No, a compiler does not catch run time errors. The reason is because by definition a run time error is not an error in the programming syntax; it is an error in program design.
3. A debug function usually contains a visual display of some data and the ability to step through the function.
4. An auto debug function is a convenient way of activating or deactivating the debug feature.

Section 4-6

1. The first step in the design of this case study program was the same as for the design of any program: to state the problem in writing.
2. The reason why a C **for** loop was selected in this program was because the loop needed a counting loop with a definite beginning and ending and increment value.
3. The beginning and ending values of the program loop were determined by calculations done on user inputs during program execution.
4. The C **return** had its required calculations done within its argument.

Answers to Self-Test—Chapter 4

1. There are five function prototypes used in the program.
2. The type of loop used in the program is a C **for** loop.
3. The initial loop condition is `counter` = `below_fr`.
4. The final loop condition is `counter` <= `above_fr`.
5. The change each time through the loop is `counter` += `freq_change`.

Answers to Odd End-of-Chapter Problems—Chapter 4

Section 4-1

1. A compound statement is a C statement that consists of more than one statement. It is set off by being enclosed between braces { }.
3. The comma operator allows you to have two sequential C statements.

Section 4-2

5. The construction of the C **while** loop is

```
while(expression)
    statement
```

7. A C **while** loop will be repeated as long as expression is TRUE (not zero).

Section 4-3

9. The construction of the C **do while** loop is

```
do
    statement
while(expression);
```

11. A C **do while** loop will be repeated as long as expression is TRUE (not a zero).

Section 4-4

13. A nested loop is one loop contained inside the other.
15. Yes, there is a problem with nesting a C **do** loop. The C **do** will always be executed at least once by the outer loop.

Section 4-5

17. A debug function usually contains a visual display of some data and the ability to step through the function.

Answers to Section Reviews—Chapter 5

Section 5-1

1. A computer's memory may be visualized as a stack or pile of memory locations, each identified by a unique number.
2. An instruction causes some computer action. The data is what is acted upon.
3. An address is a number that represents a specific memory location.
4. Immediate addressing is when the data immediately follows the instruction.
5. Direct addressing is when the instruction directs the CPU where in memory the data is located.

Section 5-2

1. A computer word is a group of bits treated as a unit.
2. The word size of a C **char** type is 1 byte (8 bits).
3. The computer represents signed numbers by using the twos complement notation system.
4. The difference between a signed and unsigned data type in C is that the MSB is not treated as a sign bit. Thus the range of values that may be represented is the same, but the magnitude of the unsigned is larger.
5. The C data type that uses the least amount of memory is the **char**. The largest amount of memory is the **long double** (which may be the same as a **double** depending upon your system).

Section 5-3

1. A pointer is a data type that represents the address of another memory location.
2. A pointer is called a pointer because it can be thought of as pointing to another memory location.
3. A pointer gets the address of another memory location by having the address assigned to it with the **&** operator. As an example, to store the address of a variable **x** in a pointer p you do the assignment p = **&x**.
4. To pass a value to a variable using a pointer, you must make sure that the pointer contains the address of the variable (see answer 3 above). Once this is done, a value can now be passed by using the * immediately preceding the pointer variable: ***p = 12**. This passes the value of 12 to the variable **x**, provided the variable p contains the address of **x**.
5. A pointer is declared by placing a space followed by the * sign immediately preceding the pointer variable name.

Section 5-4

1. The two ways of passing a value from a called function to the calling function are to use the C **return( )** or a pointer as the function formal argument.
2. The limitations of using the C **return** to pass a value back to the calling function is that only one value may be returned.

3. More than one value may be returned to the calling function by the called function through the use of pointers in the formal argument of the called function.
4. The mechanism for passing values to the called function that uses pointers in its formal arguments is to use the address (**&**) of the actual variables that will receive the values from the called function.
5. Separate functions are used in order to facilitate good program design.

Section 5-5

1. A local variable is a variable declared within a function. It is known only to that function.
2. The scope of a variable applies to what part(s) of a program in which the variable is known.
3. A global variable is declared at the beginning of the program before the function **main( )**. A local variable is declared within the function that will use it.
4. No, it is not considered good programming practice to use global variables because the value of such a variable may be changed by any function within the program.
5. Values may be passed between functions by using function arguments.

Section 5-6

1. The effect of the C keyword **const** is to cause the value assigned to the data type not to be changed (intentionally or otherwise) within the program. If the program user attempts to do this, a warning message will be given at compile time.
2. An automatic variable is a variable that is local to the function in which it was declared and whose life is equal to that of the declaring function.
3. A static variable is a variable that is local to the function in which it was declared, but whose life is equal to that of the program.
4. A register type variable is a request to the compiler to keep this variable in one of the internal registers of the system microprocessor rather than system memory.

Section 5-7

1. The address of operator in C is the ampersand sign **&**. It returns the value of the address of the variable with which it is used.
2. The indirection operator in C is the asterisk *****. It treats the value stored within it as a memory location for data.
3. The C equality symbol is ==, and the assignment symbol is =.

Section 5-8

1. A bitwise complement in C means that the binary equivalent of the value to be complemented will have each of its 1s converted to a **0** and each of its **0**s converted to a 1. The resulting binary number will then be converted back to the base of the original value.

2. The meaning of the bitwise AND operation is that the binary equivalent of the two values to be bitwise ANDed have each of their bit pairs ANDed. The resulting binary value is then converted back to the base of the original value.

3. The meaning of the bitwise OR operation is that the binary equivalent of the two values to be bitwise ORed have each of their bit pairs ORed. The resulting binary value is then converted back to the base of the original value.

4. A bitwise XOR operation is that the binary equivalent of the two values to be bitwise XORed have each of their bit pairs XORed. The resulting binary value is then converted back to the base of the original value.

5. A bitwise shift in C means converting the value to be shifted to its binary equivalent and then shifting the resulting binary number left or right the required number of bits. The new binary value is then converted back to the base of the original value.

Answers to Self-Test—Chapter 5

1. The output of each of the **printf** functions in **main( )** will be

```
The constant is 57532.
The value of memory_location_1 is 375
The contents of this_value are => b
The result is F.
```

2. There are two data values that are global for the entire program:

```
const unsigned int number_1 = 57532;
int *look_at;
```

3. There is only one data that is global for a part of the program:

```
extern char new_value;
```

Its scope is all of the functions following its declaration.

4. The value of 375 for the variable memory_location_1 is received by the statement in function_1:

```
*look_at = 375;
```

5. The statement in **main( )** that causes **memory_location_1** to get the value of 375 is

```
function_1( );
```

6. The output of the last **printf** function in **main( )** is F because the binary values to be ANDed are

$$1111_2 \quad <= 15_{10}$$
$$\underline{1111_2 \quad <= 15_{10}}$$
$$1111_2 \quad <= F_{16}$$

7. `function_1` does not require pointers in its argument because it uses a pointer within its definition. `function_2` requires the use of pointers within its parameter list because it is returning more than one value to the calling function as parameters.

8. No, `function_1` does not need to be of type **double**. The reason is because the function itself does not return a value to the calling function.

9. `function_1` "knows" the pointer `*look_at` because it is a global variable pointer.

10. `function_1` causes the value of the variable `memory_location_1` to change because it is initialized to the address of `memory_location_1`:

```
look_at = &memory_location_1;
```

Answers to Odd End-of-Chapter Problems—Chapter 5

Section 5–1

1. A way of visualizing a computer's memory as suggested in this chapter is as a pile of storage locations.

3. The process of the CPU getting an instruction from memory and then executing the instruction is called a fetch/execute cycle.

Section 5–2

5. Some of the most common word sizes used by computers are 8-bits, 16-bits, 32-bits, and 64-bits.

7. A. 0110_2 B. 1000_2

9. A. 1111_2 B. 0001_2 C. 0110_2

11. A. -6 B. -8 C. -100

Section 5–3

13. The indirection operator in C is the asterisk ($*$). When it precedes a data type, it will represent the data whose address is contained in the data type preceded by the indirection operator.

15. A. The value of `pointer` is the address of `data`.
    B. The value of `data` is 5.
    C. The value of `*pointer` is 5.

Section 5–4

17. Separate functions are used in a C program in order to facilitate good program design.

19. The **&** operator is used in the arguments of a called function in order to return values back to the calling function.

Section 5–5

21. A variable declared within a function whose life is only when the function is active is called a local or automatic variable.

23. A variable declared at the beginning of the program before **main( )** is called a global variable. Its scope is the entire active program.

Section 5-6

25. The C keyword used to ensure that an assigned value cannot be changed during program execution is **const**.
27. The name of the variable class that requests the compiler to keep the variable in one of the internal registers of the microprocessor is **register**.

Section 5-7

29. The address of **data** may be obtained by using the address of operator with the variable: **&data**
31. The C symbol used for equality is ==, for assigment =.

Section 5-8

33. A. 3_{16} B. 1_{16} C. AF_{16}
35. A. 7_{16} B. F_{16} C. F_{16}

Answers to Section Reviews—Chapter 6

Section 6-1

1. A string is an arrangement of characters.
2. You indicate a **char** string in C by the square brackets **[ ]**.
3. For a string consisting of five characters, six array elements are required. The last array element will contain the null terminator which is required in C so it knows where in memory the string ends.
4. The element number of the first character in a C string array is 0.
5. The relationship between pointers and string array elements is that a pointer may be used to access an individual character of the string array just as an individual string element may be used to access the same element.

Section 6-2

1. You can let C know how many elements an array will have by placing a number equal to the number of array elements inside the array brackets: **[N]**, where N = the number of elements.
2. The index of the first element of a C array is always 0.
3. If **int value[]** is declared then **value**, **&value**, and **&value[0]** are equal, and all contain the starting address of the array.
4. A global array is initialized; a local array is not.
5. The number of elements needed in a string array is one more than the number of characters in the string. This is needed in order to hold the null character.

Section 6-3

1. Another way of initializing an array is by using the braces { }. Each element is placed between these, separated by commas.

2. To declare an array with more than one dimension, use a set of square brackets **[ ]** for each additional dimension desired.

3. A formal array parameter that uses more than one dimension in C must have the sizes of its other dimensions stated above one dimension.

4. The method used in the programs of this section that allows arrayed variables to be added is to use a variable array index (**array[index]**) and then the **+=** operator.

5. You must ensure that each element of the array is set to **0** when using the **+=** operator with arrays.

Section 6–4

1. The basic idea behind array applications is to manipulate the value of the array index.

2. The programming method used in order to get arrayed values from the program user is to use a C **for** loop and increment the array index.

3. The method used to cause a series of entered values to be displayed in the opposite order from which they were entered is to use a C **for** loop that increments the array index on the input and a C **for** loop that decrements the array index on the output display.

4. The method used to extract a minimum value from a list of entered values is to compare these values to each other by placing the first value in a variable. The variable is then compared to the value of each array element using a C **for** loop. If the array element is smaller, a switch is made.

Section 6–5

1. A bubble sort is the process of taking two sequential quantities at a time from a consecutive list of quantities, comparing them and switching their sequence depending upon their relative value. This process is continued until no further switching is required in the consecutive list. The result will be a consecutive list of sorted quantities in ascending or descending value depending upon the requirements for a switch.

2. A sorting program must go through a list at least one time. When the list is already sorted, no switch is required, and so the program would only go through once.

3. For descending order, a switch is performed if the first number is smaller than the second. For ascending order, a switch is performed if the first number is larger than the second.

4. Four passes are required: 7 3 1 8, 3 1 7 8, 1 3 7 8, and one last pass with no switch.

Section 6–6

1. Another name for an array of characters is a string.

2. A rectangular array is an arrangement in memory where the data space used by each arrayed variable is equal.

3. A ragged array is a arrangement in memory where all of the data space used by each arrayed variable is not equal.

4. The starting address of the string is used with the C **%s** to print out a string.
5. The starting address of an array of pointers is the argument passed to a called function which requires information about a string array.

Section 6–7

1. One of the problems with the use of the **scanf( )** function is that it will not accept any characters beyond a blank space in a string of characters.
2. A better function to use in place of the **scanf( )** for string input is the **gets( )**. It will accept all input characters until a carriage return is entered.
3. The function **getchar( )** gets a single character at a time from the input.
4. The C character classifications are built-in C functions that will test to see if an input character meets certain requirements. An example would be **isalpha(ch)** where **ch** is the character being tested. This function will return a **0** if the character is not alphabetical and a non-zero value if it is.
5. You can allow the program user to enter numerical values as a string by using the built-in C function **strod( )** which will convert a string of the proper numerical format to a double precision number. Your input should first test each input character to assure that correct numerical characters have been entered and continue to prompt the user until correct data are entered.

Section 6–8

1. The purpose of the program used in this case study is to sort up to a maximum of ten strings inputted by the program user.
2. The strings are sorted according to their ASCII code.
3. The program user may input fewer than ten strings for this program. This is done by simply pressing the Enter key when prompted for another input string.
4. Top-down design was implemented in this program by first starting with the most generalized concepts and continuing with the details last.

Answers to Self-Test—Chapter 6

1. The purpose of the **char** variable ***string** in function **explain_program( )** is to point to the beginning of the string defined within the program for display in the **printf( )** function.
2. The meaning of the statement **strcmp(instring[index−1],"")** is to compare **instring[index−1]** with the **""**. This will return a **0** (meaning FALSE) if there is not a match and a non-zero (meaning TRUE) if there is a match. This is the statement that looks to see if the program user pressed only the Enter key during a prompt, thus indicating that there are no more strings to be inputted.
3. The statement is used in **main( )** as a part of the loop test that gets input strings from the program user.

4. The function `screen_scroll` operates by continuously printing newline characters. This causes existing text on the screen to be scrolled up as many lines as there are added newline characters.

5. The effect of clearing the screen by use of the `screen_scroll( )` function is to have at least 25 newline characters (there are 25 lines on the standard PC monitor).

6. The purpose of the function `flag` is to be used as a check for another pass through the sort loop. This is necessary to insure that the sorting is complete.

7. The purpose of the `getchar( )` function in `explain_program` is to cause the program to halt until the program user presses the Enter key.

8. The program does not really sort any strings. It sorts the pointers to the strings.

9. The purpose of the statement

   `if((strcmp(ptr[index],ptr[index+1])>0)&&(index<maxnumber);`

   is to determine two things: first, the relationship between two strings (is `ptr[index]` larger than `ptr[index+1]`?); second, if the string pointer is the last one in the list. If both of these are TRUE (neither one equal to zero), then a switch in string pointers will be made.

10. The purpose of the variable `int temp` in `string_sort` is to temporarily store the value of one pointer while string pointers are being switched in the bubble sort.

Answers to Odd End-of-Chapter Problems—Chapter 6

Section 6-1

1. The element number of the first character in a C string is `0`.
3. A C string is an array of characters.

Section 6-2

5. You can think of the word array as an arrangement.
7. For the array in problem 6, the number of the first array element is `[0]` and the last is `[9]`, for a total of `10` elements.

Section 6-3

9. To declare the array variable `array` as a two-dimensional array: `array[M][N]`, where M and N indicate the number of elements in each array dimension.
11. A formal array parameter that contains more than one dimension must have the size of its other dimensions declared.

Section 6-4

13. A method that could be used to cause a series of entered values to be displayed in the opposite order from which they were entered is to use a C **for** loop that increments an array index on the input, and another C **for** loop that decrements the array index on the output.

15. An easy method to use, in order to get program user input into an array, is to use a C loop and increment the array index.

Section 6-5

17. The term bubble sort is used to indicate that when values are arranged in a certain order, the order is created by having values "bubble" up to the top of the list.

Section 6-6

19. A ragged array is an arrangement in memory where the data space used by each arrayed variable is not equal.
21. The actual argument passed to a called function is the starting address of an array of pointers.

Section 6-7

23. The purpose of the **getchar( )** function is to get a single character at a time from the input.
25. A good function to use for entering strings is the **gets( )**. It will accept all input characters until a carriage return is entered.

Answers to Section Reviews—Chapter 7

Section 7-1

1. In C the keyword **enum** means enumerated.
2. An **enum** data type is used to describe a discrete set of integer values.
3. For the code: **enum** numbers {first, second, third}, the integer value of first is 0.
4. The main purpose of using a C **enum** is to make your program more readable.
5. Yes, a C **enum** constant can be set to a given integer value. An example is **enum** numbers {first = 15}.

Section 7-2

1. The C **typedef** allows you to create your own name for an existing data type.
2. The resulting new name of the data type from the following code: **typedef char** name[20] is name.
3. The main purpose of using the C **typedef** is to create synonyms for existing data types.

Section 7-3

1. The information on a check from a checking account can be treated as a structure because it contains information that is a collection of different types of data that are logically related.
2. A C structure consists of a collection of C data elements of different types that are logically related.

3. The structure of a C structure as presented in this section is the keyword **struct** followed by a beginning brace { and ending with a closing brace followed by a semicolon };. Between the braces are the structure members: **type** variable_identifier;.

4. A structure member variable is programmed for getting data and outputting data by using the member of operator (.).

5. An example of problem 4 above would be

    ```
    structure_name.member_name
    ```

Section 7-4

1. A structure tag is an identifier that names the structure type defined in the member declaration list of the structure.

2. An example of a structure tag would be

    ```
    struct tag
        {
            member-declaration-list;
        }
    ```

3. A structure can be of a variable type. This can be done by using the C **typedef** and the structure variable identifier as the name of the type.

4. A structure member is pointed to by the use of the C —> pointer to structure operator.

5. The three operations that are allowed with structures are:

    • Assign one structure to another with the assignment (=) operator.
    • Access one member of a structure (. or —>).
    • Get the structure address (using the **&** operator).

Section 7-5

1. The purpose of a C union is to be able to store one of several different data types in the same starting memory address.

2. A C union is declared in the same way a C structure is declared. The difference is that the keyword **union** is used in place of **struct**.

3. A C structure may be initialized by the program. However, it must be global or static.

4. An array of structures may be declared in the same manner as any array. The difference is that the array type has been declared as a structure with defined members.

5. An individual member of an array of structures is accessed by **struc-ture_variable[N].member_variable**, where N is the number of the array.

Section 7-6

1. A structure can be an array. In the sense of arrays, a structure may be treated as a data type.

2. A structure can contain another structure. A defined C structure may be used as a member type within another structure.

3. Yes, it is possible for an array to be a structure type that contains an array as one of its members. Since you can have a structure as an array and a structure may contain an array as one of its members, then the combination is also possible.

Section 7–7

1. A legal DOS file name has the form

   `[Drive:]FileName[.EXT]`

   Where

   `Drive` = The optional name of the drive that contains the disk to which you want to access. (Drive A: is on the left, drive B: is on the right. For a stacked drive system drive A: is on the top, and B: is on the bottom.)

   `FileName` = The name of the file (up to eight characters).

   `.EXT` = An optional extension to the file name (up to three characters).

   Both the Drive: and .EXT are optional. If Drive: is not specified, then the active drive will be used. The colon (:) following the drive letter is necessary.

2. A file pointer in C is a pointer of type **FILE**. An example would be **FILE** `*flptr;`

3. The built-in C function for opening a disk file is **fopen( )**.

4. The file pointer is assigned to the DOS file name by

   `file_pointer = ` **fopen(** `"FILENAME.EXT", "r"` **);**

5. The difference between the C file opening command for writing to or reading from a file is the character directive used as an argument in the file command. A `"w"` means write, and a `"r"` means read. Note that the double quotes are used.

6. When finished with an open file you must always close it. This is accomplished with the built-in C function `fclose(` **file_pointer** `)`.

Section 7–8

1. The four possible disk I/O conditions are

   - The disk file does not exist and you want to create it on the disk and add some information.
   - The disk file already exists and you want to get information from it.
   - The disk file already exists and you want to add more information into it while preserving the old information that was already there.
   - The disk file already exists and you want to get rid of all of the old information and add new information.

2. The different methods of reading and writing data with the C language are

   - One character at a time
   - Character strings
   - Mixed mode
   - Block (or structure)

3. The three basic C file type commands are

- "a" for append to existing file
- "r" for read an existing file
- "w" for write to an existing file (this will create the file if it does not exist).

4. The C **fscanf( )** function allows you to read a file in mixed mode.
5. The purpose of a block read or write function is to store or retrieve a block of data at a time such as a C structure.

Section 7-9

1. The two kinds of streams of data used in I/O operations are text and binary.
2. A buffer is a temporary storage place in memory.
3. The three standard files used by a C program are standard input, standard output, and standard error.
4. The redirection operators used by DOS are the < and the >. The command **A < B** means file A will get its data from file B rather than from standard input (the keyboard), and **A > B** means the data from file A will be sent to file B rather than the monitor screen.
5. A command line argument allows you to invoke commands from the DOS prompt with an executable C program.

Answers to Self-Test—Chapter 7

1. There are four structure members. They are **part[20]**, **quantity**, **price**, and **record_number**.
2. The function **read_menu( )** is of type **char**. It returns a character back to the calling function.
3. There are no global variables used in the program.
4. The purpose of the arguments in function **main( )** is to get command line input from the user. This will allow authorization for putting data into the disk files. They are called command line arguments.
5. The C **switch** in function **main( )** has the **case 'E'** placed before the **default** so that if **authorization** were FALSE, then the message in the **default** section would be displayed.
6. The functions that are unique to Turbo C are **clrscr( )** which clears the text screen and brings the curser to the upper left corner of the monitor. The other function is **gotoxy(x, y)**. This places the curser at the location indicated by x (horizontal) and y (vertical) on the monitor screen.
7. There are no **scanf( )** functions used in the program because of the potential problems inherent with this function.
8. The program will respond to either uppercase or lowercase input from the program user in response to the menu. This is because of the **toupper( )** function used on the input. This converts any character input to uppercase.

9. The member `price` must be of type `long` because the `atof( )` function converts the user input to a type `long`.
10. The information in the disk file is in binary mode. This is because the `fopen( )` function for entering and receiving data uses the `"b"` extension as part of its file directives.

Answers to Odd End-of-Chapter Problems—Chapter 7

Section 7–1

1. The **enum** (enumerated) data type is used in C to describe a discrete set of integer values.
3. No, the C **enum** does not create a new data type. Its purpose is to make your source code more readable.

Section 7–2

5. No, the C **typedef** does not create a new data type. It allows you to create your own name for an existing type.
7. The resulting new name of the data type for the given code is **structure**.

Section 7–3

9. An example of a common everyday system that utilizes the concept of a C structure is a personal check book.
11. The C member of operator identifies the structure to which the member belongs. The symbol is the period (.).

Section 7–4

13. An example of the use of a structure tag is

```
struct tag
      {
            member-declaration-list;
      }
```

15. The three operations that are allowed with structures are assigning one structure to another, accessing one member of the structure, and getting the structure address with the **&** operator.

Section 7–5

17. A C structure and a C union are both declared in the same way, the difference being that the keyword **union** is used in place of the keyword **struct**.
19. The code represents a member (identified as number) of the third element (2) of an array of structures called variable.

Section 7–6

21. Yes, a C array of structures can contain an array as one of its members. Since you can have a structure as an array and a structure may contain an array as one of its members, then the combination is also possible.

Section 7-7

23. Yes, B:FILER.02 is a legal DOS reference to a file name.
25. In C, a file pointer could be declared by:

```
FILE *ptr;
```

Section 7-8

27. The different methods of reading and writing data with the C language are: one character at a time, character strings, mixed mode, and block (or structure).
29. When the C command for writing to a disk file ("w") is given for a file that does not exist, the file will be created.

Section 7-9

31. The three standard files used by C are standard input, standard output, and standard error.
33. The physical representation of a standard file can be changed by software by using the DOS redirection operators (< and >).

Answers to Section Reviews—Chapter 8

Section 8-1

1. In order to obtain color, a color graphics card must be installed and a color monitor attached.
2. In order to change the text from 80 columns to 40 columns, an installed graphics card is required.
3. Text can have 16 different colors. All these different colors can be displayed at the same time on the screen.
4. There are eight different background colors available. The **clrscr( )** command is used to bring the full screen to the desired color.
5. There are basically two choices available to indicate the desired color, the color number and the color name. For example, **textcolor(0)** or **textcolor(BLACK)** both produce the same effect.

Section 8-2

1. The two different kinds of screens available to you on your monitor are text and graphics.
2. Text mode has all predefined images, while graphic mode allows you to define your own.
3. The one piece of hardware is an installed graphics adapter. The two pieces of software are **graphics.h** and the appropriate graphics driver.
4. The number of different colors available with color graphics depends upon the installed color graphics hardware in your system and its mode of operation.
5. A pixel is the smallest element possible on the graphics screen.

Section 8-3

1. It's important to know about your system's modes and colors so you don't mistakenly try to look for lines that are drawn in the same color as the graphics screen (you won't see them).
2. A color palette means the colors available for the current active mode.
3. There are four color palettes available for the IBM CGA color graphics card. Three colors per palette are available.
4. Yes, there are modes where color is not available. This information can be found in Table 8-3.
5. You can find out the number of modes for your system by referring to Table 8-3 or by using a program that contains the built-in Turbo C function **getgraphmode( )**.

Section 8-4

1. The effect of changing a graphic mode could be the changing of the color palette and/or the number of pixels available on the graphics screen.
2. There are four line styles in Turbo C plus the ability to have two thicknesses of these four styles. They are a solid, dashed, dotted, or center line style available in one of two thicknesses.
3. The built-in Turbo C shapes presented in this section were the rectangle and the circle. The circle's line is limited to two line thicknesses.
4. A flood fill will fill an enclosed area or the surrounding area with a given color. The difference is determined by the screen coordinates used in the flood fill function depending if they are within the area to be filled or outside of the area (not to be filled).

Section 8-5

1. Turbo C has a built-in function for creating bars because bar graphs are commonly used to present various kinds of technical information.
2. Using Turbo C built-in functions, a flat or three-dimensional bar may be displayed.
3. The options available for displaying a three-dimensional bar using Turbo C built-in functions are placing or omitting a top on the bar. This feature is available in case you wish to stack three-dimensional bars.
4. Using built-in Turbo C functions, you can distinguish one bar from the other by using the available area fill command and filling the bars with different patterns and/or different colors.
5. Some of the options available when using text in Turbo C graphics are the text style (font), text size, and text direction (horizontal or vertical).

Section 8-6

1. Scaling is a process that uses numerical methods to ensure that all of the required data appears on the graphics screen and utilizes the full pixel capability.

2. Coordinate transformation is the process of using numerical methods to cause the origin of the coordinate system to appear at any desired place on the graphics screen.

3. The values of the scaling factors are determined by the number of horizontal and vertical pixels and the minimum and maximum values of X and Y.

4. The coordinate transformation values are determined by the horizontal and vertical scaling factors and the minimum value of X and the maximum value of Y.

Section 8-7

1. The guiding principle for developing C programs is to make programs easy to understand and modify.

2. In developing identifiers for variables, you should consider using standard identifiers commonly found in specific formula relationships. The important point is to make standard formulas easily recognizable in the program.

3. Try to consider the **struct**, **switch** and user-defined **typedef** when creating your C programs. This will maken them easier to understand and modify.

Answers to Self-Test—Chapter 8

1. There are two structures in the program. Their identifiers are **wave_values** and **user_input**.

2. There are no global variables in the program.

3. There are five function prototypes in the program. They are: **explain_program( )**, **press_return( )**, **auto_initialization( )**, **sine_wave_display( )** and **program_repeat( )**.

4. The local variable is **user_input in_values**; which contains four members.

5. The function **program_repeat( )** is of type **int** because it returns a TRUE or FALSE condition to the calling function.

6. The reason why each of the strings in function **explain_program** is terminated with a **\n\r** is because the **cprintf( )** function is used which, unlike the **printf( )** function, does not give an automatic carriage return for the **\n**.

7. The **cprintf( )** function is used in function **explain_program** in order to display the text in color.

8. The user input variables are passed to **sine_wave_display** by passing the name of a structure through its argument.

9. The purpose of the C **while** in function **program_repeat( )** is to ensure that the only acceptable input response by the user is a Y or N (upper or lowercase).

Answers to Odd End-of-Chapter Problems—Chapter 8

Section 8–1

1. In order to produce text in color using Turbo 2.0 C, your system must contain a color monitor and an installed color graphics system and be an IBM PC, AT, PS/2 or true compatible.
3. The maximum number of text colors you may have is 16. Yes, they can all be displayed at the same time.

Section 8–2

5. A text screen may only display text and use text commands. A graphics screen may have individual pixels manipulated by the program user.
7. A pixel is the smallest element possible on the graphics screen.

Section 8–3

9. The built-in Turbo C function for determining the modes available on your system is

   getmoderamge(GraphDriver, LoMode, HiMode);

11. The purpose of the built-in Turbo C functions **getmaxx** and **getmaxy** is to get the maximum size of the current graphics screen.

Section 8–4

13. The palette on the graphics screen can be changed by using the built-in Turbo C **setgraphmode**(mode);.
15. Two ways of creating a rectangle using Turbo C are by using the built-in Turbo C **line**(X1,Y1,X2,Y2) function to create the four sides of the rectangle or the **rectangle**(X1,Y1,X2,Y2) function.
17. There are only two line styles for creating circles in Turbo C. They are either normal width or thick width set by the last **setlinestyle** function.

Section 8–5

19. A Turbo C fill-style represents different methods of filling in the area of a bar in order to help distinguish one bar from another.
21. There are 12 different fill-styles available for bars in Turbo C.

Section 8–6

23. Scaling is a process that uses numerical methods to ensure that all of the required data appears on the graphics screen and utilizes the full pixel capability.
25. Coordinate transformation is used so that the full range of the resulting graph generated by the mathematical function may be viewed on the graphics screen.

27. The coordinate transformation values are determined by the horizontal and vertical scaling factors and the minimum value of X and the maximum value of Y.

Section 8–7

29. Good programming style makes a source code easy to read and understand and easy to modify.
31. You should keep in mind the **struct, switch** and user-defined **typedefs** when developing C programs.

Answers to Section Reviews—Chapter 9

Section 9–1

1. The four major groupings of the registers in the 8086 μP are general purpose registers, pointer and index registers, segment registers, and flags register.
2. The difference between a register inside the μP and a memory location in the computer's memory is that a register can modify the bit pattern stored within it by shifting it or performing an arithmetic or logic operation on it. A memory location can only store the bit pattern, but never change it.
3. The microprocessor interacts with the computer's memory in two fundamental ways: (1) Selects a memory location and copies a bit pattern from one of its internal registers to that memory location. This is called a write operation. (2) Selects a memory location and copies a bit pattern from that memory location into one of its internal registers. This is called a read operation.
4. The registers inside the μP that may be treated as 8-bit or 16-bit registers are the general purpose registers.
5. A 16-bit μP can access up to 1Mb of memory by using the process of segmentation. This is where the address of a segment register is shifted left four bits and added to the address of the register containing the offset. The result is a 20-bit binary number capable of addressing a full 1Mb of memory.

Section 9–2

1. Three advantages of assembly language programs are that they are more compact, run faster, and allow more control over your computer than higher level languages.
2. Some of the disadvantages of assembly language programs are lack of portability and the ease of making programming errors.
3. A mnemonic is a short word used as a memory aid to indicate a process on a specific microprocessor.
4. When a C function is called, the first internal register placed on the stack is the instruction pointer (IP).
5. C stores locally declared variables in the top of the stack and function parameters at the bottom of the stack. (Keep in mind that the stack grows down, so the bottom has the highest memory address.)

Section 9–3

1. The advantage of confining all of your C code to a 64K segment is that you can use it on systems where memory is limited to this amount.
2. A memory model is the method used by the C system of allocating segmented memory to an executed C program.
3. In the medium memory model, far pointers are used for code but not for data. This means that code may occupy more than one 64K segment while data may not. Just the opposite is true in the Compact memory model.
4. A far pointer requires 4 bytes and may cross 64K segments. A near pointer requires 2 bytes and is confined to a single 64K segment.

Section 9–4

1. To convert C source code to assembly language source code, you need the Turbo **TCC.EXE** on your active disk along with the C file to be converted.
2. The extension normally used for an assembly language source file is **.ASM**.
3. The assembly language instruction **push bp** means to copy the contents of the base pointer register to the top of the stack.
4. **bp+4** is an address while **[bp+4]** is the value contained at that address.

Section 9–5

1. In Turbo C a pseudo variable is an identifier that corresponds to one of the 8086 family internal registers.
2. An example of the use of a pseudo variable is **_AH = 5;**
3. No, the address of the **&** operator cannot be used with a pseudo variable because such a variable represents an internal register of the μP and not a memory location.
4. Inline assembly allows you to include assembly language programming within your C source code.
5. No, pseudo variables and inline assembly are not accepted ANSI C standards. They are available as part of the Turbo C language development system.

Section 9–6

1. The purpose of a MAKE utility is to help keep your source, object, and **.EXE** files current.
2. A MAKE utility compares the date and time stamps on each of your related files to help keep them updated.
3. The use of the Turbo C TOUCH utility is to change the date and time stamp for one or more files.
4. A library utility is a separate program that allows you to create your own libraries that can then be accessed by your C programs.
5. The name of the Turbo C utility that will search files for a given string is **GREP.EXE**.

Section 9–7

1. BIOS consists of programs that are permanently resident in the computer's permanent memory while DOS consists of programs that are externally loaded into the computer's read/write memory.

2. The advantages of using BIOS routines is that they are fast and offer direct access to the computer's I/O. The disadvantage is that they are machine specific.

3. The advantage of using DOS routines is that they can make adjustments for hardware differences between manufacturers, and C programs using them are more portable between differently manufactured computer systems. The disadvantage is that they are not as fast as BIOS routines nor as direct.

4. A BIOS function is **biosequip( )** which returns an integer value that indicates the hardware and perpherals used by the computer system.

5. A DOS function would be **_dos_getdate(date)** that returns the current system date.

Section 9–8

1. An interrupt is a special control signal that causes your computer to temporarily stop its main program, perform a special task (called servicing the interrupt) and then return back to the main program.

2. An interrupt vector contains the offset and the segment address of the interrupt handler.

3. An interrupt handler is the name given to the program that services the interrupt.

4. An address is stored in the IBM PC as follows. The least significant byte is stored first. The offset part of an address is stored before the segment part of the address.

5. The interrupt vectors are stored starting at memory location 0, and for the next 1024 bytes.

Answers to Self-Test—Chapter 9

1. The meaning of **#define** DATA 0X3BC in the program is to create a descriptive identifier for the printer data port number $3BC_{16}$.

2. DB25 is the identifying name of a special 25-pin connector that interfaces your IBM PC to its printer.

3. The DB25 parallel printer connector has three ports.

4. The name and port number of the above ports are DATA port# $3BC_{16}$, CONTROL port# $3BE_{16}$, and STATUS port# $3BD_{16}$.

5. The DATA port transmits the printable data to the printer. The CONTROL port controls various printer functions, and the STATUS port reads the necessary conditions of the printer.

6. The meaning of the statement **(inport(STATUS) & busy)** is to get data from the port number represented by STATUS then do a bitwise ANDing with the number represented by **busy**. Since **busy** is defined as 0✕80, this represents testing the condition of bit 7.

7. The acknowledge bit is tested in the program by the statement:

   `(inport(STATUS) & ack)`

   Here, the identifier `ack` is defined as the value 0×40 which means bit 6 is being tested.
8. The statement `outport (CONTROL, strobe_hi)` sets bit 0 of the control port high or TRUE.

Answers to Odd End-of-Chapter Problems—Chapter 9

Section 9–1

1. The general purpose registers contained in the 8086/8088 μP are the accumulator, base register, count register, and the data register. They may be used as eight separate 8-bit registers or four separate 16-bit registers.
3. A LIFO stack means memory where the last bit of data entered into the stack will be the first bit of data read from the stack.
5. The process of segmentation allows 16-bit internal registers to access memory that contains more locations than could be addressed by only 16 bits. This is achieved by using the values in two separate registers where the contents of one register are shifted left 4 bits, then added to the contents of a second register. The resulting 20-bit binary number is now capable of addressing up to 1Mb of internal memory.

Section 9–2

7. The short word used in assembly language programming as a memory aid is called a mnemonic.
9. C places locally declared variables on the top of the stack and function parameters at the bottom.

Section 9–3

11. The differences between the medium and the compact memory models is that the medium memory model uses far pointers for code but not for data; in the compact memory model, just the opposite is the case.
13. A far pointer uses 4 bytes and may cross 64K segments. A near pointer requires 2 bytes and is confined to a single 64K segment.

Section 9–4

15. Traditionally, the files given the extension `.ASM` are assembly language source code.

Section 9–5

17. A pseudo variable is an identifier that corresponds to one of the 8086 family internal registers. A C development system that uses them is Turbo C.

19. Including assembly language source code within your C source code is called inline assembly.

Section 9–6

21. A utility, as used in a C development system, is a separate program that is designed to assist you in the development and management of your C programs and associated files.
23. A library utility helps you create your own libraries that can then be accessed by your C programs.

Section 9–7

25. The term BIOS stands for Basic Input Output System.
27. The advantages of using DOS routines is that they can make adjustments for hardware differences between manufacturers. The disadvantage is that they are not as fast or direct as BIOS routines.

Section 9–8

29. An interrupt vector contains the offset and the segment address of the interrupt handler. An interrupt handler is the actual program that services the interrupt.
31. An address is stored by the 8086 μP family as follows. The least significant byte is stored first. The offset part of an address is stored before the segment part of the address.

Index

&, 32
Accumulator, 553
Action blocks, defined, 62
Actual parameter, defined, 72
Addition, binary, 254–255
Address calculating, 557
Address, defined, 247
Addressing: direct, 249; immediate, 249
Address operator, defined, 263
AND: bitwise, 292–293; logical, 143–145
ANSI C keywords, 14
ANSI prototypes, 12
Architecture: defined, 551; segmented memory, 556
Area fills, 510–511; defined, 511
Arguments: C, 562–563; command line, 454–456; defined, 16; multiple, 76
Arithmetic operators, 25–26; defined, 25
Array applications, 332–334, 339–344
Array index, 339–340; changing sequence of, 340–342; defined, 339
Array initialization, defined, 319
Array passing, defined, 323
Arrays, 316–327; defined, 310; **int**, 317–318; inside, 318–320; multidimensional, 328–338; multidimensional structure, 424; placing your own values in, 320–321; ragged, 352; rectangular, 351; structure, 417–420; within structures, 423–424
Aspect ratio, defined, 512
Assembler, defined, 559
Assembly language, C source code to, 570–573
Assembly language code, converting C to, 559
Assembly language concepts, 558–564

Assembly language instruction, types of, 559
Assignment, defined, 121
Assignment operator, 121; compound, 27–29; defined, 23; problems from using, 290
Auto debug, 221
Auto debug function, 222–223; defined, 222
Autodetect function, making, 495–496
Autodetection, defined, 490
Automatic variables, defined, 286

Background color, changing, 486–487
Bars: creating, 513–514; filling in, 514–518
Base pointer, 554
Base register, 553
BASIC, 12
Binary addition, 254–255
Binary file, defined, 440
Binary mode, defined, 440
Binary subtraction, 256–261
BIOS: defined, 586; and DOS, 586–587; some functions of, 587–588
Bit manipulation, defined, 292
Bit mapped, defined, 520
Bit shifting, 297–298; defined, 297
Bitwise AND, 292–293; defined, 292
Bitwise OR, 295–296; defined, 295
Bitwise complement, 292–293; defined, 292
Bitwise XOR, 296–297; defined, 296
Blocks: action, 62; branch, 62; loop, 62; program, 9–10; programmer's. *See* Programmer's block; types of, 62
Block structure, defined, 61
Boolean operators, defined, 292
Branch: closed, 131–142; open, 123–131

Branch blocks, defined, 62
Bubble sort, defined, 344
Buffer, defined, 446
Buffered stream, 446–447; defined, 446
Built-in shapes, 502–512
Bytes, defined, 251

C: appearance of, 6; converting to assembly language code, 559; creating a disk file with, 425; disadvantages of, 5; elements of, 12–15; LEARN, 3; at machine level, 560; numbers used by, 14; programs. *See* C programs; QUICK, 3; reasons for using, 5–7; statements, 14–15; TURBO, 3
C arguments, 562–563
C environment, defined, 3
C functions, 63–64; inside, 70–75
C graphics, starting, 488–496
C operators. *See* Operators
C programs: identifying parts of, 19–21; needs of, 12–14; parts of, 6
C source code to assembly language, 570–573
C structure, 399–401; initializing, 415–416
C text and color, 478–487
Calling convention, defined, 562
Calling functions: defined, 78; from a called function, 81–83
Calling more than one function, 78–79
Case sensitivity: of C identifiers, 20; errors associated with, 35–36
Casting, defined, 153, 532
Characters, 310–316; checking, 366–370; defined, 14; outputting, 365–366
Character stream, 430–431; defined, 430
Character string, saving, 429–430
char type, 251–253
Circles, creating, 508–509
Closed branch, 131–142; defined, 131
Code, source, 6
Code segment, 556
Color: background, changing, 486–487; C text and, 478–487; text, 482–485; using, 493–495
Color constants, defined, 483
Color monitor, 479
Color palette, defined, 497
Columns: adding more, 334–336; defined, 330
Command line arguments, 454–456; defined, 454–456; developing, 454–456
Comma operator, defined, 202
Comments: incomplete, 37–39; nested, 37–39
Compilation, conditional, 172
Compilation error messages, 35
Compiler, defined, 3
Complement: bitwise, 292; ones, 256; twos, 256
Complementing numbers, 255–256

Completeness theorem, defined, 63
Compound assignment operators, 27–29; defined, 27
Compound statement, 125–129; defined, 15, 125, 199
Computer, inside your, 551–557
Computer memory, defined, 12
Conditional compilation, defined, 172
Conditional loop, defined, 207
Conditional operator, 167–171
Conditional statement, defined, 123
Constants, 285; color, 483; defining, 88
Conversion of data types, 152–154
Coordinate transformation, 526–528; defined, 524
Count register, 553
Creating a file, 426
Creating lines, 500–501

Data: expressing, 389–391; mixed file, 438–440; putting into created disk file, 427; reading from disk file, 427–429; saving space for, 561; types of, 14; putting into structure, 401–403
Data registers, 553
Data segment, 556
Data types: conversion of, 152–154; defined, 14; naming your own, 394–398
Date/time utilities, 584
Declaration, pointer, 264
Declarations, defined, 19
Declaring variables, 21–24
Decrement operator, 200–202
Default, defined, 479
#define, 86; forms for, 90; including, 93; saving, 95
Design, top-down, 11
Destination index, 554
detectgraph( ), 490
Dimension, defined, 316
Direct addressing, defined, 249
Disk file: creating with C, 425; creating first, 426; putting data into, 427; reading data from, 427–429
Disk input and output, 425–445
Display capabilities, 479
DOS, defined, 425
DOS calls, 588; defined, 590
double, 14
do while loop, 207–212; defined, 207; structure of, 208; using, 208–209

%e, 32
E (exponential) notation, defined, 34
Editor, defined, 3
Elements: defined, 312; location of, 312–316

enum, 389; assigning values to, 392–393
Enumerated, defined, 389
Enumerating types, 388–393
Equality operator, problems from using, 290
Equals operator, 121
Error, run time, 219
Error messages: compilation, 35; fatal, 35; types of, 35
Escape sequence, defined, 29
Expressing data, 389–391
Expressions, defined, 14
External variables, defined, 286
Extra segment, 556

%f, 31
FALSE, defined, 118
Far pointer, defined, 567
Fatal error messages, 35
Fetch/execute, defined, 247
Field, defined, 16
Field width specifiers, 29–30; defined, 30, 331
File format, 432–433
File handles, 450
File pointers, 450–451; defined, 426, 450
Files: binary, 440; header. *See* Header files; library, 3; piping, 449; random access, 450; redirecting, 448–449; sequential access, 450; standard, 448; text, 440–441
File search utilities, 585–586
Fill-styles, defined, 515
Filtering, defined, 449
Finding minimum value, 342–344
Flags register, 557; defined, 551
Float, defined, 14
Flushed, defined, 446
Fonts: changing text, 520–522; defined, 520; stroked, 520
for loop, 196–203; defined, 196; more than one statement in, 199–200; parts of, 196–198
Formal parameter, defined, 72
Formal parameter list, defined, 67
Format, file, 432–433
Format specifiers: defined, 16; for **scanf( )** function, 32; using in **printf( )** functions, 18
fseek( ), 450
Functional switching, 161–163
Function prototype, defined, 64–65
Functions: as a structure, 410–412; auto debug, 222–223; C, 63–64; calling from a called function, 81–83; calling more than one, 78–79; defined, 19, 64; functioning with, 76–86; graphing, 523–532; how to call, 85; **if** statements used to call, 129–131; inside C, 70–

75; inside the **while** loop, 207; making your own, 64; using, 63–70; what makes, 64–65

General purpose registers, 551–553; defined, 551
getc( ), 429
getchar( ) function, 365–366
gets( ) function, 363–364
Global variables, 281–284; caution with, 282–284; defined, 282
Graphic mode, defined, 488
Graphics: C, 488–496; text with, 519–520
Graphics adapter, defined, 489
Graphics driver: defined, 489; selecting, 490
Graphics modes, changing, 502–504
Graphics screen, 490–491; defined, 488
Graphics system, knowing your, 496–502
Graphing functions, 523–532

Header files: defined, 94; making your own, 94–100; using, 98–99

Identifiers: case sensitivity of, 20; case sensitivity of C, 35–36; creating your own, 19–20; defined, 19; and keywords, 21
if statement, 123–124; defined, 123; used to call a function, 129–131
if...else statement, 132–134; compound, 135–138; defined, 132
if...else if...else statement, 136–138
Immediate addressing, defined, 249
int, 14
#include<stdio.h>, 43
Incomplete comments, 37–39
Increment operator, 200–202
Index: array, 339–340; destination, 554; pointer, manipulating, 358–360; source, 554
Index registers, 554
initgraph( ), 494
Initialization, array, 319
Initializing, C structure, 415–416
Initializing variables, 23–24
Inline assembly: defined, 575; introduction to, 576–577
Inline machine code, defined, 577
Input: getting user, 31–34; record, 441–442
Instruction pointer, 554
int array, 317–318
Integer, defined, 14
Internal registers, defined, 551
Interrupt handler: creating, 601–604; defined, 592
Interrupt number, defined, 598
Interrupts, 592–604; defined, 592; functions of, 592–597; typical, 599–601
Interrupt service routine, defined, 592

Interrupt vectors, defined, 592
I/O: inside, 446; standard, 449–450; system, 450
I/O problems, 362–363

Keywords: ANSI C, 14; and C identifiers, 21; defined, 14

LEARN C, defined, 3
Library files, defined, 3
Library utility: defined, 584; using Quick C, 585; using Turbo C, 584–585
Life of a variable, defined, 281
LIFO, 555
Lines, creating, 500–501
Line styles, 504–506
Linker, defined, 3
Local variables, 280–281; defined, 280
Logic, compounding the, 149–150
Logical AND, 143–145; AND, defined, 143
Logical OR, 145–147; defined, 145
Logical NOT, defined, 147
Logic operations, 142–152; relational and, 147–149
Long strings, outputting, 364–365
Loop blocks, defined, 62
Loops: conditional, 207; **do while**. *See* **do while** loop; **for**. *See* **for** loop; nested. *See* Nested loops; nesting with different, 216–218; sentinel, 207, 211–212; **while**. *See* **while** loop
Lvalue, defined, 153

Macro, defined, 87
main( ) function, 14
MAKE utility, 578–584; defined, 578
Manipulation, bit, 292
Matrix, defined, 329
Member declaration list, defined, 406
Member of operator, defined, 401
Memory: computer, 12; how used, 251–262; internal organization of, 246–250
Memory models, 565–569; defined, 565
Memory segmentation, 556
Messages, warning, 35
Minimum value, finding, 342–344
Mixed file data, 438–440; retrieving, 439–440
Mode: binary, 440; graphic, 488; text, 440
mode function, 480–482
Modes, 497–500; defined, 493
Modulo operator, 26
Monitor, color, 479
Monochrome, defined, 479
Multidimensional arrays, 328–338
Multidimensional structure arrays, 424
Multiple arguments, defined, 76

\n, 29
Near pointer, defined, 567
Nested comments, 37–39
Nested loops, 213–218; defined, 213; structure of, 215
Nesting with different loops, 216–218
Nibble, defined, 257
Notation, e (exponential), 34
Numbers: complementing, 255–256; looking for, 370–371; interrupt, 598; representing signed, 254; used by C, 14

offset, 452
Ones complement, defined, 256
Open branch, 123–131; defined, 123
Operations: defining, 89; logical, 142–152; precedence of, 26–27, 147; read, 551; relational and logic, 147–149; write, 551
Operators: address, 263; arithmetic, 25–26; assignment, 23, 121; Boolean, 292; C, 24–29; comma, 202; compound assignment, 27–29; conditional, 167–171; decrement, 200–202; equals, 121; increment, 200–202; modulo, 26; relational, 118–122; sequential-evaluation, 202
OR: bitwise, 295–296; logical, 145–147
Output: record, 444–445; standard, 16

Parameter: actual, 72; formal, 72
Pascal, 12
Passing arrays, 321–322, 323–324
Passing strings, 326–327
Passing variables, 271–280; with pointers, 266–268, 272–274
Pipe, defined, 449
Pixels: defined, 490; lighting up, 492–493
Pointer declaration, defined, 264
Pointer index, manipulating, 358–360
Pointer registers, 554–555
Pointers, 262–269; base, 554; defined, 262; far, 567; file, 450–451; instruction, 554; near, 567; passing variables with, 266–268; pointing to, 355–356; problems in using, 289–290; stack, 554; structure, 408–410; using, 264–266
Post-decrementing, defined, 200
Post-incrementing, defined, 200
Precedence, 147; defined, 26
Pre-decrementing, defined, 200
Pre-incrementing, defined, 200
printf( ) function, 3, 15–18, 29–30; escape sequences used by, 29; using more than one format specifier in, 18; what it does, 16–18
Program, storing a, 247
Program block, 9–10; concepts of, 58–63; defined, 59; example of, 59
Programmer's block, 10–11; defined, 10

Programming, structured. *See* Structured programming
Programming utilities, 578–586
Program structure, 7–12; defined, 7
Program stubs, defined, 182
Prologue, defined, 102
Prototype, function, 64–65
Pseudo variables, 573–575; defined, 573; introduction to, 574–575; using, 575
puts( ) function, 42

QUICK C, defined, 3

Ragged array, defined, 352
Random access files, defined, 450
Rank, defined, 153
Read operation, defined, 551
Record input, 441–442
Record output, 444–445
Rectangles, creating, 506–507
Rectangular array, defined, 351
Recursion, defined, 85
Redirecting files, 448–449; defined, 448
Registers: base, 553; count, 553; data, 553; flags, 551, 557; general purpose, 551–553; index, 554; internal, 551; pointer, 554–555; segment, 551; segmentation, 556
Register variables, 287–288; defined, 287
Relational and logic operations, 147–149
Relational operators, 118–122; defined, 118
Remarks, defined, 10
return( ) function, 74
Rows, defined, 330
Run time error, defined, 219

%s, 311
Saving space for data, 561
Saving μP contents, 561
Scaling, 525–526; defined, 524
scanf( ), 363
scanf( ) function, 31–32; format specifiers for, 32
Scope of a variable, defined, 281
Screen: graphics, 488, 490–491; text, 488
Structure, C. *See* C structure
Segment: code, 556; data, 556; extra, 556; stack, 556, 557
Segmentation, memory, 556
Segmentation registers, defined, 556
Segmented memory architecture, defined, 556
Segment registers, defined, 551
Semicolon, errors in using, 36–37
Sentinel loop, 211–212; defined, 207
Sequential access files, defined, 450
Sequential-evaluation operator, defined, 202

Shapes, built-in, 502–512
Shifting, bit, 297–298
short, 14
Signed numbers, representing, 254
Single statements, defined, 15
sizeof function, 253–254
Sort, bubble, 344
Sorting, 344–350
Source code, defined, 6
Specifiers: field width, 29–30, 331; format. *See* Format specifiers; type, 22
sqrt( ) function, 81
Stack, defined, 555
Stack pointer, 554
Stack segment, 556; defined, 557
Standard files, defined, 448
Standard I/O, defined, 449–450
Standard output, defined, 16
Statements: compound, 15, 125–129; conditional, 123; defined, 14; **if**, 123–124; single, 15
Static variables, 286–287; defined, 286; storing, 563
Storage size, 251
Storing a program, 247
String-izing, defined, 99
Strings, 310–316, 324–327, 350–362; character, saving, 429–430; defined, 14; long, outputting, 364–365; passing, 326–327; storing, 310–311
Stroked font, defined, 520
Structure arrays, 417–420; defined, 417
Structured programming: advantages and disadvantages of, 11–12; vs. unstructured programming, 8
Structure members, defined, 399
Structure pointers, 408–410
Structures: arrays within, 423–424; blocking, 59–61; C, 399–401; defined, 400; function as, 410–412; introduction to, 398–404; naming a, 407–408; program, 7–12; putting data into, 401–403; ways of representing, 422–424; within structures, 422–423
Structure tag, 405–407; defined, 406
Stubs, program, 182
Subtraction, binary, 256–261
Source index, 554
switch. *See also* Switching; C, 154–167; compounding, 158–161; using, 155–158
Switching. *See also* **switch**; functional, 161–163; within switches, 164
System I/O, defined, 450

Tag, structure, 405–407
Template, defined, 405
Text, with graphics, 519–520

textbackground( ), 486–487
Text color, 482–485
textcolor( ), 482
Text files, 440–441; defined, 440
Text fonts, changing, 520–522
Text mode, defined, 440
textmode( ), 480–482
Text screen, defined, 488
Theorem, completeness, 63
Token pasting, defined, 99
Tokens, defined, 14
Top-down design, defined, 11
Tracing, defined, 219
Transformation, coordinate, 524
TRUE, defined, 118
TURBO C, defined, 3
Twos complement, defined, 256
Type casting, 153
typedef, 394–398
Type definition, defined, 394
Type specifiers, defined, 22

union, 413–417
Unstructured programming, vs. structured
    programming, 8

μP, saving contents, 561
User input, getting, 31–34
Utilities, programming, 578–586

Value passing, defined, 72
Values: finding minimum, 342–344; placing your
    own in array, 320–321
Variable class, 285–289
Variables: automatic, 286; declaring, 21–24;
    external, 286; global, 281–284; initializing, 23–
    24; local, 280–281; passing. *See* Passing
    variables; pseudo. *See* Pseudo variables;
    register, 287–288; scope of, 280–284; static,
    286–287; static, storing, 563
Vector table: defined, 598; interpreting, 598–599
void, 67

Warning messages, 35
while loop, 203–207; defined, 203; functions
    inside the, 207; structure of, 203–204
Words, defined, 251
Write operation, defined, 551

XOR, bitwise, 296–297

Microsoft® Academic Edition Order Form

To order Microsoft QuickC Compiler Academic Edition, please follow the directions below. This discount program is offered only to faculty and to students.

Registered course information:

Course name

Department

Instructor's name

School

Computer used: Make _____ Model _____

Please indicate desired disk size: ☐ 3½″ disks ☐ 5¼″ disks

Method of payment:

Please choose one of the payment options below to pay your $47.45 ($44.95 plus $2.50 shipping). *Note:* Your payment will be deposited or charged to your credit card immediately upon receipt by Microsoft.

☐ Check   ☐ VISA (13 or 16 numbers)   ☐ MasterCard (16 numbers)

☐ Money order ☐ American Express (15 numbers)

Credit card # [1] [2] [3] [4] [5] [6] [7] [8] [9] [10] [11] [12] [13] [14] [15] [16]

Print name as it appears on credit card

Expiration date ____ / ____ Signature _____

Ship my Microsoft QuickC Compiler to:

Name

Shipping address

City _____ State ____ ZIP ____

Daytime telephone (____) _____
(in case we have a question about your order)

Send this form with your payment to:

Microsoft Academic Edition Order Form
21919 20th Avenue SE
Box 3011
Bothell, WA 98041-3011

There is a one coupon per student or faculty member limitation and only one discount per coupon. This coupon is not transferable to other Microsoft products and cannot be redeemed for cash. Please allow 4–6 weeks for delivery. Offer good only in USA and Canada. Offer expires June 30, 1990.

If you have any questions about this discount offer, please call the Microsoft Information Center at (800) 426-9400.

© Copyright 1989 Microsoft Corporation. All rights reserved. Printed in USA.

Microsoft, the Microsoft logo, and QuickC are registered trademarks of Microsoft Corporation.

0689
048-399v200—QuickC (5¼″ disks)
048-395v200—QuickC (3½″ disks)

Microsoft®

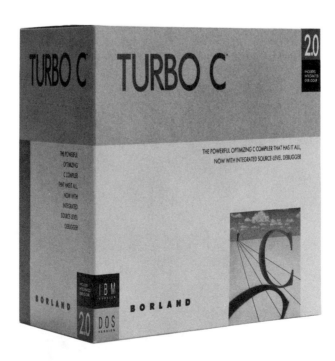

C Keywords:

auto	do	goto	signed	unsigned
break	double	if	sizeof	void
case	else	int	static	volatile
char	enum	long	struct	while
const	extern	register	switch	
continue	float	return	typedef	
default	for	short	union	

C Statements

In C, a statement consists of keywords, expressions and other statements. Statements are used to control the flow of program execution. These include statements to execute loops, transfer control (decision-making), and select other statements. What follows is a summary of C statements, listed in alphabetical order.

Assignment =

Assigns value of expression on right to variable on left.

Example:

resistance = 12;

Compound

A statement inside the body of another statement. The example of the **break**, below, illustrates a compound statement.

---

break

Causes the innermost **do**, **for**, **switch**, or **while** statement to end.

Example:

```
while(count > 0)
    {
    if(count > 5) break;
    }    /* Loop terminates when count > 5 */
```

---

continue

Used with the **do**, **for**, and **while** statements. Causes control to pass to the next iteration of these statements.

Example:

```
while(count > 0)
    {
    if(count == 5) continue;
    }    /* Loop skipped for count equal to 5. */
```

---

do statement while(expression)

Body of the **do** statement is executed one or more times until the **while** expression is FALSE (zero) and then control passes to the next statement.

Example:

```
do
{
    value = value + 1;
}
while (value < 5);
/* Loop continues as long as value < 5. */
```

for loop

```
for(expression₁ ; expression₂ ; expression₃)
  <statements>
```

Process is as follows:

expression₁ is first evaluated (done only once).

expression₂ is evaluated and if TRUE, **statements** are executed.

expression₃ is evaluated each pass through the loop.

Example:

```
for(count = 1; count < 5; count++)
{
    printf("The count is %d.\n",count);
}
/* Loop continues while count < 5. */
```

if statement

```
if(expression)
  statement₁
else
  statement₂
```

If **expression** is TRUE (non-zero), then **statement₁** is evaluated; otherwise, **statement₂** is evaluated.

Example:

```
if(value != 5)
    printf("Value is not equal to 5.");
else
    printf("Value is equal to 5.");
/* Executes first printf( ) if value is not equal to 5
   otherwise second printf( ) function is executed.   */
```

return statement

Causes termination of the execution of the function in which it is activated. Program flow returns to the calling function.

Example:

```
return(value);
/* Returns the value of value to the calling function */
```

switch statement
   Causes transfer of control to a statement within the body of the **switch**
   Example:

```
switch (value)
  {
   case 1 : {
              resistor = 10;
              printf("Resistor is 10 ohms.")
              }
            break;

   case 2 : printf("No value assigned.")
            break;

   default  : printf("Invalid selection.");
            }
            /* Value of value determines one of the three selections. */
```

while loop

```
while(expression)
    statement
```

Body of **statement** is executed zero or more times until **expression** is FALSE
(zero).
   Example:

```
while(value < 5)
    {
     value++;
     printf("Value is %d.\n",value);
    }
    /* Loop continues while value <5. */
```